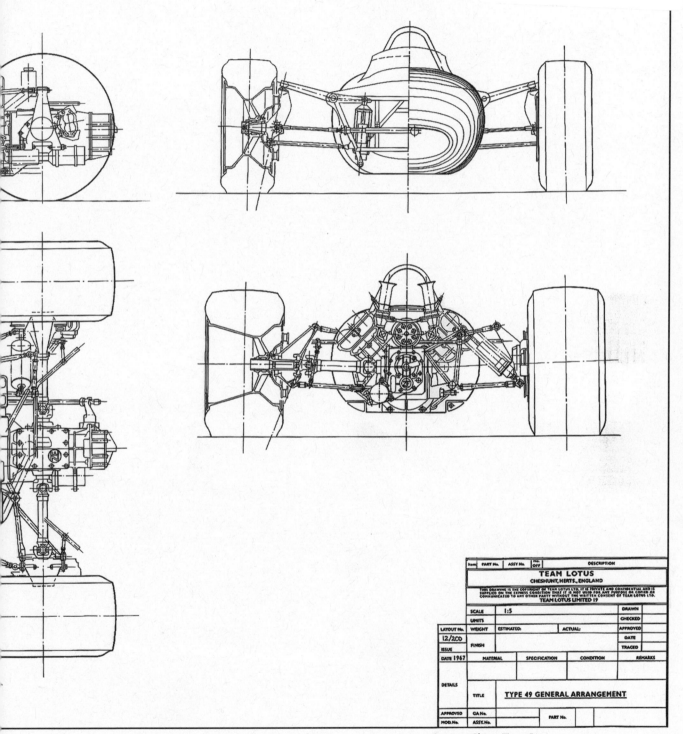

Courtesy Classic Team Lotus.

LOTUS 49

THE STORY OF A LEGEND

BY MICHAEL OLIVER

Michael Oliver 346/500

For Gill, Matthew and Dominic

Other Veloce publications -

Colour Family Album Series
Alfa Romeo by Andrea & David Sparrow
Bubblecars & Microcars by Andrea & David Sparrow
Bubblecars & Microcars, More by Andrea & David Sparrow
Citroën 2CV by Andrea & David Sparrow
Citroën DS by Andrea & David Sparrow
Fiat & Abarth 500 & 600 by Andrea & David Sparrow
Lambretta by Andrea & David Sparrow
Mini & Mini Cooper by Andrea & David Sparrow
Motor Scooters by Andrea & David Sparrow
Porsche by Andrea & David Sparrow
Triumph Sportcars by Andrea & David Sparrow
Vespa by Andrea & David Sparrow
VW Beetle by Andrea & David Sparrow
VW Beetle/Bug, Custom by Andrea & David Sparrow
VW Bus, Camper, Van & Pick-up by Andrea & David Sparrow

SpeedPro Series
How to Blueprint & Build a 4-Cylinder Engine Short Block for High Performance by Des Hammill
How to Build a V8 Engine Short Block for High Performance by Des Hammill
How to Build & Modify Sportscar/Kit car Suspension & Brakes for High Performance by Des Hammill
How to Build & Power Tune Weber DCOE & Dellorto DHLA Carburetors Second Edition by Des Hammill
How to Build & Power Tune Harley-Davidson 1340 Evolution Engines by Des Hammill
How to Build & Power Tune Distributor-type Ignition Systems by Des Hammill
How to Build, Modify & Power Tune Cylinder Heads Second Edition by Peter Burgess
How to Choose & Time Camshafts for Maximum Power by Des Hammill
How to give your MGB V8 Power Updated & Revised Edition by Roger Williams
How to Plan & Build a Fast Road Car by Daniel Stapleton
How to Power Tune BMC/BL/Rover 998cc A Series Engines by Des Hammill
How to Power Tune BMC/BL/Rover 1275cc A Series Engines by Des Hammill
How to Power Tune the MGB 4-Cylinder Engine by Peter Burgess
How to Power Tune the MG Midget & Austin-Healey Sprite by Daniel Stapleton
How to Power Tune Alfa Romeo Twin Cam Engines by Jim Kartalamakis
How to Power Tune Ford SOHC 'Pinto' & Sierra Cosworth DOHC Engines by Des Hammill

General
Alfa Romeo Giulia Coupe GT & GTA by John Tipler
Automotive Mascots: A Collectors Guide to British Marque, Corporate & Accessory Mascots by David Kay & Lynda Springate
Bentley Continental Corniche & Azure 1951-1998 by Martin Bennett
British Cars, The Complete Catalogue of 1895-1975 by Culshaw & Horrobin
British Trailer Caravans & their Manufacturers 1919-1959 by Andrew Jenkinson
British Trailer Caravans & their Manufacturers from 1960 by Andrew Jenkinson
Bugatti Type 40 by Barrie Price
Chrysler 300 - America's Most Powerful Car by Robert Ackerson
Cobra - The Real Thing! by Trevor Legate
Cortina - Ford's Best Seller by Graham Robson
Daimler SP250 'Dart' by Brian Long
Datsun/Nissan 280ZX & 300ZX by Brian Long
Datsun Z - From Fairlady to 280Z by Brian Long
Dune Buggy Handbook, A-Z of VW-based Buggies since 1964 by James Hale
Fiat & Abarth 124 Spider & Coupé by John Tipler
Fiat & Abarth 500 & 600 (revised edition) by Malcolm Bobbitt
Ford F100/F150 Pick-up by Robert Ackerson
Jim Redman - Six Times World Motorcycle Champion by Jim Redman
Grey Guide, The by Dave Thornton
Lea-Francis Story, The by Barrie Price
Lola - The Illustrated History (1957-1977) by John Starkey
Lola T70 - The Racing History & Individual Chassis Record New Edition by John Starkey
Mazda MX5/Miata 1.6 Enthusiast's Workshop Manual by Rod Grainger & Pete Shoemark
Mazda MX5/Miata 1.8 Enthusiast's Workshop Manual by Rod Grainger & Pete Shoemark
Mazda MX5 - Renaissance Sportscar by Brian Long
MGA by John Price Williams
Michael Schumacher - Ferrari Racing 1996-1998 by Braun/Schlegelmilch
Mini Cooper - The Real Thing! by John Tipler
Motor Museums of the British Isles & Replulic of Ireland by David Burke & Tom Price
Porsche 356 by Brian Long
Porsche 911R, RS & RSR New Edition by John Starkey
Porsche 914 & 914-6 by Brian Long
Prince & I, The (revised edition) by Princess Ceril Birabongse
Rolls-Royce Silver Shadow/Bentley T Series, Corniche & Camargue (rev. edition) by Malcolm Bobbitt
Rolls-Royce Silver Wraith, Dawn & Cloud/Bentley MkVI, R & S Series by Martyn Nutland
Singer Story:Cars, Commercial Vehicles, Bicycles & Motorcycles by Kevin Atkinson
Taxi! The Story of the 'London' Taxicab by Malcolm Bobbitt
Triumph TR6 by William Kimberley
Volkswagen Karmann Ghia by Malcolm Bobbitt
VW Bus Type 2, Camper, Van, Pickup by Malcolm Bobbitt
Volkswagens of the World by Simon Glen
Works Rally Mechanic by Brian Moylan

First published 1999 by Veloce Publishing Plc., 33, Trinity Street, Dorchester DT1 1TT, England.
Fax: 01305 268864/e-mail: veloce@veloce.co.uk/website: http://www.veloce.co.uk
ISBN: 1-901295-51-6/UPC: 36847-00151-3 (standard edition). ISBN 1-901295-75-3/UPC: 36847-00175-9 (British Racing Green edition). ISBN 1-901295-94-X/UPC: 36847-00194-0 (leather edition).
© 1999 Michael Oliver and Veloce Publishing Plc
All rights reserved. With the exception of quoting brief passages for the purpose of review, no part of this publication may be recorded, reproduced or transmitted by any means, including photocopying, without the written permission of Veloce Publishing Plc.
Throughout this book logos, model names and designations, etc., may have been used for the purposes of identification, illustration and decoration. Such names are the property of the trademark holder as this is not an official publication.
Readers with ideas for automotive books, or books on other transport or related hobby subjects, are invited to write to Veloce Publishing at the above address.
British Library Cataloguing in Publication Data -
A catalogue record for this book is available from the British Library.
Typesetting, design and page make-up all by Veloce on AppleMac.
Printed in the UK.

LOTUS 49

THE STORY OF A LEGEND

BY MICHAEL OLIVER

VELOCE PUBLISHING PLC
PUBLISHERS OF FINE AUTOMOTIVE BOOKS

Contents

Acknowledgements 5
Introduction 7
Foreword by Keith Duckworth OBE 8

Chapter 1	Racing Roots 9	
	Prologue - Designs on Dominance 9	
Chapter 2	A BRM Interlude 14	
Chapter 3	Doubling Up 16	
Chapter 4	The Simple Bracket 24	
Chapter 5	Consummating The Marriage 31	
Chapter 6	When The Flag Drops 36	
Chapter 7	Reality Bites 41	
Chapter 8	Beefing-up The Weak Spots 46	
Chapter 9	A Winning Streak At Last 58	

Colour Gallery 1 65

Chapter 10	From Triumph To Tragedy 82	
Chapter 11	Turbulent Times 98	
Chapter 12	Challenge For The Championship 116	
Chapter 13	Winter Works 127	
Chapter 14	Tasman Troubles 131	
Chapter 15	Wing Worries 139	
Chapter 16	Trials & Tribulations 156	
Chapter 17	Boiling Point 162	
Chapter 18	Team Spirit Returns 168	
Chapter 19	Trading Places 174	

Colour Gallery 2 177

Chapter 20	The Swan Song 195	
Chapter 21	A New Star Rises 204	
Chapter 22	The Exiles 210	
Chapter 23	The Final Analysis 222	
	The Lotus 49's Record in	
	Grand Prix Racing 222	

Grand Prix Wins in a Lotus 49
   by Driver 222
Grand Prix Starts in a Lotus 49
   by Driver 223

Appendix 1 Chassis By Chassis: The Complete
               Competition Record Of The
               Team Lotus Type 49 227

Chassis 49/R1 227
The Racing Record of Chassis 49/R1 246
Chassis 49/R2, 49B/R2 228
The Racing Record of Chassis 49/R2 246
Chassis 49/R3 230
The Racing Record of Chassis 49/R3 247
Chassis 49/R4 231
The Racing Record of Chassis 49/R4 247
Chassis 49/R5, 49B/R5 231
The Racing Record of Chassis 49/R5 248
Chassis 49B/R6 232
The Racing Record of Chassis 49/R6 248
Chassis 49B/R7 234
The Racing Record of Chassis 49/R7 249
Chassis 49T/R8, 49B/R8, 49C/R8 235
The Racing Record of Chassis 49/R8 250
Chassis 49T/R9, 49B/R9 238
The Racing Record of Chassis 49/R9 251
Chassis 49B/R10 239
The Racing Record of Chassis 49/R10 251
Chassis 49B/R11 242
The Racing Record of Chassis 49/R11 252
Chassis 49/R12 244
Replicas and Recreations 244

Appendix 2 Bibliography 253
Index   254

Acknowledgements

I would like to thank the following people, all of whom have readily given up their valuable time, in order that I could get a first-hand account of their experiences working with, racing in and racing against the Lotus 49 -

Chris Amon, Mario Andretti, Richard Attwood, Herbie Blash, Sir Jack Brabham, Bill Brack, Dougie Bridge, Dave Charlton, Tony Cleverley, Mike Costin, Gerard 'Jabby' Crombac, Bob Dance, Eddie Dennis, Keith Duckworth, Jim Endruweit, Geoff Ferris, Emerson Fittipaldi, Roy Franks, Dan Gurney, Walter Hayes, Robin Herd, Mike Hewland, Bette Hill, Ken Howes, Gordon Huckle, Colin Knight, John Love, Pete Lovely, Allan McCall, Brian Melia, John Miles, Derek Mower, Jackie Oliver, George Pfaff, Mike Pilbeam, Dale Porteous, Dick Scammell, Denzil Schultz, Trevor Seaman, Dave Sims, Bob Sparshott, John Surtees, Ron Tauranac, Rob Walker, Mike Watson, Eppie Wietzes, Leo Wybrott, Mike Young.

A huge number of people have assisted me in varying ways during this project, whether it be sharing their own records relating to chassis histories, passing on contact details, digging out old race reports or simply imparting their knowledge on a particular subject of relevance to the book. I'd particularly like to thank Clive Chapman, Bob Dance, Keith Duckworth, Doug Nye, Dick Scammell and Leo Wybrott, all of whom took the time to read through draft chapters of my manuscript. Other people whose help I could not have done without include: Andrea Bishop, Manning Buckle, John Dawson-Damer, Peter Denty, Peter Elleray, David McLaughlin, Peter Morley, Jose Phillippe, Dieter Rencken, Richard Spelberg, Jannie van Aswegen, Erich Walitsch, Janos Wimpffen, Robert Young.

If there is anyone I have omitted by mistake, please accept my apologies and my thanks in equal measure.

One of my aims in compiling this book was to try and provide a large and varied selection of photographs, both black and white and colour, and to track down shots which had never before been published. I would like to record that this would not have been possible without the co-operation of many of the people named above, a number of whom have provided me with access to photos from their personal collections. I have also drawn on the extensive resources of the Ford Motor Company's archive at Aveley Photographic and for that I would like to thank Steve Woolmington and Fran Chamberlain. I am also extremely indebted to the Ford Motor Company for the assistance it has provided in enabling me to source photos from other photographers around the world. I would like to thank the following people or organisations who have supplied photographs for the book -

Adri Bezuidenhout, Ray Berghouse (Chevron Publishing Group), Douglas Booker, Keith Booker, Classic Team Lotus, Tim Considine, Paul Cross, Ford Motor Company, Dave Friedman, Michael Lavers, Roger Leworthy, Pete Lyons, Alberto Martinez, Peter Mohacsi, Doug Mockett, David Phipps (Phipps Photographic), Euan Sarginson, Nigel Snowdon, Colin Taylor, William Taylor, Brier Thomas, Ted Walker (Ferret Fotographics), Tony Walsh, Steve Wyatt, Aldo Zana.

Every effort has been made to contact copyright holders of photographs used in this book. If anyone feels I have used one of their photographs and not given them proper acknowledgement, please contact me and I will happily rectify the situation in any subsequent edition.

Finally, I would like to say a big 'thank you' to all my family and friends, who have supported, helped and generally encouraged me throughout the time I have been researching and writing the book. Particular thanks are due to my wife, Gill, for putting up with my prolonged absence from all family-related activities and for reading through the entire manuscript many times over. I would also like to thank my father, Robert Oliver, for giving me access to his collection of books and magazines, looking things up for me at all times of the day and night and also reading endless pages of manuscript. Additionally, I am grateful to my mother, Helen Oliver, for her comments and feedback on my work. My parents were responsible for igniting my interest in motor sport, taking me to countless races as a wide-eyed child both in the UK and Europe. In fact, I think I learned to count from the numbers on the sides of racing cars! Lastly, and by no means least, thanks to my sons, Matthew and Dominic, who have had to do without a Dad being around for as much as they would have liked during the past eighteen months.

Dawn of a new era...From left, Duckworth, Chapman, Clark and Hill look quietly confident before the first time out at Zandvoort, June 1967. (Maurice Rowe)

Foreword

The first time I met Colin [Chapman] was in the mid-1950s, when I bought a Mark Six in kit form from the works at Hornsey. Then I went to work for him as a transmission engineer on the 'Queerbox', taking over from some chap by the name of Graham Hill! I didn't hang around very long as I wasn't prepared to carry on developing a gearbox that I didn't think would ever work for any length of time. So it was that I left to start Cosworth. We did very well on Coventry Climax engine tuning and then moved on to the Ford Anglia engine for Formula Junior and, I have been told, we rescued Lotus by building the engine they put into the 18s. Later, we began working with Ford, tweaking some of the production engines in their road cars. So, long before the 49 was ever thought of, we were all working together in some way.

The question of doing a Formula 1 engine had never arisen. However, when Coventry Climax gave up making Grand Prix engines, Colin asked me if I could design one for him. As I had already decided to do a 1600cc four-valve Formula 2 engine, it seemed fairly reasonable, provided it went alright, to consider virtually doubling it up to produce a three litre engine for Formula 1. That's how the DFV came about, basically.

We had surprisingly few problems with the DFV, really. The main one was the gear train torsional problem and that took a long time to solve. By 1972-73 it was getting almost impossible to run and the engine was nearly abandoned. It was at that stage that I designed my quilled hub and that rescued the whole of the valve gear and it carried on. I'm fairly certain that, even now, all gear-driven Formula 1 engines embody a fairly similar piece of kit.

The 49 was a beautiful little car. It looked right from the start– nice, neat and small. I suppose I can claim to have played a fairly major part in defining the shape of the monocoque. I talked very early on with Colin and Maurice Phillippe about the preferred engine format and how we could integrate it with the chassis.

I am delighted that, at last, someone has chosen to record the full story - from a 'behind-the-scenes' perspective - of the Lotus 49 because it was such an important milestone in Grand Prix racing. Michael has clearly gone to considerable lengths to unearth the story of the car and I think it shows. It's a sobering thought that I am the only surviving member of the quartet pictured opposite. The fact that other key figures such as Maurice Phillippe and Jochen Rindt are no longer about must have made his task even more difficult.

When you look at it, Ford took a bit of a gamble with the whole Lotus 49/DFV thing. After all, it could have gone pear-shaped and they would have looked rather daft. Fortunately for me it didn't! Now Cosworth is part of Ford and they are back at the sharp end of the grid, too. With Dick Scammell having come back to oversee its integration into Ford, it is only a shame that Colin and Team Lotus are no longer to be found on the grid to complete the full circle.

Keith Duckworth OBE
Northampton, England

Colin Chapman: The man who made it all happen at Team Lotus. Persuasive, charming, sometimes caustic and often intolerant of the shortcomings of others. But above all, a conceptual genius. (Ford Motor Company)

Introduction

When I first set out to write this book I knew exactly the kind of thing I wanted to produce. I had just finished reading Andrew Ferguson's *Team Lotus, The Indianapolis Years* and it was one of the most amusing, informative and fascinating motor racing books I had read. If ever I was going to write a book, then that was the standard to which I would aspire.

Andrew's book was written from the unique perspective of having been there and worked with all the key characters in the story. By talking to as many of the people involved in the Lotus 49 project as possible, I have sought to gain a similar insight and I hope that this is conveyed in what I have written. In many cases I have left it to the people involved to tell the story in their own words: they do so far better than I could.

I've deliberately sought to emphasise the experiences and recollections of individuals involved with the Lotus 49 – in other words, the human side of the story. I'm afraid I'm one of those people whose eyes glaze over at the sight of endless pages of technical jargon. Consequently, I have tried to keep the use of complex terminology to a minimum, although when one is discussing the specification of the car and the ins and outs of the Ford Cosworth DFV engine, it is impossible not to use it in places. I hope I have found a good balance ...

Sadly, the passage of time has robbed us of the experiences and recollections of many of the people most closely associated with the Lotus 49, including drivers Jim Clark, Graham Hill, Jo Siffert and Jochen Rindt, the combined design talents of Colin Chapman and Maurice Phillippe and, of course, Andrew Ferguson, who was so active in creating and maintaining Team Lotus records, as well as providing the inspiration to write this book.

The exceptional level of co-operation I have received from everybody I have dealt with during this project, be they mechanics, designers, team managers, engineers or photographers has, I hope, enabled me to paint an accurate picture of the way it was. I have certainly enjoyed researching and writing this book – I hope you enjoy reading it.

Michael Oliver
Witney, England

Chapter 1
Racing Roots

Prologue - Designs on Dominance

Like many of Colin Chapman's best creations, the Lotus 49 began life over dinner and on the back of a napkin. His companion on this occasion was Walter Hayes, then Director of Public Affairs for Ford of Great Britain. As Hayes sat back and listened, puffing away on his pipe, Chapman sketched out the shape of the car that would transform Grand Prix racing for ever. His vision followed the tried and tested formula which had already brought him tremendous success in the outgoing 1.5-litre Formula. The car he envisaged was to be light and compact, with the engine acting as a load-bearing structure carrying the rear suspension. Most importantly, it would be his first ever truly integrated Grand Prix car, where the car was built for the engine and the engine was built for the car. The only problem was linking up with a manufacturer prepared to fund such an engine, which was where Hayes and Ford came in.

The result of that meeting between Hayes and Chapman one night in 1965 was that Hayes, having been convinced by Chapman that Ford was the right company for the job, put forward a proposal for the company to back a Formula 1 engine programme - and the rest is history. The Ford-funded Cosworth DFV engine went on to score a total of 155 Grand Prix victories during a 17 year period which was ended only by the onset of the all-powerful turbos. Thus, the DFV was not driven out by another, more superior normally aspirated engine but by an altogether more powerful format which, to this day, DFV designer Keith Duckworth maintains was outside the spirit, if not the letter, of the regulations at the time.

The Lotus 49 was the car which brought the Cosworth DFV and the name of Ford, into the Grand Prix arena. It was a car which gave the DFV the means by which it could prove itself the best engine in Formula 1, both in 1967, when it won four races, and in 1968, when it won both the Tasman series and the World Championship. Remarkably, given the pace of technical development at the time, the 49 remained a competitive proposition during 1969 and into 1970 when, in 'C' specification, the car scored what was probably its most memorable victory at the Monaco Grand Prix in the hands of Jochen Rindt. Not bad for a 'simple' chassis design which had been introduced as a stop gap to enable any development niggles to be ironed out in the DFV engine …

***

To put the development of the Lotus 49 and the Ford Cosworth DFV engine into context, it is worth considering the circumstances under which the project came together and the background of the key personnel involved. The roots of the Lotus 49 and the partnership with Ford and Cosworth stretch back many years, to the early days of Team Lotus as a Grand Prix entrant. The relationships established during this period go a long way towards explaining the ease with which the project fell into place nearly a decade later. What is more, the paths of the main protagonists involved with the 49 crossed again and again in the years leading up to 1967. In the autumn of 1957, a young Duckworth, fresh out of university with a mechanical engineering degree, was weighing up whether to join either Napier or Rolls-Royce as an apprentice, when he was offered a job by Colin Chapman at Lotus, on a salary of £600 a year. He had already worked for Lotus in the previous summer holiday, as an understudy to Graham Hill in the gearbox shop. Towards the end of 1957, Hill was about to leave to pursue a full-time career in racing, and Duckworth was offered the grandly titled position of gearbox development engineer, which he duly accepted.

The job was something of a poisoned chalice, for his first task was to try and build some reliability into the new five-speed Lotus gearbox, nicknamed the 'Queerbox'. While the young engineer quickly arrived at solutions which he believed would solve the problems, Chapman was less willing to make the proposed changes, as Duckworth explains. 'The dog flanks needed to be flat and they were very short, hence they rounded off. Then you were really trying to balance two balls against each other, and they do tend to fling out, and they used to jump out of gear. You had only to widen the gears by about 0.1 inch each to have made their life useful instead of hopeless. But he [Chapman] said that they couldn't afford to throw away the pile of gears that they had and so, therefore, they had to press on with them.' After ten frustrating months, and displaying the same lack of compromise that would stand him in such good stead in the future, Duckworth quit Lotus: 'I wasn't prepared to carry on developing a gearbox that I didn't think would ever work for any length of time. By that stage, I was discussing with Mike [Costin] whether we could start a business, messing about with racing cars and engines.'

Duckworth had struck up a good friendship with Costin, who was responsible for day-to-day technical matters on both the production and racing side at Lotus. It was a case of mutual admiration of each other's strengths – Duckworth was a pure thinker, rarely used to working under time constraints, whereas Costin was an instinctive engineer who with Lotus had learned to think on his feet and improvise in the time available.

Such was their understanding of one another that by the summer of 1958, they had decided to go into business together. Costin had become increasingly concerned by Chapman's way of doing business, as well as the tremendous demands placed on him in terms of working hours, which meant that he was seeing less of his family than he wanted to. However, the difficulties of starting up in business led him to conclude that, for the time being at least, his interests and those of his family would be best served if he stayed at Lotus, as Duckworth recalls: 'I had no ties whereas Mike had a wife and children. He was offered the job of technical director [of Lotus] on a three year contract, which forbade him becoming involved in any 'extra-mural' activities. So he had to drop out from being a director and a founder shareholder of the company.' Consequently, Duckworth set about establishing the new business on his own. And so it was that Cosworth Engineering Limited - Cosworth being an amalgam of the two partners' surnames - was born.

It wasn't until late 1962 that Duckworth and Costin finally got to join forces in the business officially. By this point Cosworth was doing very well on Coventry Climax engine tuning and had moved into tuning the Ford Anglia engine - the 105E - for Formula Junior. This had come about largely as a result of its involvement in supplying engines for the new rear-engined Lotus 18 Formula Junior car – a contract it had won after a testing 'shoot-out' against the Speedwell A35 BMC engine. Ironically, the choice came down to the engines of Duckworth or his predecessor at Lotus, for Graham Hill was a director of Speedwell! There is little doubt that Costin played an important role in helping the decision go Cosworth's way as well as smoothing the waters between Duckworth and Chapman, as the former recalls: 'I think Mike was the one who was instrumental in converting Colin to the use of an engine by us and overcoming the difficulties over my leaving. There was obviously some slight strain for a period due to the fact that in 1958 I had told him what I thought of his establishment.'

Consequently, a mildly-tuned Cosworth-Ford engine made its bow on Boxing Day 1959 in the back of a Lotus 18. However, this was a cobbled together last-minute effort and it was only in the Easter Monday meeting at Goodwood the following year that a full-blown engine made its appearance in a car driven by an emerging young Scottish driver by the name of ... Jim Clark. Once more, paths were crossing that would finally converge in 1967. This victory, and the subsequent domination of the Ford-Cosworth 105E, laid the foundations for the financial stability of Cosworth and established Lotus as a leading British racing car constructor.

For 1964, two new classes of single-seater racing, the 1-litre Formula 3 and a new format Formula 2, were introduced to replace Formula Junior. With the MAE (Modified Anglia Engine) and SCA (Single Camshaft, A-Series) engines, Cosworth continued to develop its reputation within the motor sport fraternity. Just as importantly, its wizardry with Ford engines had not gone unnoticed and, by late 1962, the manufacturer was actively working with Cosworth on refining and developing the Lotus-Ford twin-cam engine that would be used in the Ford Lotus Cortina, to be launched the following year. Even at this stage the three key technical partners in the Lotus 49 project were working together, albeit in an area totally unconnected with Grand Prix racing.

The SCA engine was the first to feature a cylinder head designed by Duckworth from scratch, although it retained the old-faithful five bearing Ford cylinder block that had formed the basis of the company's initial successes. Although the single overhead camshaft design had several deficiencies (particularly in the combustion department) which still make Duckworth wince today, it was good enough to get the job done and won the 1964 and 1965 Formula 2 Championship titles.

It was only when the sophisticated Honda four valve engine emerged in the back of the works Brabham in 1966 that serious competition threatened to leave the SCA behind. Duckworth was (and is!) not a great believer in development saying: 'Development is only really required due to the ignorance of designers!' Instead, he resolved to design and build a completely new engine for the upcoming 1600cc Formula 2, the FVA (Four Valve, A-Series), which was a twin-cam, four-valve unit to be introduced for the 1967 season: 'We decided we were going to do a 1600cc Formula 2 engine, again using the same Ford block as our 1000cc (SCA). And we decided that it was going to be a four valve engine - I'd had enough of my problems with the single overhead cam and difficult combustion.'

During the final year of the 1-litre Formula 2 format, Honda's engine had overshadowed the SCA with a twin overhead camshaft design which had made use of four valves per cylinder. Although there were successful precedents for this design dating back as far as 1912, it was a concept that very few manufacturers had managed to make work effectively. Even the much-admired Coventry Climax concern had struggled to obtain as much power from such a layout in their engines as from their ultra-successful two-valve layout. It wasn't until 1965 with their 32-valve V8, which took Jim Clark to the Formula 1 World Championship title, that they really proved the four valves per cylinder concept.

The Honda engine was clearly considerably more powerful than the SCA, something which Duckworth put down to their attention to detail but also resolved to pursue and refine himself, as his partner Costin recounts: 'In engineering, there is nothing that hasn't been done before. If you want a four valve engine, go back to 1898. When I was an apprentice, a chap I was an apprentice with had a four valve 600cc Ariel motorbike. When the Japanese did it, they started out with the wrong pent roof type of four valve arrangement. What Keith did was to flatten it ... the included angle of the valves was the difference between all the other four valve engines and our DFV/FVA. And that was Keith's design thinking, from nothing. Not from what was done in 1898, but because that was what was required. He analysed what was wrong with the Japanese four valve just by looking at it externally and saying "No. When we do it, we'll do it this way." That was just brilliant.'

As Costin pointed out, the difference in Duckworth's approach was related to the valve angle or, more correctly, the included angle of the valves. This is defined as the angle between the axes of the inlet and exhaust valves in a cylinder. Historically, racing engines had featured valve angles of anything between 60 and 100 degrees, with many settling on 90. The benefit of large included angles was reduced valve travel into the cylinder. However, Duckworth queried the validity of the established hemispherical (literally the shape of half a sphere) head arrangement favoured by many designers and opted instead for a so-called 'pent-roof' combustion chamber – that is, one which is more like the shape of the underside of a roof on a house - and a much narrower valve angle of 40 degrees. The inlet valves were on one side of the chamber roof, while opposite were the exhaust valves, with a single spark plug, in the middle, between them.

Although Duckworth had, at one stage, considered having inlet and exhaust valves diametrically opposed he abandoned this as not giving satisfactory turbulence for combustion as well as having many complications, not least of which would be the fact that there would have been an inlet and exhaust manifold to be accommodated each side of each cylinder head. This was something which both Repco and BMW would later discover for themselves, to their cost, whereas Duckworth's ability to think out solutions before committing resources to building engines ensured that diametrically opposed valves got no further than his drawing board.

Duckworth's conceptual analysis concerning narrow angle four valve heads changed the way Grand Prix engines were designed and he claims that, to this day, many present day engines utilise the same principles of air flow and flow management that he established over 30 years ago. Given that Honda had managed to produce 150bhp from its four valve 1 litre engine, Duckworth, quite understandably, reasoned that an output in excess of 200bhp ought to be easily achievable from a 1.6 litre power unit.

Walter Hayes: The visionary who enabled the whole Lotus 49 project to get off the ground. (Ford Motor Company)

In February 1965, five months before Duckworth began design work on the successor to the SCA, the principal 'customer' engine supplier in Formula 1, Coventry Climax, shocked the Grand Prix world when it announced its decision to quit motor racing. The company's board of directors had decided, for cost reasons, not to develop and supply a new engine for the 3-litre Formula which had been announced by the FIA, and which was to take effect from January 1st 1966. This decision had major ramifications for Chapman's Team Lotus, which had won one World Championship using Coventry Climax engines in 1963 and would go on to win that year's title as well, both with the by now established Jim Clark at the wheel. It left the team just ten months before the first race of the new Formula to find an alternative engine supply.

There were very few options open to Chapman, since it just wasn't possible to go out and buy a Grand Prix engine 'off the shelf' at that time. The alternatives seemed to be a link-up with an existing manufacturer, such as BRM or Maserati, or to develop a totally new engine, which obviously required significant financial investment. Although the former was probably a more realistic objective given the time span involved, the latter was by far the most attractive to someone who was always looking for that unique advantage over his rivals. Given the links between Lotus and Cosworth, it was hardly surprising therefore that Duckworth was one of the first people that Chapman approached. 'Colin came along and said did I want to have a go at a Formula 1 engine and how much did I think I needed to design and develop it? And, in our great scientific costings of that era, I asked my co-directors, Mike, Bill Brown and Ben Rood, how £100,000 [about £1.1 million at late 1990s values] sounded. Yes, we should be able to do it for that. So Colin said he'd try and see whether he could raise £100,000 from somewhere.'

Armed with this figure, Chapman went out in search of funding. His first stop was the industry body, the Society of Motor Manufacturers and Traders (SMMT). His approach generated nothing more than faint amusement, despite the fact that he was offering an opportunity to showcase the talents of the British motor industry to the world at large. The only serious offer of funding had come from David Brown, boss of Aston Martin, but, as Chapman recounted to Doug Nye several years later: 'He was very interested but ... virtually wanted to buy Cosworth.' This, naturally, did not appeal to Duckworth and his partners. Approaches to other so-called interested parties also drew a blank and things were looking pretty desperate by the time Chapman had his dinner with Walter Hayes.

At that time Ford was enjoying a tremendous period of success in motor sport, as Hayes explains: 'The sixties were a quite remarkable time for Ford in motor racing. I was running all Ford's motor sport outside of North America and Ford US was engaged in what they called a 'Total Performance' programme. The Cortinas were winning all the rallies that were going and doing well in saloon car races. We had the wonderful victories at Indianapolis and later on we had tremendous triumphs at Le Mans. So the world was young, people were young and the excitement for car ownership was still very much alive and therefore motor sport was what a sensible manufacturer did. And we really had an almost unalloyed period of success.

'There was a coincidence to it because in my earlier career I had edited a newspaper, now defunct, called the *Sunday Despatch*, and when I was casting around for a motoring correspondent with a difference, I persuaded Colin Chapman to write for us. He was not the world's best correspondent but we got to know each other and we became very good friends. And when I went to Ford in 1962 I could see that the friendship was going to have some greater meaning because we were both in the same line of business.'

Hayes' move to Ford was at the behest of the then Ford Chairman, Sir Patrick Hennessy, who had come under pressure to recruit some expertise from outside the company to join the Ford of Britain board of directors. Hayes had been appointed with a wide-ranging brief, including publicity and advertising, government affairs and, later on, motor sport.

It was the Coventry Climax pull-out that proved the catalyst for Ford's entry into the top flight, according to Hayes. 'The really seminal event which caused the change [in Ford policy regarding Formula 1] was the fact that Coventry Climax really felt that they had to come out of the sport. It was not a huge company and it was making too many demands upon them, so all of a sudden Formula 1 teams didn't have an engine. And the way Formula 1 developed was that it was, perhaps, not too difficult to build a chassis, or to design cars, or to put very good teams together, but they could none of them afford to do their own engine. Even if they had the expertise - Brabham was a bit of an exception - by and large they just couldn't afford an engine, they didn't have the technology for an engine.

'In 1965 I went out to dinner one night with Chapman and he did his usual thing and got hold of a paper napkin and started talking about an engine and how it could be done and he'd thought it out. In other words he'd build a car up to the engine and then the engine would bolt onto the chassis [actually a concept that had been tried by other teams before] and it would all be very compact and of course, starting from a piece of paper with Keith, we could design a very simple and compact engine.'

To Hayes, with Ford having conquered all other challenges, suddenly all the right elements appeared to be in place to go and do Formula 1 properly. 'All sorts of people had tried to tempt me to recommend Ford going into Formula 1 earlier. But it seemed to me that you started with your production cars and then you did the other cars that seemed to have relevance to you. In other words Formula 1 really was the icing on the cake. It was the pinnacle and if you are going to climb Everest, go and climb a few other mountains first. Also it was really going to be expensive and, while I had a motor racing budget, which was very satisfactory, spending on Formula 1 meant not spending on other kinds of the sport, which you might think were more important.'

Hayes agreed that Duckworth was a natural choice to design the engine. 'When it came to "Could we do an engine?", the "Who would do the engine?" was much less difficult because, by that time, we both could see that Duckworth was a genius.' Hayes also readily admits that being familiar with the people with whom he was going to be working was a big influence on his decision to go ahead. 'What really persuaded me that we should go was not just the idea of the engine but the whole thing seemed right. There was Colin, whom I trusted, and there were Keith and Costin, whom I trusted. It seemed to me that we could go and buy the very best people who would do their very best to bring it off. I went to Cheshunt with the American Ford executives to sign the Indianapolis contract with Chapman and I was able to observe very closely what Chapman did. There was a perfect example of somebody who had been asked to take on something that had never been done before. Nobody still realises or appreciates that there were 300,000 open mouths around that great oval when this funny little car came onto the track. They'd all had front-engined Offenhausers and so on and it absolutely revolutionised American racing. And I had seen what we had done with the Lotus-Cortinas. It was a sort of seminal car for Ford at the time. So, if somebody called Herbert Bottomley had come to me and said that Ford should do a Grand Prix engine, the chances are that we wouldn't have done it!'

Formula 1 was the last bastion of motor sport that Ford sought to conquer during the 1960s. Having won Indianapolis in 1965, and sewn the seeds for Le Mans success which would be reaped handsomely during 1966, the company sought to right this situation. According to Hayes, the international appeal of Grand Prix racing was a major attraction to the company because it created many marketing opportunities. 'The fact was, there is no question that - from the European point of view - Formula 1 was the pinnacle. Later on, when we won the Manufacturers' Championship with the GT40, if you had asked people to name the other races we won apart from Le Mans, they wouldn't have been able to tell you. And of course, another nice thing about it was – and it was more important than you think – that Grand Prix racing was very much an international arena. You had Austrian, French, Italian drivers and they were all terribly valuable to us in the countries where we were doing business. So, the internationalism of the sport was appealing. You could go on being terribly successful at Le Mans but what would that be? Once a year, everybody would say "Ford wins again".'

Following his meal with Hayes, Chapman was invited to dinner with Harley Copp, a motor sport enthusiast from Detroit in the US who at that time had just been appointed Vice President of Engineering at Ford of Britain. It was on this occasion that Chapman is reported to have blurted out to Copp: 'Look, you'd be missing out on the best investment you will ever make ... for £100,000 you can't go wrong.' As those who knew him will testify, Chapman in full swing was an extremely charming and persuasive character, and the simple logic of his argument won Copp over too.

The next stage was for Hayes to obtain formal approval for his proposal. 'I went to the British board - of which I was a member - and said I thought it was a good idea. I then went over for an annual motor racing review with the Policy Committee which is the top management group at Ford Motor Company in the States. In a sense, the programme had already been given the go-ahead by the board in Britain and really we were seeking a rubber stamp. But it was very important to get the approval of the Committee. At the end Henry Ford said to me "What is this engine going to do?" And I said "I think it is going to win some GPs and, if we are very lucky, we could even win the World Championship." He said "Well, all we can say is the best of luck to you."

'So I came back and we started the programme in mid-65. I swore everybody to secrecy because I thought that people would laugh at me because, here was Ford doing another V8. It was a time when people were talking about V12s and H16s but I had been in motor sport long enough then to know that the secret of motor sport is simplicity. If you can un-complicate things you are much more likely to be reliable and you are much more likely to win. Therefore I was, by upbringing and intellect, a V8 man. But I was terrified to tell anybody that was what I thought because it would have seemed so silly to people at that time.'

Since Duckworth had already begun work on the FVA (Four Valve, A-Series) and he had never before designed a complete engine (let alone a Formula 1 engine) from a clean sheet of paper, it was agreed that they would proceed with the design and development of the FVA and, if that proved successful, the Formula 1 engine would follow. This also provided for an early indication of the potential power output of the Formula 1 engine, since it would be virtually the Formula 2 unit doubled up, as Duckworth recounts: 'I was designing the FVA at the time. And the argument was, that if that went well enough and gave 200bhp or thereabouts, it seemed fairly reasonable that we could get over 400bhp from an F1 engine.' As the standard-setting single overhead camshaft Repco engine of the time was developing around 360bhp, this hopefully meant that the DFV would arrive with a significant power advantage over its rivals from the outset, even before any development work had been done to increase output.

Although a verbal agreement would probably have been sufficient for Hayes and Duckworth, legal eagles at Ford insisted on the two parties having some sort of formal contract. This caused a few problems for the Cosworth man, who had something of a reputation for being difficult where contracts were concerned. 'In the end, Ford said "Seeing as we don't seem to be able to write anything that you'd sign, perhaps you could come up with something and we can see whether we can sign it!" So I wrote out a contract which I thought gave them what they wanted, and gave me what I wanted, which was the rights to carry on selling it at my own discretion. And also, avoiding them being able to test engines and verify and check things and keeping the intellectual property rights to myself rather than to Ford.'

The contract eventually agreed to by all parties (and signed by Duckworth on June 23rd 1966) stated that Cosworth was to get £25,000 on March 1st 1966 (the date when it was originally drawn up), a further £50,000 on January 1st 1967 and a final instalment of £25,000 on January 1st 1968, with the initial £25,000 being intended to cover the design of the FVA engine. It provided for Cosworth to make available five engines by January 1st 1968 and that the engines would be maintained until the end of 1968. The choice of team was not difficult, with Team Lotus effectively having exclusive use of the engine for 1967, although the agreement actually stated that the choice was at Ford's discretion, while stating that Cosworth would be 'available in an advisory capacity if required.' This created the option for Ford to make the engine available to other teams should it wish to do so which, as it turned out, was a very prudent clause which probably prevented Grand Prix racing from being destroyed by an all-dominating Lotus-Ford partnership.

- First Person -
Walter Hayes on Colin Chapman

I could see that some people might think he was difficult at times. It was part of the nature of the man. He had such a clear conviction of the rightness of what he was doing that he did not want to be deflected. The point about Chapman, and where you had to fight to control him if you could, was that he would get a wonderful idea and then he would put it into action and before that idea had been properly worked out, he had a better idea. Then, before that idea could be worked out, he'd got an even better idea.

He was impatient with the pace of car development and wanted to get ahead all the time. So he could actually allow himself to be diverted. You can see it in the history of Lotus, because they always had one smashing season where they won everything and then a season when everything seemed to come apart. And then another sensational season and then it seemed to come apart again. It wasn't that he couldn't do it, it was that, in a sense, he was always pushing forward so fast that he didn't stop sometimes to get what he was doing right.

Chapter 2
A BRM Interlude

While all the arrangements were being concluded for the tie-up with Ford and Cosworth, there was the more pressing problem of which engine Team Lotus would use for the interim season of 1966 before the Ford contract came on stream. Since the options were pretty limited, the BRM H16 engine looked to be the most promising choice at the time, particularly as there was the prospect of using a bored-out version for Indy as well. As a result, an agreement was struck between the rival concerns at Cheshunt (shortly to become Hethel) and Bourne. To take the H16 unit, designer Maurice Phillippe produced a simple chassis, the Type 43, along with a virtually identical Type 42, intended for use with the 4.2 litre Indy derivative of the H16.

Phillippe had replaced Len Terry at the end of 1965, the latter having recommended him for the job on his departure. Like many others involved with Lotus over the years, Phillippe had worked for the aircraft manufacturer De Havilland during the 1950s and so had considerable experience of monocoque structures and hanging engines on them. In the mid-1950s, long before they were used in Formula 1, he had built an 1172cc special called the MPS featuring a stressed-skin full monocoque chassis, which he began racing in 1955. When the new Formula Junior came along, he built a car called the Delta in conjunction with Terry and another fellow De Havilland employee, Brian Hart. Eventually, he left De Havilland to join Ford as a project manager on Anglia 1200 development, all the time continuing to race at club level with considerable success. After he was offered the Lotus job, his first task had been to convert the Lotus 39 chassis – which had been designed to take the still-born flat-16 Coventry Climax engine – to carry a Climax FPF for use in the 1966 Tasman series.

The 43 tub was, in effect, the precursor to the 49 chassis, since it would employ the same concept of using the engine as a stressed member (on which the rear suspension was hung) attached to the monocoque by four bolts. In looks, its ancestry was clearly visible since it was quite similar to the successful Len Terry-designed Types 29, 34 and 38 Indy chassis, except that it did not have pontoons extending rear of the main bulkhead to support the engine. Since it was only required as an interim car for a single season, this was a relatively low-budget effort using as much existing Lotus componentry as possible.

Even before the season had started, it was frustratingly evident that Lotus and Clark would be in no position to defend their 1965 World Championship title. Due to development and reliability problems, the BRM engine did not appear in the back of a Lotus until the second round of the Championship, the Belgian Grand Prix at Spa-Francorchamps. Clark resorted to using a Lotus 33 - with a 1.5 litre 32 valve Climax FWMW V8 engine bored out to 2 litres - in the first round at Monaco and his team-mate, Mike Spence, a BRM V8 suitably expanded in capacity to 2.1 litres. The slightly larger capacity of Spence's car was a consequence of the fact that the BRM V8 could be stretched to 2070cc, whereas the Climax unit could not go out to more than 2 litres.

At Spa, the prototype Lotus 43-BRM, in the hands of Pete Arundell, failed to start after engine problems in practice, while at the next race, the French Grand Prix at Reims, it at least managed a handful of laps before retiring. The car (or rather the engine) was 'rested' for the next three rounds in Britain, the Netherlands and Germany while BRM engineers attempted to iron out various shortcomings. It reappeared for the Italian Grand Prix at Monza in September, this time with Clark at the wheel, having his first taste of three-litre Formula 1 power. The Scot promptly put his car on the front row of the grid and ran strongly in the race only to be thwarted by more niggling problems until eventually the engine blew with two laps to go.

In the next race, the US Grand Prix at Watkins Glen, Clark's legendary mechanical sympathy ensured that his car – rather gallingly for the BRM works team, using their spare engine – got to the finish and, with virtually all serious opposition eliminating itself, did so in first place! That this proved to be the only victory for the H16 engine in a two year career says a lot for Clark's ability, and rather less for that of the engine to run reliably for a sustained period of time … Predictably, the same engine lasted less than 50 miles of practice in Clark's car for the next and final round of the 1966 World Championship in Mexico, and gear selection problems eliminated the Scot from the race.

The Lotus 43 made its final bow in the South African Grand Prix in January 1967, when two examples were wheeled out for Clark and new driver, Graham Hill. This race marked the first appearance of Team Lotus's new star-studded line-up, the first time since the 1950s when two World Champion drivers or two absolutely topline drivers had lined up alongside one another in the same team. The Englishman had been lured back to the team from BRM by the promise of great things to come and with the help of some extra money from Ford as a carrot. This was due to Walter Hayes' desire to have two top drivers at the wheel of Ford-powered car, as he recalls: 'I really didn't want to be in the position where we had a number one driver and a number eight driver or anything like that, so I said that I would very much like to get Graham Hill and so we got him.' Ironically, even in his first race for Lotus, Hill was still maintaining his links with BRM by being powered by one of their engines.

The combination of Clark and Hill was one of immense experience. However, they were quite different characters, both away from the track and behind the wheel. Clark had been born into a reasonably affluent family and was a complete natural when it came to racing. Hill had been born into a relatively humble background and had literally pulled himself up by his bootstraps, including the spell as gearbox engineer at Lotus. During that time, and afterwards, he made quite an impression on Mike Costin: 'It was fascinating seeing Graham develop himself. I remember him saying to me, in the very early days "I reckon I could become World Champion". Now it was all very well for the likes of Hawthorn and Collins and Moss to start off in motor racing and have their Dad or somebody buy them a car, but Graham started without any of that. And I remember saying to him "And the best of luck, mate". And yet, he did it.' Although they were different, what the two drivers had in common was that they were extremely quick, were consistent, knew how to win races, and were both former World Champions. This was exactly the combination, which Walter Hayes sought in order to give Ford the possibility of maximum exposure results-wise.

Neither car finished the race in South Africa and it marked a sorry end to the link-up between two of the great Grand Prix teams of the early 1960s. Since the bulky 43s were not particularly renowned for their agility and responsiveness, Lotus reverted back to the more nimble 2-litre cars for the next race at Monaco. Originally, this was intended to have been the event where the 49s would make their debut. However, initial problems with the DFV engine put back the schedule somewhat, so that they would not make their first appearance until the following month in the Dutch Grand Prix at Zandvoort. As a result, Clark appeared in the Climax V8-powered car he had driven on occasion during 1966, while Hill continued to maintain his BRM links by driving a similar chassis but with the BRM 2.1 litre V8. While Clark set fastest lap and then retired, Hill soldiered on to finish a worthy second, continuing the run of good results in the Principality that he had enjoyed during the 1960s. Both drivers were particularly pleased when this race was over, since for the next race they would have their hands on their new secret weapon, and not a moment too soon.

The visual similarities between the Lotus 43 and the 49 are clear in this picture. In essence the 49 was a slimmed-down version of its predecessor. (Goddard Picture Library)

The complicated, unreliable and bulky BRM H16 engine. Installed in the back of a Lotus 43, it scored its only Grand Prix victory with Clark behind the wheel at Watkins Glen in 1966. (Goddard Picture Library)

Chapter 3
Doubling Up

Mike Costin undertook most of the testing of the FVA concept in the back of a Cosworth-owned Brabham. The unit he ran was actually a 1500cc FVB, exactly half of a DFV. (Cosworth Technology Library)

By the time the contract between Cosworth and Ford was actually signed and sealed, the first part of the agreement had already been completed, as Duckworth recalls with a laugh: 'The contract wasn't actually signed until halfway through 1966, by which stage we actually had the FVA running!' It first ran on a test bed in February 1966 and, by the summer of 1966, Mike Costin was pounding the tracks with an engine most people presumed to be an FVA installed in the back of a Brabham Formula 2 car. In fact, what appeared was not quite a 'full' FVA, as Costin recalls: 'It wasn't an FVA, it was an FVB 1498cc. The FVA was a 1600cc engine built for 1600cc Formula 2. But we built that one [the FVB] and that is the one I raced in our Brabham because it was half a 3 litre. That was the original 'FV' of the DFV'.

The new engine reached its 200bhp power target easily the first time it ran and proved immensely reliable. The blend of the cast iron Ford Cortina block with a light alloy cylinder head, double overhead camshafts driven by spur gears from the crankshaft and 16 valves proved a potent combination. Deliveries of the FVA to customers commenced early in 1967 and the engine became the one to beat as soon as the new Formula 2 format began in March of that year, with most customer units sold that year developing in excess of 220bhp.

Back in the spring of 1966, with the FVA running well on the test-bed and about to begin its programme of track testing, Duckworth's thoughts turned to the Formula 1 engine. Thoughts being the operative word, for he literally designed the complete engine in his head before committing pencil to paper. This was to be the first engine he had designed which would not base itself in some way on a proprietary Ford cylinder block, although of course his design was heavily influenced by the success of the Ford Cortina-blocked FVA engine, as well as many other proven aspects of that design.

Given his experience with the FVA and considering the regulations in force at that time, Duckworth was adamant that an eight-cylinder engine was the optimum format and not one with 12 or 16 cylinders. In time, of course, he was to be proved absolutely right: 'Colin had asked me if I thought I could make a 12 down to weight and get the fuel in the right place, for 200 mile races. I said that I didn't think it looked possible, but that we could make an eight down to weight. I knew how much power I could get because it was effectively two fours as regards the breathing and the tuning of the intakes and the exhausts. If you make a flat crank, it's the same as two fours. We knew a lot about fours, therefore we knew that we could get into the required area of power doing it that way. So I wasn't that happy to make a 12, but I was quite happy to make an eight. Also, I thought that, carrying your fuel for the whole race, the extra power of the 12 would be cancelled out by its extra weight and extra fuel consumption and the difficulties of designing a sensible car. And, as I am a 'total concept' man, I was there!

'The most important things in racing are the tyres

and a car's all-up weight is what kills them. If you can make a light car then you can use softer rubber with lots more grip without overheating and excessive wear. This allows you to see heavier and more powerful cars off. The original 3-litre Formula rules were for 500kg cars with no refuelling stops, which made a V8 the logical choice. A little later, on 'safety' grounds, weight was increased to 550kg, which made the choice of a V8 a little less wise. As soon as refuelling and tyre changing is allowed, more cylinders for more power becomes the correct choice. So to decide on how many cylinders you should use requires an analysis of the whole picture, all the regulations.'

The V8 unit was a monumental undertaking by anyone's standards, let alone for someone who was designing their first engine from 'scratch', as Duckworth freely admits: 'That was the first 'clean sheet of paper' engine I'd done, designing the block and the heads - well, all the bits - from square one. And there is a major element then, of determining what was the minimum length or the bore centres for a satisfactory crank, satisfactory valve gear, satisfactory liner arrangements and everything, and you have got to proceed with schemes for all these to determine the most suitable one.'

Although the original idea had been to 'double up' the FVA to create a Formula 1 engine, it wasn't quite as simple as it sounds. For one thing there was the difference in capacity per cylinder (400cc for the FVA versus 375cc for the DFV). So just what was carried over from the FVA to the DFV? Duckworth takes up the story: 'Well, we learned that our four valve general system worked. From the start, we were obviously going to make the V8 a single plane crank, because then it meant that you had two four cylinders and therefore the exhaust systems that we had on the FVA, we just had two of them on the DFV. Also the tuning followed the same principles, the firing orders down the banks were the same as two fours. So the exhaust tuning was there and the intake lengths and things like that were there as the rpm we were talking about when we started was very similar.

'Unfortunately Formula 2 was 1600cc, whereas Formula 1 was three litres, therefore two 1500ccs. I was keen on trying to keep the pistons flat at that time and, therefore, in order to do the flat piston and to maintain the compression ratio with the reduction in swept volume, I closed the valve angle up from 40 degrees included on the FVA to 32 on the Formula 1. The valves and springs and all the valve gear bits were the same, the bore was the same but the stroke was reduced. However, other than bits of valve gear and principles, there wasn't much carry over.'

The first task was to provide Lotus and its suppliers with basic details about the size and shape of the front and rear faces of the engine, as Duckworth recounts: 'We had a few meetings on principles. I, very early on, produced the gearbox arrangement and the rear face of the engine. When I'd decided what the front should be, I told them, defining the shape of their rear monocoque. I also said how I thought my flexible plates should go into a slot and the skins should come round there and a plate, with a few rivets around it. So I was fairly responsible for the way in which the engine and chassis were integrated.

'The first drawings to be done were the back end of the engine. Colin proposed to use a ZF box, because he was keen on ZF boxes whereas I was quite keen on Hewland boxes. I thought that the ability to change all the individual gears in the Hewland was a great advantage over having a collection of boxes with different crown wheel and pinions to give you different ratios. So DFV1 is the drawing of the back face of the engine and the starter, to allow ZF to try and design a box.' Walter Hayes was also in favour of the ZF box due to the previous relationship that Ford had developed with the company. 'We had started the [Ford] GT40 with a Colotti. And then we went to ZF and they were very slow in developing, but we did eventually get a terribly good ZF box for the GT40. Therefore, ZF might have seemed the natural choice because it had proved a box that could run well with big Ford V8s.'

Cosworth records show that this first DFV drawing, 'Flywheel housing and starter details – rear view' was dated 15th June 1966, or just under 12 months before the DFV would make its race debut. Subsequent drawings were also aimed at helping Lotus finalise their suspension concepts and the marriage of tub to engine, with the second one being 'Rear suspension pick-up points' and the third 'Front bottom engine mounting'. Both these drawings are dated 8th July 1966 and were the result of consultation between Duckworth, Chapman and Phillippe at a meeting held in June at Cosworth. At this stage, there were no detailed design drawings of the rest of the engine. It was not until late August that Duckworth began design work in earnest.

Working from home, which was not far away from the Cosworth factory but far enough to allow him the peace and quiet he needed to think his ideas through, Duckworth literally worked all day every day for several months, looking at all the possible alternatives for different aspects of the engine. Sometimes, a little nudge was needed from his fellow director and partner to keep the

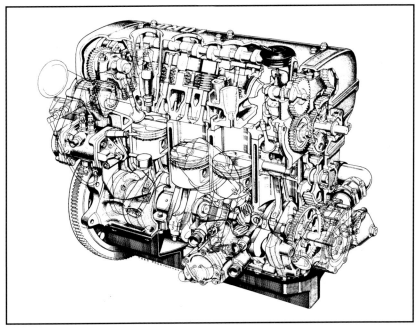

This superb cutaway drawing by Theo Page shows the 1600cc FVA engine, the first to be produced under the Cosworth-Ford partnership brokered by Walter Hayes. With four valves per cylinder, and double overhead camshafts driven by a series of gears from the crankshaft, it was the forerunner of the DFV. (Ford Motor Company)

project on track, as Costin recalls. 'The main input I had at the design stage was to try to make him make up his mind on the bolt spacing for the cylinder heads. Because he'd got, as Keith always has, about six different schemes with all the pros and cons, and I remember saying to him. "Hey look, you've got to come to a decision because you're going to hold up everything".'

Costin perhaps tends to understate the roles played by him and other key personnel, such as Ben Rood, at Cosworth during the days when the DFV was being designed, as Duckworth is at pains to point out. 'I am innovative and numerate, whereas Mike is an incredible intuitive engineer and he just knows whether something will work. I've caused everybody who is close to me to think very hard how things work. I believe you should never do sums before you understand what the problem is, you should just sit and try and work out what is happening.

'The Mikes and Bens of this world were very quick to pick up the analysis and thinking about problems. Ben on machines, particularly. Again, Ben was a machine shop man, and very intuitive. He just knows G-forces and can feel what's happening. Mike knows how much metal will bend, twist, shape under what load and knows whether things will work. And we had the communication system that over the phone I could say, "I've thought of a way of doing this. Picture this: have you got a pencil? Well, we've got this here, that there, and so on" and Mike would draw what he thought I was saying. And then say "What I've got is this, that and the other" and I'd say "Do you think it's any good as an idea?" and he would perhaps just check on another bit and say whether he liked it or he didn't think it was very good.

'If he didn't think it was very good, or thought there was something wrong with it, he would say. He wouldn't in a lot of cases, explain why he didn't like it but I reckon he was right about 96% of the time and if he didn't think something would work or something didn't sound right I would then go through it all and generally speaking, sooner or later, I'd find something wrong with it! Therefore, at all stages, if I was doing something that was innovative or not standard, I would put it to him as to whether he thought that it was sound or not and get his answer. We are wonderful joint operators because we are sharp enough to see that you spend a fair amount of time wrong in this life - or up your whatsit as we normally refer to it - and we can both point out when each other is "up it!"'

Although, quite fairly, the vast majority of the credit for the DFV engine goes to Duckworth, there were three other men working with him on the detail design, and without whom he could not have completed the job in the time he did. 'I did the schemes - and they were fairly rough - of the sump and the auxiliaries. Peter Stemp detailed the sump and all the auxiliaries were developed from my rough schemes and detailed by Mike Hall. I did a lot of the final drawings. The block and head drawings - the complicated major bits - were my final detail drawings. Most of the other drawings were schemed in a fairly detailed manner by me, but were untidy and therefore they were re-drawn and dimensioned and everything else by Roy Jones. He was the detail draughtsman, really, and drawing office manager as well.' The other member of the team who had an involvement in drawing the engine was Paul Sherman-Erpe, who drew several parts relating to the oil system.

Although he naturally wanted an engine that would enable him to keep the car right down to the weight limit, Chapman had no input on the detail specifications of the engine. In fact, as those who worked with him testify, his interest in and knowledge of engines was limited relative to other parts of the car, as Walter Hayes explains. 'The interesting thing about Chapman was that I never ever really felt he was a sensational engineer. But nobody had more ideas than he had. When he first sketched cars for you, he would draw suspensions and other things in some detail but, wherever the engine was he'd just draw a box and write "engine". And I don't believe he ever really had very much detailed technical knowledge of engines, but he was a superb innovator.' This is an assessment supported by Mike Costin. 'No, he wasn't an engine man. This is borne out by all the cars in which he put engines on their side and didn't make adequate provision for oil scavenging or breathing ... not very nice cars any of those!'

However, Chapman was a truly great concept man, as Duckworth points out. 'Certainly the most brilliant conceptual engineer that I've known. He wasn't a detail designer and never had any ideas of limits and fits. And for mechanical bits, lots of things require clearances to be about right to work, and you've got to make pieces to within reasonable limits to make certain the greatest fits and differences will still operate. And he never seemed to take the disciplines of detail design on board.'

Mike Costin believes that Chapman's other great strength was his ability to persuade people that what he was saying was right and to persuade them to do things, often against their better judgement. 'The one quality Colin had, at which he was better than anybody, is that he had the ability to convince you that you should do what he wanted. He could convince you that you should carry on working, or start working, or do something, without your questioning it. Not on technical terms, but by engendering enthusiasm.'

As well as being very persuasive, Chapman had a unique ability to divert the course of a debate if he felt he was losing the argument, something which in his early dealings with the Lotus boss, used to infuriate Duckworth, who invariably took the 'bait' hook, line and sinker: 'He could marshal his arguments in a superhuman fashion. You'd be having a discussion about something and he would say, "Well this follows" and I would look doubtful, without having been able to sift through my mind why I didn't like what he said. At that stage, realising that he might be out-argued, or that there was something fallacious about what he'd said, he would change the subject and throw in a red herring and I would lose the main argument while chasing this useless red herring. And I'd find that I'd agreed to something that I shouldn't have done. I gradually learnt that all I could do was to say "Yes Colin, I've heard your arguments, I need time to think about them. I'll let you know my view in the morning". And, with that proviso, there wasn't a problem.'

Chapman and Duckworth were both acutely aware of the benefits of not carrying any excess weight, as the latter relates: 'I think he was totally wedded to getting cars down to the weight limits. We'd had discussions very early on about this. We were both people who subscribed to the view that lightness is next to Godliness and you must be down to weight. Racing is about the use of rubber - and your rubber is killed by the weight on it - apart from the poke and a few other things, so you need as light a car as possible. Therefore, if you have got a weight limit you should be on it. And it did strike me that a V8 was lighter, smaller and - with the knowledge that there was at the time - you needed to get the fuel on the C of G [Centre of Gravity] of the car. Therefore a short engine looked valuable for that because, with the width of the V8, you weren't sensibly able to get [fuel] tanks down the side of the engine, nor were you able to get any sensible structure round a V8. So the decision was made that we would make it cantilevered off the back of the monocoque and all the suspension be just directly coupled onto the back of the engine, and that the loads would go through the engine.'

Having made the decision as to the broad size and shape of the engine and the fact that it would act as a stressed member, the next problem which Duckworth faced was to come up with a realistic method of mounting the engine: 'I was left with this as a problem and gradually, thinking about it and sketching what size the engine was, it struck me that the total shape of the engine was reasonably round looking, or at least the pick-up points were. Having tried to design beams to go across to take the loads into the block, I didn't like those and they were heavy. I gradually convinced myself that you could probably hang the rocker covers onto the extensions of the monocoque in line with the skin and then, picking it up on the bottom at the sump as well, that gave you the biggest base possible and conformed to the outside shape of the monocoque.' This way the two bolts at the bottom into the sump took the shear, while the thin steel plates from the rocker covers allowed for the small amount of heat expansion of the engine.

A piece of solid aluminium was positioned between the bottom of the monocoque and the front of the sump. It was bolted to the sump of the engine and then had two holes for the forward-facing bolts which went into the threaded holes in the bottom of the 49's tub. The reason for the presence of this component was so that, in the event of an accident, this would absorb the loads created and possibly break, rather than breaking the sump, which would write off the entire engine casing. It was a neat and very effective solution, which never ceased to amaze Duckworth. 'We spent all the first year looking at this fragile-looking piece, machined from solid aluminium, on the grounds that we thought it would break. It never did break and it stayed there for years [even] when the cars got more pokey and heavier and had more downforce: it went on for ever did this bracket, which was designed as a "break here" - seldom did anything crack, nor did the bolts into the front of the sump.'

Duckworth's engine layout effectively dictated the final shape of the rear of the monocoque: 'When I had got the height of the engine, and where the rocker covers were, and where my fish plates would go and pick up on two bolts, that defined the shape of the back of the tub. So, whilst the H16 was bolted in, that had lots of structural members to try and get from the substantial pieces of the engine into the chassis, I didn't like that when I looked at the weight of my members and the space they took up. So I arrived at my flat engine at the front to go along with the flat sheet of the back of the monocoque. I would claim to be a general design engineer, rather than specifically an engine designer. At that time, I had little idea about anything, I suppose, so was very bold! It was before my confidence had got dented!'

Having settled on the overall nature of the engine package, Duckworth then set about detailing specific aspects of it and liaising with Chapman and Phillippe on things like where the rear suspension would pick up on the engine. 'They came back and said they wanted to hang the suspension units off the head and that they would like members going here, there and everywhere and holes here! So that defined the holes at the bottom of the bellhousing face and across the top to pick up supports for the wishbones. Then they wanted to come forwards with some tubes along the engine to give it fore and aft stability. The gearbox had to be able to be changed, it had to be able to come off easily. Obviously, the driveshafts on the gearbox were five or so inches behind the back face of the engine and, therefore, the pick-ups for the rear suspension were hung back five inches behind the engine, it was all hung backwards. We co-operated on the design of the rear of the cylinder head to pick up some of their members and then I found some suitable spaces along the sump line to give it the fore and aft stability - some diagonals coming from the pick-up points forwards. So there was co-operation on that kind of thing, but they had to go where I thought we could get them, having already put too many things in their way! So some of them looked a bit "hung on"!'

With the engine such a vital part of the structure, it was important for Duckworth to produce a design that would be strong enough to withstand the various forces that would be imposed upon it: 'We decided very early on that with all the torsional loads and the fore and aft loads to the engine going in through the rocker cover, the design of the cam carrier and the rocker cover and all that - with too many bolts on it, looked like a Rolls-Royce engine! - had to be more structural. In fact, it was grossly over-engineered when you see how everybody now hangs the buggers in by any means without any due care and attention! But that's experience, whereas I was trying to do it by doing the odd sum as to what I thought the loads were in chassis and what they were likely to be when hitting kerbs gently, and so on. So I made a fairly good structural attempt at the head and the cam carrier and I had these mountings on the back with some reasonable holes in so that the suspension loads could go into those as well.'

As if the stresses and strains of virtually single-handedly designing a new Grand Prix engine from scratch were not enough, Duckworth also took responsibility for designing the engine components to facilitate their manufacture in limited volumes: 'I designed all the fixtures to machine the bits! During the time that I was designing bits [for the engine] I was also designing them for the machinery on which they would be made – our existing machinery – so that they would be economical to produce. I tended to choose several ways of tackling a problem and would quite often throw out a solution which was inherently the best, but which would require such accurate methods of manufacture that I couldn't see us ever making more than one of them! Therefore, I would choose a solution that, whilst not so ideal from a functional point of view, was much easier to manufacture. That way, we had a chance of making a few of them!'

Eventually, the time came to transfer everything from drawing board to three-dimensional form. 'I oversaw the first building' says Duckworth. 'I can remember marking out the original block castings. There is a picture of George Jackson and I looking at the first cylinder block. And I was deeply involved with the manufacturing as to how we would make everything.' The building of the original DFV was done by George Duckett. He was the builder of Cosworth's prototype engines and had built the first FVA. The second engine builder Cosworth took on, after Bill Pratt, Duckett moved up from Edmonton when the company relocated to Northampton. As he recalls, it was a slow process. 'I started building it towards the end of 1966 and actually finished it in early 1967. The first one took an awful long time to be built because we were waiting for each component as it arrived.'

It was around this time that Ford put its foot down and insisted that Autolite spark plugs (Ford Motor Company's in-house brand) be used instead of Cosworth's preferred choice, Champion. Brian Melia was a Ford rally driver at the time but, as the rally budget had been slashed to pay for the DFV, he was at something of a loose end and this is how he came to be involved with Autolite, as he explains. 'Henry Taylor [then Competitions Manager of Ford in the UK] realised that Ford were going to hand the Cosworth plug kudos to Champion because Ford of Britain didn't have anything to match what Champion could offer in terms of service and backup. The Autolite company in the States had a full range of far superior plugs to Champion, but of course they were there and nobody over here knew one end of a race plug from the other.

'Taylor suddenly put two and two together and thought, 'well, this guy's an engineer, he's going to be sat on his backside at Boreham doing nothing for a year'. I suppose it went through Walter Hayes and Si Clifton, MD at Autolite. They pointed out to Autolite that they were going to have their noses put in the mud, Ford's brand new engine and they are going to have to use Champion. So they sold the idea to Si Clifton and within a matter of days they said "Go and report to Si Clifton and discuss what you think should happen", and they basically sat me down at a desk and said "Write your own budget" which is unheard of at Ford. Because of course nobody at Autolite knew what the hell it was about.'

Such decisions at the top level of Ford did not go down well at Cosworth, as Melia recalls. 'Cosworth were a little bit anti-Autolite. All their previous projects had been very much helped by Champion and they were very *au fait* with the Champion personnel and the range of products, and so on. I didn't really say to them "You will use Autolite", obviously it came down through management. But I was *persona non grata* at Cosworth in a way because I represented the companies that were imposing unacceptable conditions!'

The key characteristics of the engine that emerged were that it was a 90-degree V8 of 2993cc capacity, with a single (flat) plane crank. The flat plane crank was chosen because it offered advantages in terms of a more simple exhaust system and allowed the lessons learned with the FVA's layout to be carried over to the DFV. A pent roof combustion chamber with four valves per cylinder was used, as per the FVA, with a valve-included angle of 32 degrees, shallower than the FVA. Double overhead camshafts - therefore four camshafts in all - were driven via a series of spur gears from a half speed compound gear in the Vee of the engine which was driven by another compound running off the crankshaft gear. The half speed shaft also provided drive, via a triangular belt drive, for ancillaries such as water pumps (two, one on each side), fuel pump, oil pressure pump and two oil scavenge pumps. It also drove yet another shaft for the distributor, alternator and fuel injection metering unit, all of which were in the Vee of the engine.

It was with bated breath that the first DFV was set up in the test room at St James Mill Road. However, as Duckworth recollects, the assembled throng need not have worried: 'It was late March or early April. It gave enough power as soon as we ran it. It gave 408bhp on the first run or something like that. So, on that, we were in business, weren't we? At a modest RPM too ...'

From a clean sheet of paper to running engine had seen a period of less than twelve months elapse. However, although the initial running on the test bed showed that the engine had comfortably exceeded its horsepower target, according to Duckett there were several problems that manifested themselves early on. 'The first time we got it on the test bed, it blew up! The valve springs broke.

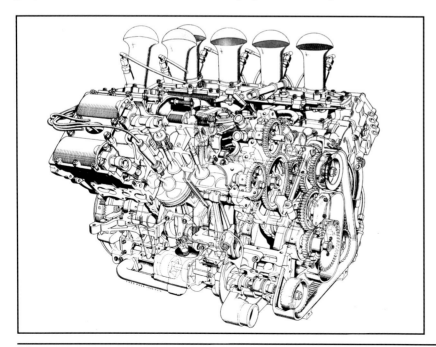

Another of Theo Page's beautifully detailed cutaway drawings, this time showing the DFV. The front face of the engine is to the right, with the complex train of gears running from the crankshaft nose to the camshafts and the triangular belt drive providing drive to the auxiliaries at the side of the engine. (Ford Motor Company)

I think they were motorcycle valve springs from the [BSA] Gold Star, which we only used for a very brief period. After that we got the springs made for us by a German company. We used to have a standard Ford 1500 or 1600 engine and we'd build it with these springs instead of the standard Ford springs and run it on the test bed for about ten minutes. If the springs lasted, we would take them and use them in the DFV! And we had to do that until we'd got 32 springs for each engine. So we had to run the engine four times to get the required number of springs. We only ran it for a short period - these springs used to break in the first few seconds and if they didn't break in the first few seconds they were good!'

As Duckworth recounts, other problems encountered in early tests after the valve springs were sorted out were serious enough to postpone the debut of the engine, which had originally been targeted for the Monaco Grand Prix in May: 'What we had to do was to stop it spewing oil out and then, a bit later, spewing teeth out. I hadn't made reasonable provision for getting oil and air down from the cam boxes into the sump and air up to the heads from the crankcase and, basically, we filled the heads with oil. We also needed to improve the methods of getting air back into the head because oil and air came down from the head but we couldn't return any air, and it all went wrong!'

The answer to the problem of oil in the cam boxes, as Costin recalls with a laugh, was neither high-tech nor elegant. 'Having identified the problem, Keith had gone through it all and said "Now, we know we're in the s**t, we know why we're in the s**t, but we haven't got the room to get out of it. Therefore you could only tackle it by some "bear's arse" piece of kit!' With very little room to play with, Duckworth was forced to adopt a rather inelegant solution. 'The first engines had these big pipes up the side and into little tin boxes fabricated on the top [of the cam covers]. Mike and I were in at weekends, welding up and testing engines. That was the thing that stopped us appearing one race before we did.'

Duckett recalls another ingenious piece of improvisation that helped to separate oil and air in the early engines. 'The oil flow through the crankcase, the scavenging, was a problem in the early days. The more oil you've got in the engine, the more drag you've got. And you had to separate the oil from the air. So we put pan scrubbers in the sump! We opened them up into a sausage, like a tube and laid them in the sump, where the pump drew the air out of the sump and it broke the bubbles down so that it separated the air and the oil. Later on we had a centrifuge [to separate oil and air] but the original ones didn't have that.'

The subject of [spark] plugs and their performance soon cropped up, although initially this was caused by them being in short supply, as Brian Melia recalls. 'They were having problems with plugs. I went up to look at them, and found there were only three sets of plugs in the country. They were American made, and nobody had bothered to do anything other than give them a couple of sets of plugs. All the test work had been done on two or three sets of plugs that were very secondhand by the time I got involved. The production was very small in the States, so it wasn't a question that they'd got hundreds on

the shelf. It took me a while to fill the pipeline, to have plugs coming over.'

Hayes had intimated the existence of Ford plans to embark on a Formula 1 engine programme to Denis Holmes, then motoring correspondent of the *Daily Mail* at the Geneva motor show in March 1966. Apart from this, nobody knew anything more about the engine, suggesting that Hayes' intentional shroud of secrecy surrounding the project had worked. 'I had wanted to delay the announcement as long as possible to avoid the huge guffaws that would arise when people discovered it was a V8. Although that sounds silly to say now, at that stage none of the motor racing journalists were going to think anything [of a V8].' The stunned silence among the assembled throng of pressmen when the engine was first revealed suggested he had been right to do so. However, by that stage, as Hayes says, it was only a matter of weeks before they would find out if it was any good or not. 'I didn't really mind too much. After all, we were racing in June, so they didn't have too long to guffaw really.'

The Ford Cosworth DFV, as launched to the press in Ford's Regent Street, London, offices at the end of April 1967. With its aluminium front cover off, the triangular belt drive to the auxiliaries on either side of the engine is clearly visible. (Cosworth Technology Library)

The key men at Cosworth poring over an early DFV, identifiable by the absence of aluminium boxes on the cam covers: (left to right) Bill Brown, Keith Duckworth, Mike Costin and Ben Rood. (Cosworth Technology Library)

These rather ugly fabricated tin boxes and piping needed to be added to the early engines to alleviate breathing difficulties. By 1968, the problems had been solved and they disappeared. The odd routing of the pipes was to get round the top of the cam covers and behind the exhausts.
(Cosworth Technology Library)

Remarkably, the DFV was wider than it was long. This shot, of a late-1967 specification engine with the ignition box between the inlet trumpets, clearly demonstrates its width, as well as the 90-degree format of the engine and the triangular belt drive to the auxiliaries.
(Cosworth Technology Library)

Duckworth inspects the exhaust exits of one of the early DFVs. With the left-hand cam cover off, the double overhead camshafts, with holes for the plugs between them, are clearly visible.
(Cosworth Technology Library)

Doubling Up

In order to marry up the two banks of four cylinders, they were offset, with the left-hand bank placed further forward than the right-hand. As a result, the aluminium plates that slotted into the tub on each side had to be different lengths. On the far right is the water header tank or swirl pot. The pipe running out of it to the bottom of the shot takes water back to the radiator and the chrome pipe entering it comes from the head (there were two - one from each head). (Ford Motor Company)

The official launch of the Ford Cosworth DFV engine took place on Tuesday 25th April 1967 at Ford's showrooms in Regent Street, London, with Hayes unveiling it in the company of Ford Chairman Sir Patrick Hennessy, Colin Chapman, Keith Duckworth and Graham Hill. It was a momentous occasion for all concerned. Immediately after the launch, this engine, numbered 702, for 1967, engine number two, was handed over to Team Lotus, who had the first Type 49 chassis awaiting its installation.

- First Person -
Walter Hayes on Keith Duckworth

He could appear to be a rather wry and cynical chap and could be hugely sarcastic if he wanted to be. He has always spoken in a rather measured way, you can see that before Duckworth talks he thinks. But the finest thing about him was his extraordinary truthfulness because, if you'd say to him "We think we'd like to do this" he'd pause and say, "Well, it won't work". And there was this kind of absolute truth about the guy all the way through.

When he retired, I sent him a picture of Isambard Kingdom Brunel and I wrote on the bottom 'To a better engineer' and that's what Duckworth has always been. He had this extraordinary ability to see what was wrong with things, which I think is as much a part of being an engineer as seeing what is right."

Chapter 4
The Simple Bracket

Right-hand man: Maurice Phillippe (centre) was the designer/draughtsman responsible for interpreting Chapman's concepts (as well as coming up with many ideas of his own) and turning them into practical realities. Here he is seen talking to Jim Clark (right) and Firestone representative Brian Hayward, while Chapman is talking to mechanic Bob Dance in the background. (Jose Phillippe Collection)

What emerged from the Team Lotus works two weeks after the delivery of the first DFV was not a radical car but one which, by necessity, was simple in nature. The reason for this was that the engine was, as yet, unproven, and Chapman wanted a reliable, straightforward chassis which would give him no problems – a 'bracket' which would allow the team to focus all their energies on sorting out the DFV. Sound logic, but, as it happened, this did not turn out to be true, for the chassis let the team down far more times than it should have done during 1967.

The Lotus 49 was the first Formula 1 car conceived by Colin Chapman which did not involve compromises attributable to trying to mould his design around an existing engine. Early on in the project he had a very good idea of the kind of car he would like to build but, of course, this very much depended on the final engine format that Duckworth, in consultation with Chapman, chose. While Chapman was responsible for the overall concept of the car, the detailing work was left to his design chief, Maurice Phillippe.

In June 1966, with the design of the Type 43 and the Indy programme out of the way, the attentions of Phillippe and Chapman had turned to their 1967 Formula 1 challenger. Following the meetings that month with Keith Duckworth which had produced his template for the front of the engine and its overall profile, the two Lotus men then embarked on a series of discussions about the new car before committing pencil to paper. These discussions mostly took place while travelling, either in Chapman's Elan or aircraft, on their way to races or meetings with Cosworth and ZF. To illustrate the demands placed on their time by other projects for 1967, they were also designing a new Formula 2 car (the Type 48), new Lotus Cortinas and considering various Indy alternatives! This was in addition to having to prepare for the move to new headquarters at Hethel aerodrome in Norfolk - a world apart from the existing site at Cheshunt - and all the disruption that would bring with it …

The first official drawing for the Type 49 is dated July 14th 1966. It is titled 'Proposed gearbox layout for 1967 F1 (discussion only)'. However, it seems certain that the layout was finalised soon after this, for time had to be given to allow ZF to get started on detail design of what would be a completely new gearbox in order that the company could meet the production schedule. Although he produced a number of sketches in the intervening period to help firm up ideas in his head, it was not until early in December 1966 that Phillippe sat down again at his drawing board to begin full-scale design work on the Lotus 49. After drawing general layouts and detailed body lines, a scale model for wind tunnel tests was drawn up, but it is not clear whether such a model was ever actually produced. Geoff Ferris, who did the

The Simple Bracket

drawings for the model, recalls that they used the MIRA tunnel for the 49 but is not sure whether this might have been later on in its life.

The layouts also set out the fuel capacity of the car, which was originally planned to be 42.25 gallons, consisting of two side tanks each of 14.75 gallons and a seat tank of 12.75 gallons. Range at six miles per gallon was calculated as 254 miles, while at five miles to the gallon it fell to 211 miles. On a fuel-related topic, the main problem connected with the chassis design was the late decision to locate the electric fuel pump in the chassis, rather than as an integral part of the engine, as Duckworth had hoped. This was due to lack of space in the engine. However, the solution – embedding the pump in the rear left-hand side of the tub – proved to be a neat one, although it was to cause problems later on.

Most of the drawings produced in December and January concentrated on the front and rear suspension, probably the most crucial details other than the tub and the engine/gearbox. According to published accounts at the time, the first orders for parts were placed on February 9th 1967 (three months before the car's planned debut in the Monaco Grand Prix) while the manufacture of the tub - by Team Lotus fabricators including Roy Franks and Colin Knight - began on February 17th.

The build programme was co-ordinated by Chief Mechanic Dick Scammell, supported by Leo Wybrott and the two mechanics assigned to what would be Graham Hill's car, Dougie Bridge and Dale Porteous. Scammell had joined Team Lotus in 1960, going straight on to working on the Lotus 18 Formula 1 cars and had progressed up through the mechanics' ranks in the Formula 1 team in the intervening years. He was renowned as an efficient organiser, a good progress chaser who ensured that the mechanics always had the parts they needed at the time they needed them, and it was largely this role that he fulfilled with the 49. 'Maurice did all the design and I managed the programme. I liaised with the people who made the chassis, who were in the same shop as the mechanics and with Cosworth, and pulled the whole thing together in the end I suppose; and with a group of people put the car together.'

Considered and quietly-spoken, New Zealander Wybrott had joined the Formula 1 team in 1964, with responsibility for gearboxes/transmissions and had gradually moved more into the role of general mechanic, working with Mike Spence in 1965 and then Jim Clark in 1966. Dougie Bridge had joined Lotus on the production line around 1958/59, again working on transmissions, then moved to working on the development of the Lotus Elan and from there onto making racing parts for monocoques. In early 1963 he joined the race team and went straight into Formula 1, working alongside the likes of Jim Endruweit and David Lazenby. Porteous, another New Zealand 'import', had joined the Team at the end of 1966, so 1967 was to be his first full season.

In essence, what emerged from the Potash Lane works at Hethel once again bore a strong family resemblance to previous Lotus cars from the pen of both Phillippe himself and his predecessor Len Terry. However, because the engine it had to accommodate was substantially smaller and lighter than the BRM H16 unit,

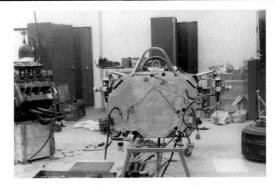

Chassis 49/R1 approaching completion in the Team Lotus works at Hethel, awaiting the first DFV. The two holes for the bottom engine mounts can be seen at the bottom of the bulkhead, with the two slots for the top mounts at the top left and the top right of the bulkhead. (Dick Scammell Collection)

Another view of 49/R1, showing the mounting brackets for the upper and lower rear radius arms, which were to cause so much trouble during 1967. The pipe which carried water to the engine can be see running in the channel at the bottom of the tub, nearest the camera, while oil pipes were run down the other side. (Dick Scammell Collection)

Mounted on the dash behind the Lotus trademark red steering wheel were, from left to right, a combined oil pressure/temperature gauge, rev counter, water temperature gauge, fuel pump and ignition switches, fuel pressure gauge and ammeter. (Dick Scammell Collection)

Initially, ventilated discs front and rear were adopted, although these were later replaced by solid units. Also visible are the four-pot calipers, which provided stopping power, the fabricated steel rocker arms, short, stubby rocker arm fairings and lower front wishbone mounting points. (Dick Scammell Collection)

the whole chassis was scaled down in size compared to the Type 43.

The basic monocoque was skinned in 18 gauge Alclad aluminium, riveted to mild steel bulkheads. It commenced at the rear mounting point for the front suspension rocker arms and terminated behind the driver's back, with a flat face to match that of the DFV

Lotus 49

The sub-frame which was built into the front of the tub provided the mounting points for the front part of the rocker arms and the lower front wishbones, while smaller forward-facing tubes attached to it held the radiator in place. Also visible are: the pipe running from the 'bird-bath' oil tank above the driver's knees to the main tank; the bungee cord holding the oil tank in place; and the end of the water pipe (blanked off) which would be connected to the radiator. (Dick Scammell Collection)

engine. A footbox riveted to the front of the tub housed the pedals and there was a subframe attached to this which acted as the forward mounting point for the suspension cast alloy rocker arms, the rear mounting point being located protruding outwards from the front of the tub within a stubby fairing. Inboard Armstrong coil/damper units were adopted, with an anti-roll bar completing the set-up. The subframe also carried tubes holding a Serck combined oil/water radiator, while a triangular oil tank (the shape being designed to channel air out sideways as it exited the radiator) was mounted between the radiator and the tub and held fast by nothing more sophisticated than a bungee cord! Water travelled to the engine via a 'step' at the base of the right side of the tub and came back via a return pipe mounted in the air-stream on the left side. Both 'outbound and return' oil pipes were channelled down the lower left side in the 'step' of the tub. An auxiliary 'bird-bath' oil tank was located above the driver's legs in the scuttle in case oil consumption of the new DFV engine was heavy.

Independent suspension was adopted all round. The lower half of the front suspension consisted of a split wishbone arrangement, with the lateral link mounted at the base of the subframe and the radius arm locating on a mounting point within the tub, the two items being pinned to each other close to the upright. The upper half followed the traditional Lotus pattern of fabricated steel rocker arms, the front of which also located in the front of the subframe with the rear being fixed to the tub. These rocker arms were attached to the inboard coil/damper units. Cast magnesium uprights were designed in such a way that the ventilated Girling discs used were positioned out in the airstream in order to assist with cooling, while four pot calipers provided stopping power. Wheels were 15 inch diameter Lotus cast magnesium of eight inch wide design, the deep rim necessitating the use of a special tool to reach the wheel nut, which was also designed at the same time as the rest of the Type 49 components! Steering was Alford & Alder rack and pinion type while Firestone tyres completed the package.

Fuel storage arrangements had been altered slightly since the original layout drawings were completed. The fuel was now housed in two 15 gallon pontoons on each side of the driver's body and feet which fed into a 10 gallon seat tank behind the driver, situated between the seat-back and rear bulkhead where it met the engine. This gave a total capacity of 40 gallons. A non-return valve allowed the main tank to be filled from the side

The 'hi-tech' method of mounting the triangular oil tank with 'bungee' cord is clearly visible! The tank was shaped like this to help air exit from either side of the nose. The oil temperature sensor can be seen at the bottom right of the tank. Front uprights were shaped so that the brake discs could be out in the airflow. (Dick Scammell Collection)

The Simple Bracket

Having just arrived at team Lotus, the DFV is admired by (left to right) Dick Scammell, Dougie Bridge and Leo Wybrott. On the left hand-side of the engine (the front-face) are the four holes for attaching the aluminium plate which held the engine to the tub, while the bolts on the rear face are ready to accept the ZF box. On the side of the engine is, to the right, the oil pump with a small diameter tube leading into the main chamber. To the left of the oil pump is one of the water pumps, with a pipe leading into the head. (Dick Scammell Collection)

tanks under acceleration and stopped the fuel making the return journey under braking. Fuel was added via a small round filler mounted in the scuttle above the driver's knees, where there was also a small round access hatch forward of the fuel filler, with a NACA duct in it to provide some cooling to the cockpit. However, the hatch's principal purpose was to enable mechanics to adjust the brake balance. On both sides of the tub towards the front, there were hatches permitting access to the most forward parts of the side pontoon fuel bag tanks. Inside the cockpit there were panels with indentations to give elbow room for the drivers which, when removed, also offered access to the side fuel pontoons, while the seat tank was reached via a round hatch in the seat back.

The steering wheel was offset to the left to provide gearchange space on the right-hand side, while the battery was located beneath the driver's knees. A simple dashboard display included a rev counter, a combined oil temperature and pressure gauge, a fuel pressure gauge, a water temperature gauge, an ammeter to show that the alternator was charging, and two switches, one for the ignition and the other for the fuel pump. This was protected from the elements by a Lotus twin-piece air deflector windscreen. The whole ensemble was completed by the trademark Lotus red leather steering wheel, while behind the driver's head was a rather low and ineffectual-looking two-piece roll hoop.

As Keith Duckworth described earlier, the Ford Cosworth DFV engine was attached to the tub via four mounting points, consisting of two bolts running from the sump into the base of the monocoque and two thin steel plates which slid into a recess just beneath the main

The new DFV sits on its trolley awaiting installation. In the background, mechanics prepare Jim Clark's 2-litre Climax V8-engined Type 33, just back from the Tasman series, in readiness for the Monaco Grand Prix. (Dick Scammell Collection)

Lotus 49

Chapman wanted to remain faithful to ZF, who produced the 5DS12 box for the 49, while Cosworth thought Hewland would offer greater flexibility. In the end, Cosworth were proved right! The slab-sided design of the box also suffered from flexing, causing many crown wheel and pinion failures. (Dick Scammell Collection)

skin of the tub at about the height of the driver's shoulders. Rear suspension was hung on tubular subframes that were bolted on to the DFV engine at the rear. These subframes quickly became known variously amongst the mechanics as the 'fir trees' or 'pine trees' because they had so many protrusions at different angles! Forward and rearward loads were taken by upper and lower radius rods that attached at one end via neat brackets to the rear corner of the tub, and at the other to the cast rear upright. The upright held the rear disc brakes and was attached at the bottom by unequal length wishbones to the 'fir tree' and at the top via a link to the upper part of the 'fir tree.' This part of the fir tree was also the mounting point for the top end of the coil spring/damper unit, the bottom end being mounted at the base of the upright. In all, there were some fairly significant loads going through the fir tree. As it transpired, this was to prove something of an Achilles Heel on the car during 1967.

Power was relayed to the wheels via the ZF 5DS 12 syncromesh gearbox, Chapman staying loyal to the German concern despite Duckworth's preference for a Hewland box, and the mechanic's misgivings about the unit's inflexibility in terms of ratios due to being a fixed ratio box. The quick-change drop-gears on the tail of a Hewland or Colotti-type gearbox – which enabled final-drive ratios to be juggled speedily and easily 'in the field' – were not available on the ZF box. Instead, the entire unit had to be stripped and rebuilt to change ratios, something clearly not suited to dusty or sandy paddocks where the mechanics invariably had to work. Former Team Lotus Racing Manager Jim Endruweit explains that to avoid such a situation, the team used to carry at least two spare gearboxes per car, each with different ratios. 'We used to have what was considered to be the

This upside-down shot shows an early mock-up DFV supplied to assist Team Lotus in solving installation problems before a 'live' engine came along. The water pipe from the radiator can be seen coming in from the left. Water was pumped into the head from both sides, reaching the other half of the engine via a pipe in the base of the sump (visible in this picture). Next to the water pump is the oil scavenge pump. The small diameter pipe leading out of it returned oil to the other side of the engine through another crosspiece in the base of the sump and thence via a pipe down the left side of the car to the radiator for cooling. (Dick Scammell Collection)

The Simple Bracket

most suitable ratio in the cars and then a ratio up and a ratio down.' The box also had a mechanism on it for preventing missed changes, which meant that drivers could not go 'across' the box by accident. Final drive was via BRD tubular halfshafts, which took power from the differential through ZF's latest sliding splines. This neat arrangement enabled the splines to take up travel caused by suspension movement. The car's ignition system was mounted on a tray on top of the gearbox alongside an early type of Lucas electronic rev limiter, designed to keep the engines from going above 9200 rpm.

The entire car had only one piece of bodywork, this

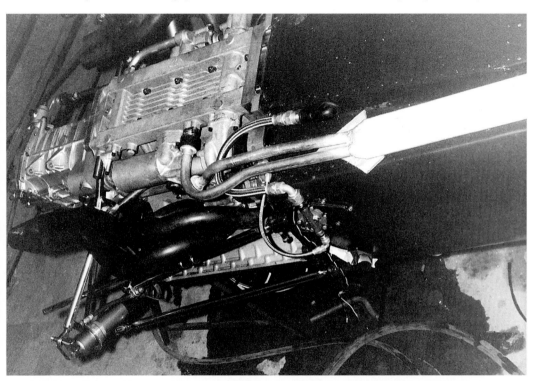

Looking at the other side of the engine (the left if looking at it from the rear) the pipe returning oil to the radiator can be seen looping out of the base of the engine, while the pipe bringing oil in from the oil tank is alongside, feeding into the pressure pump. Also visible is the feed from the seat fuel tank to the fuel pump, which was embedded in the side of the tub. The pipe exiting the pump runs to the metering unit. (Dick Scammell Collection)

Oil returned from the engine down a pipe into the combined oil/water radiator (nearest the camera) while cooled water was sent to the engine along the pipe on the other side of the tub. The pipe carrying oil to the engine can be seen exiting the bottom of the oil tank. The sculpted aluminium ducting was intended to assist airflow exiting the radiator but was abandoned as it made access to the oil tank and removing the radiator too complicated. (Dick Scammell Collection)

being the glassfibre nose cone shrouding the water/oil radiator and oil tank. The end result was a very sleek, purposeful racer, resplendent in Lotus's traditional shade of British Racing Green with a yellow stripe down the front.

As a final comment on the build programme, the original Type 49 costings make fascinating reading -

Figure 1: Type 49 Costing (Formula 1)

Design	2,407}	
Jigs & patterns	1,980}	(a)
Overheads	1,400}	*These charges remain*
Parts discarded	530}	*regardless of number*
	£6,317}	*of cars built*
Parts to build	2,070}	
Labour	1,003}	(b)
Sundries	436}	*These charges accrue*
Gearbox	474}	*for each car built*
Overheads	400}	
	£4,383}	

Assuming that our normal commitment is 3 cars

$$\text{Cost of each of 3 cars} = \frac{6,317 + 3 \times 4,383}{3}$$

This system absorbs all the 'a' expenses over three cars, any additional cars would cost only 'b'

= £6,489
add engine
======

So each of the three 49s would cost just under £6,500 to design and build in 1967, excluding the cost of the engine (which was provided free of charge by Ford). At today's values, this is around £67,000 (US$107,000). The total cost of the project was £19,466 (or £201,000/US$322,000 in the late 90s). This was the equivalent of around one fifth of Ford's total investment in the DFV engine and was split between fixed costs of £6317 (£65,000/US$104,000) and a per car cost of £4,383 (around £45,000/US$72,500) which would remain the same regardless of how many additional cars were built. A small price to pay indeed for a world-class Formula 1 car …

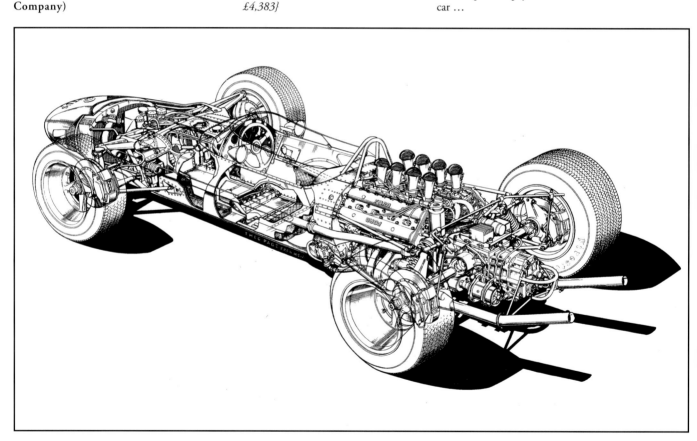

Legend revealed: Theo Page's cutaway shows the simplicity and compact nature of the 49. (Ford Motor Company)

Chapter 5
Consummating The Marriage

It was on Monday May 8th 1967, the day after the car's intended debut race at Monaco, that Mike Costin had the honour of being the first person ever to drive a Lotus 49 and a DFV engine. These first runs were very much 'systems checks' and consisted of nothing more taxing than going up and down the runway at Hethel. At such an early stage it was deemed prudent not to involve any Team Lotus drivers and, in any case, Graham Hill was more than occupied during the month of May in trying to qualify his recalcitrant Lotus 42F for the Indy 500. As well as Costin, Chief Mechanic Dick Scammell drove the car that day. Amazingly, despite being present at Hethel that day, Colin Chapman was not there to witness the first run of his brainchild.

The first problem to materialise was with the clutch, something that came as no surprise to Costin: 'We blew up the clutch, so I sent all the bits back to AP [Automotive Products]. They'd virtually insisted that we used this new twin clutch plate of theirs. I'd said "This is stupid, it is never going to survive". But we put it in, and when it fell apart after just a few times up and down the straight, I put it in a Jiffy bag and sent it back with a compliments slip saying "I told you so!" Then we had no clutch, so we dashed over to where the works Lotus Cortinas were, swiped a clutch off one, put it on and that was the clutch used on the DFV for the next 10 years. It was the Cosworth twin-plate assembly with the peg drives, which we used to knock up and sell - a good bit of kit, lasted for donkeys' years!'

As well as trouble with the clutch, the first day's running also revealed a potentially worrying handling problem, as Costin recounts. 'When I was driving at Hethel, I couldn't keep the bloody thing straight and I was saying to Dick "Look, this is utterly undriveable!" and they were all taking the p**s, saying "Oh, you've never driven anything with this much power!" I said "Look, this is set up so that it's got toe-out on the back-end or something, it's hopeless". When the clutch blew

up and we took the car back into the workshop and took the clutch out, I said "Here, let's have some bits of string and some straight edges". What did I find? Two and a half inches toe-out! Now, the car would've been undriveable with half an inch toe out ... So we fixed that, and then of course it was fine ...'

After flying back to the Cosworth works overnight in Chapman's plane for a few bits and pieces, Costin resumed testing duties the following day, this time at Snetterton and once again without Chapman in attendance. In those days, the old circuit was still in use, using a loop that left the current circuit at Sear corner, down a straight to a hairpin and back to rejoin at the Esses. To those familiar with the circuit today, the old track is still visible, being the site for a regular market. This section of the track was concrete and rather uneven, due to the fact that there were gaps between each concrete section.

Very early on in the day's testing, a serious problem emerged concerning the brackets on which the upper

On the first day the 49 runs in front of the press (May 23rd 1967), Dick Scammell rolls out of the race shop as the mechanics look proudly on. It had already tested in secret as early as May 9th but the mounting points for the radius arms had proved too weak and had to be reinforced by a 'top-hat' structure inside the tub, a process which took some time to complete and delayed testing of the car. (Ford Motor Company)

Lotus 49

This rear view shows the tubular sub-frame, to which the top links, coil spring/damper units and unequal length wishbones were attached, and which was to give so much trouble during 1967. Chunky tubular BRD drive shafts would later be replaced by solid torsion shafts. (Ford Motor Company)

Dapper Graham Hill and designer Phillippe share a joke as they admire the new 49 on the runway at Hethel. Hill was always immaculately dressed, charming, approachable, a confident (and hilarious) public speaker and a real 'man of the people'. He and Clark, also always well turned-out but a little less approachable, were a sponsor's dream. (Jose Phillippe Collection)

radius rods - taking acceleration and braking loads - were mounted. Costin takes up the story: 'Accelerating out of the hairpin through the gears, every time you went over one of the joints between the pieces of concrete, you could just get on the verge of wheelspin - it would go over one and just spin a bit, and I thought "That's really going it." After a few laps seeing what the temperatures and the pressures were like and how it was going, and looking at it and saying "Thank gawd it's still in one piece, no rods coming out the side!" somebody noticed these brackets were kinked in.

'Dick said, "That's you putting bloody wheels up kerbs" because the chicane was there and I would put the odd wheel up the kerb now and again. And I said, "If the bloody car can't take that, you want to take it back now and start on it again, because it'll never survive the first

Hill and Chapman look very pleased with themselves as they pose before press photographers with their new 'weapon'. (Ford Motor Company)

The Lotus 49 was a sleek-looking, simple design. As the years went by, it gained a variety of aerodynamic appendages and these tended to spoil its basic good looks. (Ford Motor Company)

Consummating The Marriage

The key protagonists in the Lotus 49 project: from left, Maurice Phillippe, Keith Duckworth, Graham Hill, Mike Costin and Colin Chapman. Only Jim Clark, in tax exile, is missing. (Ford Motor Company)

practice". But in my own mind, I knew when it had happened because the biggest thump that those radius rods were getting was over those concrete joints, because I'd got as much throttle on as the car would take and you could feel every one. It was the top [radius rod] that caused the trouble all year and very unkindly, as well. Maurice had drawn a very nice bracket, but it was slightly offset and the bloody thing kinked. Dick [Scammell] will tell you how many times it failed, because he was there at every race, making it stronger and stronger and stronger!'

In fact the problem was so serious that all testing had to be suspended while an additional internal member was fitted under the skin to provide extra support to the front edge of the bracket where it kept kinking in to the tub. This, as Dick Scammell recalls, was not such an easy job given that they did not want to have to dismantle the tub and start building all over again: 'That was a big piece of work. We had to put some "top hat" sections inside the monocoque, which, with the car finished, was really quite difficult to get access to.' Although, initially, this seemed to have cured the problem, the radius rod mounting brackets - particularly at the bottom - would be a source of much aggravation throughout the life of the Mark I Type 49s, with the result that the design for mounting the lower radius rod was radically changed for subsequent variants. In addition, there were also problems with the front suspension radius arm pick-up points, which were being pushed in and required strengthening.

As well as chassis problems, several niggles became evident with the engine, particularly with the Lucas OPUS (Oscillating Pick-Up System) transistorised ignition system, which as Duckworth says was: 'Usually

known as 'Opeless, which it nearly was! The OPUS gave us misfires in quite a few ways and we had to do a lot of development on the system. As it got hot a coil got bent. It was made out of nylon and we had to strap it to a post. We had a problem with the bits which weren't made well enough. The drum warped on the OPUS trigger and, if it got too far away from the other bit that would cause a miss, or it could rise or lower a bit. In the end we had to re-machine the Lucas bits that we used in our own distributor base within better limits, to make it work.'

Having reinforced the radius rod brackets, which took quite some time because of the difficulties involved, the team felt sufficiently confident to bring in Hill to try

His head dwarfing the rather puny-looking roll hoop, Hill sets off down the runway at Hethel for his first taste of the Lotus 49 and DFV power. (Ford Motor Company)

Lotus 49

'The Guvnor' tries hard not to mow down any photographers as he takes his own creation for a blast down the runaway at Hethel. (Ford Motor Company)

Ready to go: With the formalities of posing for the press out of the way, Hill prepares for his first run on a track in a 49, at the Snetterton circuit. Looking on are (left to right) mechanics Dale Porteous and Dougie Bridge, Chief Mechanic Dick Scammell, Cosworth men Keith Duckworth and Mike Costin and Team Lotus Chief Designer, Maurice Phillippe. (Phipps Photographic)

Consummating The Marriage

On track at last: Hill gets to grips with the 49 at Snetterton. (Dick Scammell Collection)

The verdict: apparently Hill is complaining to Chapman and Duckworth about the abrupt power delivery of the DFV engine, although neither of them looks particularly concerned! (Phipps Photographic)

and get some more serious development testing under their belt. Consequently, on May 23rd he was given his first opportunity to take the wheel, after posing for official Ford photographs showing the car and the design team involved. After a disappointing Indy, the Englishman was itching to get his hands on the new car. Teammate Clark was unable to take any part in the testing programme due to having been forced into tax exile, sharing a flat in Paris with journalist Gerard 'Jabby' Crombac. Instead, all he could do was to look longingly at pictures of the car and read press reports, as the Frenchman recalls: 'We knew how it would look and we had seen a picture of Graham testing it, but that was all.'

After testing sessions, there was the inevitable Chapman job-list that had to be got through before the cars could be despatched for Zandvoort. Later, Maurice Phillippe recounted that there were 84 items on the list, which was apparently issued three days before the team was due to depart!

Meanwhile, the build programme on the second car was progressing. Despite the fact that the first car was completed almost four weeks before the Dutch Grand Prix, it would still be a tight thing to finish chassis 49/R2 for Zandvoort because of the re-inforcements necessitated by the kinking of the tub. Although it was completed in time, the car hadn't even turned a wheel by the time it was loaded up on to the transporter …

Chapter 6
When The Flag Drops ...

Not only had his car never turned a wheel but also Jim Clark had never even sat in it until the Thursday night before the Dutch Grand Prix. He travelled up from Paris with Crombac in his Elan and, when they arrived in Zandvoort, they went straight to the garage where Team Lotus were based to have a look at the car. The first official practice session began next day, on Friday morning. Consequently, there was no time for making major set-up changes, it was just a case of making the best of the situation, as his mechanic Allan McCall recalls: 'There might have been a little fiddle around, but basically he drove the car exactly as Chapman presented it.' While this might have been an almost insurmountable problem for many drivers of his generation, Clark's unique ability to adapt stood him in good stead.

In the first session, he was eighth fastest, while team-mate Hill spent most of the session having the pedals adjusted to his liking and wound up 14th. After a change of gearbox on both cars during the lunch break - to use an alternative final drive ratio - both drivers began to get on the pace, moving up to fifth and sixth in the first afternoon session. However, it was in the final Friday afternoon session that Hill really stretched the legs of his 49, when he put in a time nearly two seconds quicker than

The packaging of the DFV was very neat and compact, with only the ignition components (on top of ZF box) and fuel pump (sunk into tub just forward of lower rear radius rod mounting point) not integral parts of the engine. The installation of the engine into the chassis was also very simple, with only four bolts holding them together. Here Clark enthuses about the 49 in the pits at Zandvoort during the first day's practice. (Ford Motor Company)

On the second day, Clark brought his 49 into the pits, complaining that there was something wrong with his car and refused to go out until the mechanics had found it...which they did. (Ford Motor Company)

Looking down on Clark's 49/R2 from above the pits at Zandvoort reveals its sleek lines and just how short the DFV was in relation to the rest of the car. Chapman is squatting at the left rear wheel, while at the front a Goodyear man checks out the opposition. (Cosworth Technology Library)

his previous best. Only Dan Gurney, in the V12 Eagle-Weslake, could live with his pace, his being the only other car with an engine developing anything like the sort of horsepower being pushed out by the DFV. Meanwhile Clark was struggling with a car that didn't feel right to him. Although he could not put his finger on what might be wrong, he was certain something was wrong. When the mechanics checked the car over, they could find nothing other than a small amount of play in the front wheel bearings.

A big crowd turned out to witness the final practice session on Saturday afternoon and it was a truly epic battle, with Hill and Gurney trading fastest laps. The Eagle driver was convinced that his time of 1.25.1 would be sufficient for pole, but in the dying moments Hill went out and took half a second off this: his 1.24.6 being fully 6.2 seconds inside the existing Formula 1 lap record for the circuit! The Lotus-Ford combination had well and truly arrived ...

Clark's session was ruined by a continuation of his previous day's problem. He had only run a few laps before he came in it was found that a ball-race in the right-rear hub had broken up and split the hub carrier. More than thirty years on, in his book *Jim Clark* (Dove Publishing, 1997), Eric Dymock quoted from an interview with Clark later that season, in which the Scot described his conviction that something was wrong and his determination not to continue until the problem was found: 'The night before I told them there was something wrong with the car, but they didn't find it. The next day I was sure there was something wrong with the car and I just brought it in and I just wouldn't bloody well drive it until they'd found it. They found it all right. You could hardly see it but it was there.' To be able to feel or sense the ball-race breaking up before mechanics could detect it was quite amazing and was another demonstration of Clark's tremendous empathy with and sensitivity to the cars he drove.

After a new hub assembly had been fitted, there was barely any of the session remaining and he was unable to better his Friday time. The result was that, while Hill would line up on the front row alongside Gurney and Jack Brabham, Clark was back down on the outside of the third row, only eighth fastest. It transpired that the problem on Clark's car had been caused by the bearings not being properly spaced apart. As a result, Dick Scammell recalls that an all-nighter was necessary in order to rectify this on both cars. 'The rear hubs were

Lotus 49

Judgement Day: seconds after the start at Zandvoort, Hill is fractionally ahead of fellow front row members Jack Brabham (far right) and Dan Gurney (centre), with Clark (just above Gurney's head) already making progress from the third row. Following Hill are Rindt and Rodriguez (Cooper-Maserati), Amon (Ferrari), Surtees (Honda) and Stewart (BRM). Just visible in the haze is the race official who was still standing on the grid when the flag was dropped. (Cosworth Technology Library)

wrong and Keith [Duckworth] stayed up with us all night adjusting the spaces between the two bearings in the rear uprights by hand with files to get the preload right, and it did take all night!'

The outcome of that race, on Sunday June 4th 1967, is etched into the book of motor racing legend. Race day dawned overcast but dry, although the track was wet from overnight rainfall, and a strong wind was making the sand swirl across the dunes in which the track is set. At the start Hill out-dragged his front-row companions and led them into the first corner. He simply streaked away from the rest of the field more in the style of his team-mate, so that by the end of the first lap a clear gap was visible as he flashed past the pits, his advantage measured at 1.2 seconds on the stopwatches. However, he did not continue to pull away, instead maintaining a

The second corner of the first lap, and already Hill has pulled out quite a lead over his pursuers, Brabham (Repco-Brabham), Rindt (Cooper-Maserati), Gurney (Eagle-Weslake), Amon (Ferrari) and team-mate Clark. (The Dave Friedman Photo Collection)

After a subdued start (by his standards, at least) Clark moved steadily through the field and into the lead after Hill's retirement. Here he leads Rindt, Hulme (Repco-Brabham) and Amon. (The Dave Friedman Photo Collection)

steady two second gap over the chasing pack, which comprised the Repco-Brabhams of Brabham and Hulme, Rindt's Cooper-Maserati, Gurney in the Eagle, Clark (who at this stage was still getting used to his car) and Amon's Ferrari.

Alas, this fairy tale beginning was destined not to last. On lap 11, just as he was coming onto the start and finish straight, Hill felt something go in the engine, switched off and coasted in to the pits. It transpired that some teeth had broken off of one of the cam-drive gears and his race was run. However, salvation was at hand for, by lap 15 Clark, now more comfortable with his car, had moved up to second position behind Brabham, and on the next lap he passed the Australian with ease. He led for the rest of the race, driving well within himself and the limits of the car, but still pulling away in a controlled manner from Brabham who vainly, doggedly and gallantly pursued his rival. At the chequered flag, Clark was 27 seconds ahead of 'Black Jack', who had eased off somewhat towards the finish once it was clear he was not going to catch the Lotus. To cap a great day, Clark ended up with the fastest lap of 1.28.8. The fact that it was 'only' two seconds inside the existing lap record and some four seconds slower than Hill's pole time illustrated that he had actually been driving well within himself.

It was a dream debut, something which Lotus and Ford had obviously hoped for but dared not expect. Hayes' decision to take Ford into the highly competitive arena of Formula 1 was instantly vindicated. After only one race, half of his prophecy about the likely achievements of the DFV had been fulfilled. The car's victory generated publicity for the company across the world, the value of which alone probably exceeded its total investment in the project. As the national anthem played and Clark stood on the rostrum flanked by a proud Keith Duckworth and equally proud Colin Chapman, it was clear that the writing was well and truly on the wall for rival teams and engine manufacturers ...

This was a fact that had not escaped Tony Rudd, BRM's engine man at the time (who would later go on to work at Lotus). He went up to Chapman after the race, shook his hand and said "You've really set the cat amongst the pigeons now!" This clearly indicated his immense regard for the new performance standard which Lotus and Cosworth had set.

Not wishing to place undue pressure on the Lotus team for their first appearance with the new car and engine combination, Walter Hayes and Harley Copp had adopted a low-profile presence at the meeting, flying out in a light aircraft from Southend to Schipol airport, where they were met by a car and then on to Zandvoort. After the race they had slipped quietly away again, their only celebration being to stop on the way back to the airport and buy two bottles of beer!

In the Team Lotus camp, the mood was ecstatic. Jabby Crombac recalls Clark's post-race comments: 'He was delighted, very happy. He knew it was a winner. He wasn't sure he would win but he was hoping to. After practice he said it was a potential winner. But he thought it was really too early to say, and that the car would have some development problems so the win came as a surprise.' His drive at Zandvoort was also the first evidence of a change in race strategy on the part of Clark. There is a general consensus of opinion, among those who knew him that - with the introduction of the Lotus 49, which he regarded as a fine but brittle piece of machinery - his attitude towards racing changed. Instead of demoralising his opponents by crushing them (although it was still a pretty comprehensive victory), he chose to do merely enough to win and not to push the car to its limits. This ensured that the still-new DFV engine was not put under too much strain, but also that the unproven chassis was given an easier ride.

At first sight, this strategy seemed to have enabled Clark to bring the car home in 'as-new' condition, for it appeared to have been a trouble-free, all-conquering debut for both car and engine. In fact, his 'softly, softly approach' was probably the only thing that enabled him

Clark, looking pleased with the outcome of his day's work, completes his victory lap. (Ford Motor Company)

On the Zandvoort rostrum, Clark tells Duckworth and Chapman how he did it. (Ford Motor Company)

Lotus 49

to complete the full race distance since, unbeknown to casual onlookers, the car had been in far from perfect fettle at the finish. As well as clutch trouble which he experienced during the race, and fading brakes, it transpired that his engine had also suffered from almost the same problem as Hill's with the timing gear.

In a typically realistic and down to earth assessment of the situation, Duckworth admits that it was little more than good fortune (and, although unsaid, probably also Clark's tremendous mechanical sympathy) that his engine made it to the end of the race: 'We were lucky to win first time out. Graham's was an older engine [no. 702, compared to Clark's slightly later 703] and that lost two teeth next to each other and lost the drive to a cam, therefore it ground to a halt. On the way back to the garage after the race, I could hear a severe rattle coming from Jim's car. It turned out that it also had two teeth missing but fortunately there was one standing in between and therefore the drive was maintained, but it was obviously fairly touch and go with finishing ...'

Looking back now, Chief Mechanic Dick Scammell remembers Zandvoort primarily as an exhausting weekend. 'We got a huge blast [from Chapman] for arriving late, although we'd been working all sorts of hours. We then ran the cars and they were all right. But because we'd had to work all night before the race I remember it as a very hard race meeting. Obviously it was a pleasure to have got it all there in the end, and it was nice to win. But that was all a bit taken over by sheer relief that the meeting was over. I really hadn't got the energy by that time to get terribly excited!'

- First Person -
Allan McCall on Jim Clark and Graham Hill

"Absolutely chalk and cheese, but Graham could be just as fast on a given day as Jimmy, there's no doubt about that. I've got great admiration for Graham and his performance but some people can operate on a different plane and Jimmy was one of those. I imagine Nuvolari and Fangio were the same.

He carried so much more speed into, and through, a corner. Jimmy never used brakes really, he was just magic: he could damn near run a season on a set of brake-pads and Graham couldn't get through a weekend! He could go quicker, easier, than the other guys. He had his way of rushing out in the races, mainly because he always thought it was dangerous to race with people - his thing was "get away from the maniacs!""

Clark and Hill: different ... but as fast as each other on their day. (The Dave Friedman Photo Collection)

Chapter 7
Reality Bites

Duckworth had just two weeks to investigate the timing gear problems and, as there were only two engines at the time, this meant that Team Lotus could not do any testing during this period either as the engines had to be returned to Cosworth for stripdown and thorough examination. Thus, they rolled up at the Spa-Francorchamps circuit in Belgium with the cars virtually unchanged from Zandvoort.

One area where the team had sought to make improvements was in the delivery of power from the DFV, which Clark had described as like suddenly having two engines instead of one. According to Duckworth 'It had a big hole at about 5000 rpm, or so, and then came out of it rather sharply. When it went from being a hole to power on, there was a colossal horsepower change or torque change in very few rpm and it made it a handful.' As a result, Chapman conceived - and Maurice Phillippe drew - a neat little progressive throttle pedal linkage that would help Clark cope with this problem. Although it was not ready for Spa, it was fitted for the next race in France and used until Duckworth and his engineers produced a cam linkage at the throttle slides that had the same effect.

Additionally, the 'bird-bath' oil tank above the driver's legs was discarded for this race, since the DFV had turned out to consume relatively small quantities of oil during a race. In fact, its principal contribution at Zandvoort had been to leak, dripping oil over Clark's overalls!

The 8.76 mile triangular circuit, nestling within the hilly Ardennes region, was ultra-fast, with the first gear La Source hairpin being the only respite from a near flat-out blind around the rest of the track. It was therefore ideally suited to the most powerful engines in the field - the Ford Cosworth DFVs in the two Lotus cars and the V12 Weslake in the Eagle.

Conditions were good for Friday practice, which was held in the late afternoon between 4pm and 6pm. Clark emerged fastest from this session, his time of 3.29.0 fully nine seconds inside the 1966 pole time set by John Surtees in a Ferrari! However, only 2.2 seconds slower was Gurney's Eagle, while Hill was a further 1.7 seconds adrift in third.

During practice both 49s appeared with bib spoilers around the nose cone, the intention being to counter any negative lift that might be induced by the high speeds attained on the circuit. However, although they were effective, they made the back end unstable since there was nothing to balance the downforce created at the front of the car, so both drivers elected to have them removed for the race. The future significance of such aerodynamic

In the Spa paddock, Duckworth discusses engine matters with Harley Copp, Ford of Britain's Vice-President of Engineering, while Chapman (right) does his best to look disinterested. (Ford Motor Company)

Lotus 49

The master at work: Clark, seen in practice, tackles Spa's daunting Eau Rouge. (Ford Motor Company)

Below: Clark got a flyer at the start of the Belgian race and here he leads Rindt, Stewart (BRM-H16), the Ferraris of Parkes (car no. 3) and Amon, Surtees (Honda), Rodriguez, Brabham and the eventual winner, Dan Gurney. Hidden behind the cloud of smoke over the grid, mechanics are working frantically to get Hill's car started. (Ford Motor Company)

Hill exiting the La Source hairpin during practice. Note holes in nosecone, where bib spoilers were tried briefly the previous day, black adhesive tape to stop air-deflector windscreen lifting at high speed and the round water pipe running along the bottom channel of the tub, which contrasts with the original flattened design still on Clark's car. (Ford Motor Company)

aids had not been truly understood, as Clark's mechanic Allan McCall describes: 'It definitely worked, but at that stage none of us understood aerodynamics, we weren't thinking that way. It was going to cure a little bit of understeer but Jimmy was one of those drivers who drove with an awful lot of understeer, so it wasn't a problem for him.'

During Saturday practice, Clark further reduced his time to a stunning 3.28.1, while neither Gurney (who equalled his Friday time) nor Hill, who was experiencing engine maladies, were able to improve. However, none of their rivals improved sufficiently to knock them off the front row, so the sharp end of the grid remained unchanged. The day of the race was dry, hot and sunny. After the warm-up lap, the cars formed up on a dummy grid prior to rolling forward to their start positions. At this stage, disaster struck, for Hill's engine turned on the starter, coughed and then died and he had to watch helplessly as the rest of the field moved forward.

Although in his autobiography *Life at the Limit* (William Kimber, 1969), Hill put the likely cause of the problem as being that the terminals on the battery were not done up, Keith Duckworth says the real cause was that, in the stress of the start, he failed to follow the correct starting procedure: 'In order to guarantee starting of the engine you had to crank the engine and then turn the ignition on. If you turned the ignition on and then cranked it, it would do quarter of a turn and fire, backwards, and not start and another quarter turn, and, if you kept the starter engaged, it cooked itself and then wouldn't turn the engine and there was no hope. It was thought that the battery was flat, whereas in fact it was no longer able to turn the engine. But, by the time they'd changed the battery it had taken so long that the starter motor had cooled. I think that perhaps a little lecture was given on how you start it ... and it was then started, but he was a long way behind.'

While the mechanics were working away frantically to change the battery on Hill's car, the rest of the field was already well into its first lap. After 1 1/4 minutes, he was on his way again but it was not long before the leaders came through for the first time. Clark was already pulling away steadily from his rivals, led by Jackie Stewart in the H16 BRM. By the end of the second lap, his lead was up to five seconds. However, Hill was back in the pits again, this time for good with terminal clutch problems.

As Clark sailed serenely on his way, it seemed like a repeat of Zandvoort. By the seventh lap (quarter distance) he was a whopping 17 seconds ahead, a lead which was increasing at around two seconds a lap. However, at the end of the 12th lap, Clark suddenly pulled into the pits with the core blown out of one of his spark plugs. It took some two minutes for the mechanics to change the offending plug, so that when he rejoined, he was well down, in eighth place. Three laps later and he was in with the same problem, this time dropping him to a lap behind the leader, who by now was Dan Gurney in the Eagle. Due to retirements, the Scot was back up to seventh place by three-quarters distance and when Pedro Rodriguez went out four laps later with a broken piston, he was back in the points.

This is how it stayed, a disappointing outcome given the promise shown in practice. Clark didn't even have the consolation of setting the fastest lap of the race, which went to Gurney with a time of 3.31.9, only seven tenths slower than his best practice time. However, in the circumstances, Clark did well to finish at all, Dick Scammell's notes recording that he finished with no clutch and 1st, 3rd and 5th gears only! Fortunately at Spa, which is nearly all top gear running bar the hairpin, this was less of a handicap than it might have been elsewhere.

Although after the race the finger of suspicion was originally pointed at the Autolite plugs, which had not been Duckworth and Cosworth's preferred choice for the DFV engine, it transpired that the mechanics had been over-tightening the plugs, as Ford's Autolite representative at the time, Brian Melia, recalls: 'Cosworth's attitude was "Well, you will tell us to use these crap Autolites, they quite often come loose in the engines on the test beds." What happened in Belgium, and what was happening on the Cosworth test bed, was that they were over-tightening the bloomin' things. They hadn't got that experience of 10mm spark plugs, so they were just winding them into things and stretching the body. What happens then is that you get a slight leak of gas from the cylinder to the atmosphere past that seal, which pushes more heat up than should go up through the body. Eventually the body expands and actually blows the ceramic out of the body of the spark plugs, which was what happened at Spa. Our engineers looked at it and rightly diagnosed what the problem was and all we did was go out and buy about 20 torque wrenches with a 10mm hexagon and we handed those out and then we never had any more bother. Of course, this didn't exactly enamour me to Cosworth. I'm walking into the 'bees knees' of engineering in Britain and trying to tell them how to tighten spark plugs ...' Just to add insult to injury, the wrenches were pre-set and could not be altered without special keys, all of which the Autolite man retained firmly in his own possession!

Even now, Allan McCall rejects the charge that they were 'bolt-stretchers' when it came to the plugs, saying: 'They tried to blame us for over-tightening them because no-one wanted to accept the fact that the centre porcelain just blew out of the spark plugs'. However, the fact remains that, once the torque wrenches were issued, the problem did not recur.

Following the Spa race, a decision was taken to revise the stiffening brace which had originally been put in at the front of the tub on either side to provide support for the front radius arm mounting. This had been the cause of problems in early testing and was still causing problems, so it was decided to replace it with a D-shaped baffle made out of 18-gauge aluminium sheet. This would be riveted onto the inside of the side skins from the top of the scuttle, just forward of the windscreen, and ran down in an arc to the base of the tub, providing support for the radius rod mounting. This facet of the design was incorporated as new into the third tub, which was being built at the time, but had to be retrospectively fitted into the first two cars later in the season. This work necessitated cutting the fronts off the tubs, adding in the baffle and then stitching on a new replacement panel forwards

The lap almost completed, Clark enters the La Source hairpin with Eau Rouge stretching away in the distance. (The Dave Friedman Photo Collection)

of the baffle to the front bulkhead. The other effect of this arrangement was to split the side pontoon fuel cells into four tanks instead of two. Not only did this lose some much-needed fuel capacity but it also made it devilishly awkward for the mechanics to install, remove or connect up the tiny tank between the baffle and the front bulkhead!

The 1967 French Grand Prix was held on the Bugatti circuit at Le Mans, which for the rest of the year was home to a racing drivers' school. The characteristics of the circuit could not be more different to those of Spa - a tight, twisty, tortuous track running mostly through the car parks used for the legendary 24 hour race and utilising just the start and finish straight and pits complex and the run under the Dunlop bridge down to the Esses from the full La Sarthe track. As a result, Denis Jenkinson and other journalists quickly dubbed it 'The Grand Prix of the Car Parks', while the drivers dismissed it as a 'Mickey Mouse' circuit.

The weekend got off to an ominous start when the Team Lotus transporter was delayed at Dieppe by a customs official who at best could be described as overzealous and, at worst, downright bloody-minded! The upshot was that the cars arrived too late for Friday practice, Hill and Clark being left to twiddle their thumbs while Jack Brabham and Denny Hulme set the early pace in their Repco-Brabhams, closely followed by Spa victor Gurney.

When the cars did eventually arrive, it could be seen that small detail changes had been made to both 49s for this race, including the front section of the air-deflector windscreen being widened, the introduction of the progressive throttle linkage to help the drivers cope with the surge of power from the DFV and the addition of a brake balance knob on the dashboard to enable in-cockpit adjustment of the balance from front to rear as desired. At the rear of the car, the clutch had been beefed up to prevent a repeat of the Spa problems, while tubular structures had been added to the rear of the ZF box. This was so that the car conformed to regulations that stated that exhaust pipes should not protrude more than 10 inches beyond the rear of the car. At Spa and Zandvoort, the cars had technically been illegal, although no-one had bothered to protest them at the time (thank goodness - think how that would have ruined the fairy story debut of car and engine!) but someone had obviously had a quiet word in Chapman's ear because here they were on the cars ...

The second and final practice session on Saturday was of only 90 minutes duration, in which the two Lotus drivers had to learn the circuit and secure a decent position on the grid. Frustratingly, both cars were af-

Reality Bites

flicted with a persistent misfire due to a fuel injection problem and it was only in the dying minutes of the session that Hill's car cured itself, enabling the Englishman to turn in a stunning last-gasp effort to snatch pole from under the nose of Jack Brabham, by a tenth of a second! Clark's car never ran cleanly the whole session, but he still ended up fourth on the grid! Late that night, after some further tweaking by the mechanics back at the garage, Clark took his car for a quick blast down the main road to check everything was running OK. Fortunately for him, most self-respecting over-zealous officials were tucked up in bed by this time, so he managed to avoid a brush with the local noise abatement officials and, anyway, this was Le Mans ...

A clash with the Tour de France bicycle race, which was passing nearby, contributed to the paltry 20,000 crowd that turned up to watch the race and, in the giant grandstands opposite the pits, they looked quite lost. A rain shower before the race had cleared the air somewhat by the time the field took to the track, a sharp contrast to the hot, muggy conditions that had prevailed during the two days of practice. As the flag dropped, Hill jumped into the lead. Across the line for the first time, he headed Gurney, Brabham and Clark. After some initial jostling which saw Brabham wrest the lead from Hill, Clark pushed his way to the front of the field on lap five and Hill followed him through two laps later. A hesitation while lapping Guy Ligier's lumbering V12 Cooper-Maserati saw Hill nip past his team-mate on lap 10 and this was the way it was to stay for a further four laps until suddenly Clark came round alone. Hill had pulled off the circuit with no final drive and the Scot led for only a further nine laps before he, too, pulled off into the pits with the gearbox clanking ominously. Another promising showing had been spoilt by an early exit for both cars and reality was beginning to bite after the dream start at Zandvoort.

Clark sits impassively in his car while, under the watchful eye of race officials, Dick Scammell and Allan McCall change the plugs. To the left of the car Colin Chapman, lap chart in hand, inspects one of the offending items. Behind him, leaning over the barriers looking concerned, is mechanic Gordon Huckle, while to the right Graham Hill keeps a watching brief as Keith Duckworth inspects another plug. Autolite's representative Brian Melia is obscured behind Chapman. (The Dave Friedman Photo Collection)

Chapter 8
Beefing-up The Weak Spots

The tyre marks over the 'Hill' sticker show where the right front wheel came back into the side of the tub on impact. The huge dent and split in the side skin clearly illustrate why the car could not be rebuilt in time for the Grand Prix. (Goddard Picture Library)

An early demise in the French Grand Prix was precisely the outcome that Mike Costin had predicted before his partner left for the race. 'I told Keith "Look, you're wasting your time, why on earth do you want to go to Le Mans? They won't last more than three laps!" because we'd seen what was wrong with the gearbox. We looked at it, and it was slab-sided. Now, here's ZF: marvellous people, I mean beautiful gears and everything, but they'd designed a gearbox that had got flat sides. The sides of the gearboxes – which take all the push-off loads from the pinion pushing the crownwheels sideways – have got to be like a cone to feed loads back into the main casing. The 49 was as flat as a kipper, therefore the sides would pant [pulsate] and crownwheels and pinions just knackered themselves, quick as a flash.

They were changing them more often than their bloody underpants! So when they went to a "Mickey-Mouse" circuit like that little Le Mans one, with first and second gear every other corner – forget it!'

Even before the race was over, Chapman had sat down and sketched out what he thought was needed to prevent a repetition of the problem, which had been brought about in the first place by a late decision to replace the planned sliding joints in the drive shafts with sliding splines. This entailed using larger ball-races, which in turn meant larger diameter holes in the side plates, making them even less rigid and more prone to flexing under load. This flexing allowed the pinion to come out of full mesh with the crownwheel, with the result that the teeth on both chattered and broke away.

The offending gearboxes were removed immediately after the race and Chapman, Duckworth and their wives flew straight to the ZF works at Friedrichshafen in Germany, while Scammell followed by road with the gearboxes. As the Cosworth designer remembers: 'I went with Colin and everybody to see ZF about fixing the gearbox, getting it strengthened-up. I would like to record that Mike and I, when we first saw the gearbox just said "That'll never work, that'll never last very long" and it didn't! After our visit, ZF did a very good job and put long bolts through it and strapped big heavy sideplates on it and made a great job.'

The solution arrived at involved the addition of cast iron side plates on each side of the differential housing, plus a new top plate with extra lugs, through which bolts were run to hold both side plates firmly together. It was a neat, albeit heavy, solution to the problem which must have gone against all of the Lotus boss's principles. As Duckworth recalls with a chortle: 'Anything was a weight penalty to Chapman, he was such a purist!' Nonetheless, it was a modification that must have been effective, for no further trouble was experienced in this area.

The British Grand Prix weekend was the setting for

Beefing-up The Weak Spots

yet another Team Lotus epic. The cars sported their revised gearboxes with bolt-on side plates, the ever-efficient ZF having delivered six of the new units in time for practice. They also featured a revised clutch and a different type of progressive throttle slide linkage to that used in France as part of a rolling programme of constant improvement.

Such changes were introduced despite the fact that the team was unable to go testing, due to the lack of a spare engine. In effect, the practice sessions and races at Grand Prix meetings had doubled as the team's testing. Even if there had been an engine, there was no spare car, so going testing would have run the risk of one of the race cars being damaged. At this stage, the third tub had been completed by the fabricators (once famously called 'Those Communists' by Chapman!) back at Hethel, but was awaiting all its key components. So far, both drivers had been lucky, staying on track and managing to keep out of trouble during the first three meetings they had participated in, so there had been no real urgency to the build programme.

Practice took place on Thursday and Friday, the race being scheduled in those days on a Saturday to avoid the Sabbath. Two sessions were held on the Thursday and, in both, the Lotuses yet again had trouble with persistent misfiring. Despite this, Clark managed to turn in the third quickest time behind the omnipresent Brabhams. Overnight, the source of the misfire was traced to a tiny bleed hole in the bypass system of the fuel injection system, which had been passing too much fuel. Blocking it immediately cured the problem, and so it was no surprise when Clark went out on Friday and immediately put his car on pole, some four seconds inside the 1966 Formula 1 lap record and two seconds quicker than Jackie Stewart's pole-time for the International Trophy in April.

After some further adjustments Clark went out again for another seven-lap stint and reduced his time by half a second, to 1.25.3. He was joined on the front row by Hill, who had worked his way down to 1.26.0, and the two Brabhams. However, towards the end of the session, disaster struck Hill. Unable to improve his time further in a car that felt a bit 'twitchy', he headed for the pits. As he came down the pits approach road at 60-70mph the lower rear radius arm mounting plate on the left side of the car detached itself, turning the car sharp right into the pit retaining wall. The impact ripped off the right front wheel, nose cone and radiator and left Hill sitting in a cloud of steam and smoke. More significantly, there was a huge dent in the tub where the wheel had been pushed back on impact and the chassis was badly deformed. After an inspection in the paddock, it was clear that the car could not be repaired in time for the race, so Chapman took the decision to fly nearly all the mechanics back to Hethel and attempt to build up the bare tub of 49/R3 into a new car overnight.

It turned out that the accident had been caused by a faulty weld so not only did they have to modify the bare tub at Hethel before building it up, but Allan McCall also had to improvise his own modifications to Clark's 49/R2 at the garage in Brackley which the team used as its base. McCall remembers, 'I got towed down to Brackley and shoved in the Ford garage by Jim Endruweit and left there. Overnight I cobbled up a repair on Jimmy's car. I was left there by myself and I had access to somebody in the stores, bits and pieces like that. Anyway I redrilled the bracket that was the same as the one that came off the monocoque of Graham's car and caused him to crash, and I put in some 10/32 nuts and bolts to hold the radius arm bracket on the chassis and I also cobbled up some other things around the car.'

Meanwhile, New Zealander Dale Porteous, in his first year with the team, was given a double headache as the one charged responsible for getting the battered 'donor' car 49/R1 back to base: 'I had to drive the transporter back. Everyone else flew back. I'd never driven back from Silverstone and I didn't know the way! The guy with me said he knew the way but we were going

Chapman, Duckworth and Phillippe look on as the mechanics, assisted by Hill, put the finishing touches to R3 in the Silverstone paddock. (Phipps Photographic)

into London instead! Finally, we got going in the right direction but when we got to Cambridge the truck wouldn't go above idle, it had black smoke coming out of it. We stopped and phoned the workshop to say that the truck was playing up because they were all waiting for us to get the car back. Eventually it fixed itself!'

What followed was a true all-nighter in the finest Lotus tradition. It was a case of 'all hands on deck', with anyone from the works who was available being drafted in to help. Dick Scammell recalls that weekend with a particular sense of accomplishment. 'It was quite a momentous weekend. We went back to factory and, fortunately, we had a chassis there which was sort of finished, it was finished as far as the sheet metal workers were concerned but it didn't have any of the pipes in or anything. We stripped the crashed car and built the other one overnight, it was quite an effort.' There wasn't even a spare fibreglass nose, so a makeshift solution was adopted, as Porteous recalls: 'It had an aluminium nose. It was one of the early bucks that Jim the aluminium guy had made to make the fibreglass ones out of. And we used that one because we didn't have anything else to put on it ...' The slightly incongruous looking car was topped off with an earlier air-deflector windscreen.

It was touch and go whether they would get back in time for the race. Even as the car was being wheeled onto the grid the mechanics were still working on it. Hill had also flown back to Hethel and, on the morning of the race, had gone to the works to check that everything fitted him and to get the pedals right. It was here that his fastidious nature in terms of writing down all of his settings paid off, for he was able to have it set up in exactly the same way as the car he had crashed the previous day. All in all, it was a magnificent effort. Colin Chapman described it as 'sixteen of us doing three weeks' work overnight' and he was not exaggerating.

The car that emerged was subtly different to the first two 49s produced. In place of the braces originally specified, the strengthening baffles described earlier had been fitted, and longer rocker arm fairings had been added. These modifications were to spread braking loads over a wider area after it had been found that the shorter, stubby, fairings had caused the tub to kink. Finally, a larger, oval-shaped hatch had been fitted into the scuttle, combining access to the pedal area and the fuel filler cap. These modifications to the access hatch were deemed necessary because of the tremendous problems the mechanics were having gaining access to brake balance adjustment, as the diminutive Allan McCall describes: 'The reason those cars got converted was that I was the only one who could adjust the balance bar and I had to lay on my back to do it. I had to be picked up and lowered into the monocoque upside down with the spanners in my hand! Chapman came along and saw this happening and yelled at Maurice [Phillippe] "Come and look at this! Get Allan out of the car, put Maurice in, make him do this. For Christ's sake, Maurice ..." and there was a great big to-do! Chapman did one of his lovely acts and the result was that we were given more access.'

There was one last scare for the team before they could get down to the serious business of racing, as Allan McCall explains: 'After that all-nighter, they got to the circuit and realised they had forgotten me! I was still in Brackley with Jimmy's car. So someone came out and towed me in - I spent half the time on the grass verge because it was wall-to-wall traffic. And I got in there in just enough time to put fuel in the car and get it onto the grid ...'

So it was that, against the odds, two Lotuses took their positions on the front row of the grid for the 3pm start. As the flag dropped, the field, led by Clark and Hill, streaked down to Copse leaving a cloud of dust and tyre smoke to settle in their wake. At the end of the first lap, Clark led comfortably from his team-mate, who was still settling into his new car. As a result of this, Brabham nipped past him on the following lap to head a five car gaggle also comprising Amon's Ferrari, Gurney in the Eagle and Hulme's Brabham. Once he felt comfortable, Hill really began to motor, passing Brabham on lap nine and closing up on his team-mate, who had literally been cruising round at the front, doing nothing more than was necessary to maintain an advantage over his pursuers.

Blast off!: Clark and Hill, the latter complete with odd-looking aluminium prototype nose cone, surge forward off the grid at Silverstone. (Ford Motor Company)

Beefing-up The Weak Spots

After becoming accustomed to his new car, Hill passed Clark and led until lap 55 of 80 but it was not to be his day... (Ford Motor Company)

The absence of team orders was illustrated by the fact that, after running in formation for five or six laps, Hill slipped past Clark to assume the lead. This was the way it stayed until, on lap 55 out of 80, Hill slowed dramatically, after getting a bit wayward at Becketts. It was reported that his left-rear wheel was canted in at a decidedly unorthodox angle and he toured in dejectedly to the pits. The mechanics quickly identified a missing Allen screw from the top transverse link (which, in the rush to rebuild the car overnight, hadn't been tightened up properly), replaced it inside 60 seconds and sent Hill out again. However, the combined effect of this stop and his slow drive round to the pits meant that he had lost two laps. He stormed back into the fray, passing the BRMs of Hobbs and Irwin and was closing on the Honda of sixth-placed Surtees at the rate of three seconds a lap when his engine expired as he went past the pits to complete his 64th lap.

Clark continued serenely in the lead, clearly driving well within his own and the car's capabilities. By the time he took the chequered flag, only four cars were on the same lap, with Championship leader Hulme following him home 13 seconds adrift and Amon getting the better of 'Black Jack' in the closing stages for third. Indicative of the supremacy of the Lotus 49s was that Clark's fastest race lap of 1.28.0 was a second slower than the fastest lap of the race, set by Hulme at 1.27.0, yet he was never challenged.

Something which may have restrained his pace was that, before the start, Clark had made a rare decision to make an adjustment to the set-up of his car. This was unusual because, according to McCall, he did not normally proffer suggestions for changes, particularly at such a late stage: 'He was having a bit of an understeer problem at Woodcote, which in those days was a flat out bit of a big old corner and it was decided to put another one eighth of a degree of negative [camber] on the outside [left hand] front wheel.' However, despite the ease of his victory, it obviously wasn't to the Scot's satisfaction: 'Straight after the race I went to Jimmy, he put his hand on my shoulder and the very first words he said to me as he got out of the car were: "Putting that negative on the front was a mistake, take it out again." That and changing

the height of the roll bar at the next race in Germany were, basically, the two adjustments made to that car that year! All Jimmy ever wanted was a car that repeated. He could adapt himself to anything, provided he knew what the recipe was.'

Victory was just reward for the overnight efforts of the team, although Hill's mechanics in particular must have despaired at his poor run of luck in the reliability stakes. The result lifted Clark up to joint second in the World Championship title race, nine points behind the canny Denny Hulme, who was building a strong Championship challenge on the back of his consistency and the reliability of the Repco-Brabham package.

For the next race in Germany, the programme of rolling improvements and modifications to both engine and car continued, with changes to both engine and chassis. On the engine front, the package was enhanced by the re-siting of the ignition box, as Duckworth describes: 'It wasn't part of the engine, it was a loose bit. That was why we built it "in-Vee" between the throttle slides and put it in a box and turned it upside down. We

The cause of Hill losing the lead at Silverstone was an Allen screw, which held the top rear suspension link, dropping out. This caused the left rear wheel to lean in at a drunken angle. (Ford Motor Company)

Lotus 49

Clark oversteers round Copse on his way to a popular victory in the British Grand Prix. (Ford Motor Company)

moved it in order to get an engine that was one lump.' In addition to this change, further small refinements had been made to the throttle linkage in order to improve power delivery.

On the car front, to try and prevent a repetition of the problem experienced in the British race by Hill, a fixed stud with a locknut, which was also wired-up, replaced the Allen screw fixing of the rear suspension. Meanwhile, 49/R1 had been rebuilt following its Silverstone accident to the same specification as 49/R3 and would be taken to Germany as the spare car. The rebuild involved the addition of the strengthening baffles in the forward tub in place of the original braces, the addition of longer rocker arm fairings to spread their loads over a wider area, plus fitting the larger access hatch on the top of the scuttle above the driver's knees.

It was most likely on this occasion that an indentation was also added to the bulkhead, in line with the nose of the crankshaft. The reason for doing this is explained by Dick Scammell. 'As the fuel and the air got hot in the tanks they expanded and this back wall [the rear bulkhead] was pushed out. I can remember seeing front covers off DFVs where the bolt on the end of the crank had machined its way straight through the back wall, where I think it had been pushed back by the fuel. If it did that, then the next thing it was going to machine itself through was the tank …' This was another feature that was built into R3 from new, but that had to be added retrospectively to both R1 and R2. Indeed, the author understands that all original specification Lotus 49s (*e.g.* chassis numbers 1-5) had this indentation and it is still present today on all three surviving cars from this batch.

While all this was going on, Chapman was hatching plans to improve the car's oil system and its weight distribution by siting the oil tank in 'saddle' fashion over the ZF gearbox. However, clearly the various problems with the 49 and the hectic pace of life in the Team Lotus race shop precluded such a dramatic modification, for this modification would not appear on a 49 in a race for another eight months …

The German Grand Prix meeting was to be yet another eventful weekend for Team Lotus. However, it started innocuously enough. During the first practice session on the Friday morning, Hill, who had decided to stick with 49/R3, tinkered with the settings of his car and only used the 'warm-up' loop, so did not record a timed lap. This loop comprised the pits straight, the South Curve, the straight running back behind the pits and a link road back to the pits straight. It meant that drivers could satisfy themselves that their car was running well

In an era when drivers didn't wear seatbelts, it was possible for them to turn around and 'supervise' the work being carried out, from the cockpit. This is Clark in practice at the German Grand Prix. Note the ignition system, housed from this race onwards in a box between the inlet trumpets. (Ted Walker/Ferret Fotographics)

Beefing-up The Weak Spots

before setting out on a full 14 mile lap, and they didn't have to do a whole lap before embarking on a timed one.

Surprisingly, Clark, in his regular 49/R2, was almost twenty seconds off the pace set by Surtees in the Honda. The reason for this was that his car was suffering from brake trouble. It turned out that the original specification ventilated discs had been cooling a little too effectively, and were glazing the pads. As a result, solid brake discs were fitted to his car for the afternoon session with harder pads. As they were not ventilated, these solid discs were considerably narrower, at 3/8ths of an inch compared to the ventilated ones. At the same time, Hill's 49/R1 was left with the ventilated discs, but softer pads were fitted in an effort to find an alternative solution.

Half way round his first full lap of practice on Friday afternoon, Hill's car ground to a halt because there was no oil in the gearbox, so he did not really get a chance to decide whether the brakes were improved or not! The spare car was therefore pressed into service, Hill finally getting to record a lap time, although by the time he got back to the pits there was not much of the session left. In the meantime, Clark was still not happy with his car's brakes, but had improved by nearly 24 seconds to 8.19.8. However, this still placed him only seventh fastest. Virtually the whole field had been humbled by the young Belgian driver, Jacky Ickx, at the wheel of Ken Tyrrell's Formula 2 Matra (a select group of Formula 2 cars having been invited to participate in a race, to be run simultaneously with the Grand Prix, to bolster the size of the field). Capitalising on the agility of his chassis and the usable power of the Cosworth FVA, he had turned in a quite incredible 8.14.0, with only Denny Hulme's time of 8.13.5 preventing the humiliation of the entire Grand Prix field!

For the final session on Saturday morning, Hill stuck with 49/R3, this car by now having been fitted with solid brake discs as per Clark's car. Having yet to complete a lap, and on his third different braking configuration of the weekend, he set off with Chapman's warnings about the brakes ringing in his ears. However, it was not long before this car was *hors de combat* too as a result of a hefty shunt on the twisty run down the hill towards Adenau Bridge. In *Life at the Limit*, Hill described the incident thus: 'I was having some trouble with the brakes - they didn't feel quite right; Jimmy, too, was experiencing trouble. Anyway, I got to a point where I just didn't get slowed down enough - I went up one of the banks and did a great big wall of death act.

'Stewart, who was following me, said he saw me come right out of my seat - we weren't wearing seat belts in those days. The car came back down on the road again going backwards and I was able to step out of the car a trifle breathless. Jackie Stewart stopped his car and came back to see if I was all right - he looked more shaken than I felt; he was quite white. He had witnessed the whole horrifying incident; he had seen my car go right up the bank backwards with me almost standing up in the cockpit. Of course it made a bit of a mess of the car and so we had to go and get the other one for the next day.'

'A bit of mess' was an understatement! The ZF box had been ripped off the back of the car, the left-side wheels and suspension had been removed and a radius arm had punctured the tub on the left-hand side at the front. After retrieving the battered wreck, the weary mechanics unloaded 49/R1 from the transporter and began preparing that for the race. It was at this point, almost at the end of the session, that it was realised that, with his accident-prone weekend, Hill had not even managed to complete the mandatory five laps to enable him to start the race, so Clark's car was co-opted, much to his consternation, to enable the Englishman to do a very slow tour, under express orders to bring it back in one piece at all costs! This he managed to do, much to the relief of everyone concerned.

Fortunately, by this time Clark had already completed his practice laps. The team had been working hard on sorting out his car, primarily the brakes, but also with another rare set-up change, as McCall recounts: 'At the beginning of practice, he went out in the car. Now they had these skinny little roll-bars for adjustment. You had the rear roll bar that was in the middle adjustment. He came in and said "Maybe we should ..." and there was a big discussion with Chapman - in those days we sat down and waited for orders and listened to discussions. It was agreed that, maybe, a little bit more oversteer wouldn't hurt the car. So they shifted the rear roll-bar up half an inch. Think of this in relation to an adjustment in today's world when it's all absolute micro-adjustments. Anyway we shifted the bar up half an inch, Jimmy went out and did a lap, came straight in and said "Put it back". And then he went out and qualified, and he was very nearly the first under eight minutes.'

Once the car was to his satisfaction, Clark put in some all-out laps. His first effort was 8.08, which would have been enough to earn him pole anyway, but the next time round he managed to improve further to 8.04.1, tantalisingly close to breaking the magic eight minute barrier, and nearly ten seconds faster than anyone else had managed around the circuit all weekend.

As a portent of things to come, the mechanics found that the front rocker arms on both cars had begun to collapse. In a clear demonstration of the favouritism, which most within the team at the time acknowledge always existed on Chapman's part, he was happy to put sturdier but heavier rocker arms on Hill's car for the race, but would not countenance such a move in the case of the Scot. As Allan McCall says: 'It collapsed the rockers in practice, as it did in the race. Jimmy's car had 20 gauge, although we did have 16 gauge [therefore heavier] ones with us. Graham's collapsed too, so they put 16 gauge on his car but they wouldn't put them on Jimmy's! Chapman did one of his calculating things where he said "No, they are 300% too strong, blah, blah, blah ..." so I got a new pair of 20 gauge ones which we put on the car and which collapsed on the first lap! They just crinkled up and collapsed and actually burst the radiator as it scraped on the ground.' In fact, this was not the full story, but then we are jumping ahead of ourselves ...

It was during this weekend that Walter Hayes suggested to Colin Chapman that, contrary to their original agreement, which had promised exclusivity up until the end of the 1968 season, he felt that the DFV engine should be made available to other teams due to its overwhelming superiority over other engines on the

Lotus 49

The young Canadian, Eppie Wietzes, was the first driver other than Clark or Hill to race a Lotus 49, when he started his home Grand Prix at Mosport. (Phipps Photographic)

Grand Prix grid. Hayes says, 'I thought that we were going to ruin motor racing if we had such dominance. So I talked to Chapman and said "I think that we are going to have make this engine freely available to other people". And he, to his great credit said "Yes, I suppose so". So, from the beginning of 1968 onwards the engine became progressively available to different teams and we had an absolutely wonderful period of motor racing. There's no knowing what the 49 would have done if we had maintained its exclusivity where engines were concerned …'

Race day was dry and warm, and a huge crowd gathered in expectation of a tremendous battle. At the start, Clark used the power of his DFV engine to jump into his customary lead, while in the middle of the pack a cloud of tyre smoke at the entrance to the South Curve indicated that someone had spun. When it cleared, Hill, who had started from the middle of the grid, could be seen gathering it all together again after having been forced onto the grass, which caused him to lock a wheel and spin. Amazingly, everyone missed him, so he was able to continue at the tail end of the field.

By the end of the first lap Clark had pulled out a small advantage, and had broken the existing lap record from a standing start. However, even at this stage his car was handling oddly: it transpired he had picked up a slow puncture in the right rear tyre and was having to work hard simply to stay ahead of his pursuers, Hulme in the Brabham and Gurney's Eagle. Clark simply put it down to a combination of full fuel load and new tyres, but, by the fourth lap, he could no longer stay ahead and toured in slowly to retire. When he arrived in the pits, his right front wheel was leaning inwards due to the rocker arm having collapsed and his radiator was holed as a result of the front of the car scraping on the ground. It seemed that, while the collapses in practice had shown the rockers were weak anyway, the changes in ride height caused by the punctured tyre at the rear may have placed further strain on the right front rocker. Either way, there was nothing that could be done quickly to get Clark back in the race and his car was pushed away.

For subsequent races, heavier gauge rocker arms with a stiffening brace down the middle were introduced as standard to try and prevent a re-occurence of the problems but, as McCall recalls, 'Even then, Jimmy got 18 gauge rockers instead of 16 gauge, because Chapman wouldn't go all the way! That was the way he worked - brilliant but sometimes he stuck to a problem a little too long.'

Hill's nightmare weekend continued, with a loose front wheel nut causing him a heart-stopping moment out on one of the faster parts of the circuit. After a stop to have it tightened and the other wheel nuts checked over as well, he rejoined, but after eight laps his race was run when he drew up in the pits with the Allen screw which bolted the 'fir-tree' to the block missing and the whole assembly leaning inwards at a crazy angle.

These episodes did little to dispel the impression that Team Lotus built cars that were very fast but which were not strong enough to get the job done. However, Keith Duckworth, like Chapman a staunch advocate of being down to the weight limit, defends the Lotus boss's design philosophy. 'He was trying to design light cars to win races. If you start off light and it breaks then you will strengthen up the pieces that break. An awful lot of bits that you might have thought would have broken, won't break. And that is the only way of getting a light car. If you start off by designing everything conservatively and nothing fails, there is no way that you are ever going to lighten your car. You never lighten things under those conditions and therefore you don't learn quickly enough. People talk about being able to stress cars but the only problem is that you don't know how hard your driver is going to hit kerbs or you can't work out what the loads from hitting a kerb on any occasion are going to be. If you design it such that it will hit brick walls in all directions all the time, you'll never win a race.

Beefing-up The Weak Spots

'Therefore, you actually have to hang it out if you want to make a race winner. Then you've got to look at pieces sufficiently often so that your failures should really be fatigue failures from a number of cycles and not be an instant, ultimate failure. If your mechanics and your system of looking at your cars and preparing them is any good, you should be able to catch most things at the stage that, either some stretch in the material - it has collapsed a bit - has occurred, or there is a crack forming and you can catch the crack before it is a disastrous failure. So, the only real way is to design light and then to have loads of inspections and crack detection and things like that.'

The Formula 2 race which was run in conjunction with the Grand Prix was won by a young Jackie Oliver in the Lotus Components works car, who had a steady and consistent race and inherited victory when the rapid Ickx's Matra retired. The maturity of his performance did not go unnoticed by Colin Chapman, who marked him down as someone of promise for the future.

Between this race and the next round of the Championship, the Canadian Grand Prix, Chapman and Phillippe were working on another innovation – the nose fin. It is not clear who might have come up with the idea but, given Phillippe's background of working with wing sections at De Havilland, it seems likely that he would have been a keen advocate of such a development. While Jim Hall's Chaparral 2F sports cars had run from the beginning of 1967 with a revolutionary suspension-mounted rear aerofoil, there was no sign yet of any Formula 1 teams attempting to adopt this technology. However, both Brabham and Lotus had tried different types of spoilers fitted to the noses of their cars at Spa and it seems that the Lotus design team wanted to take this idea a stage further.

Drawings produced by Phillippe on August 21st 1967 show a design for an 'Airfoil assembly' which is basically a fin attached to the side of the nose cone, but much higher up than a bib spoiler. Amazingly, like the oil saddle tank, this was another development that would have to wait until the following year before being seen on a Lotus racing car. Perhaps, given the undoubted superiority of Team Lotus at that point in the 1967 season, Chapman felt it was unnecessary to introduce these modifications and better to keep them up his sleeve for 1968. Furthermore, it was not unknown for such innovations to be rendered obsolete between seasons due to sudden rule changes, so this could well have been a defensive move aimed at avoiding such a scenario …

After the dramas of the previous two races, it was with some trepidation that the whole team (and particularly Hill!) travelled to Canada, wondering what sort of a weekend would unfold. This was the inaugural Canadian Grand Prix, the race being held at the Mosport circuit, outside Toronto. As was often his practice if he had a spare car available, Chapman had rented out a car to Chuck Rathgeb Jr, team manager of Comstock Racing Team, for a promising local driver, Eppie Wietzes, to drive. The link with Wietzes came about because Ford of Canada were co-sponsors of Comstock Racing. At the time, Wietzes was lying second in the Canadian Racing Championship at the wheel of a Ford GT40. Although he had plenty of experience behind the wheel of 'big-banger' sports cars with torquey V8 engines, he had certainly never sat at the wheel of anything with the sort of power-to-weight ratio of a Lotus 49, so this was going to be something of a baptism of fire.

Only small modifications had been made to the Team Lotus cars for this race, with different filters of the kind used in the Climax V8 days added to the fuel system to try to improve the persistent misfire maladies experienced at previous races. A further small revision had been made to the fixing bolt for the subframe holding the rear suspension, again with the aim of stopping it from becoming detached at any time other than when it was being disassembled by the mechanics! Additionally, solid brake discs were fitted as standard on all three cars following the success of the set-up first tried at the Nürburgring. Clark was in his usual chassis 49/R2, while Hill chose to revert back to 49/R3 - which had been repaired back at Hethel following its German GP practice crash - leaving Wietzes to take over the original prototype, 49/R1.

Due to the unfamiliarity of the drivers with the Mosport track, extra practice sessions were laid on, with unofficial sessions on Thursday afternoon, Friday morning and Saturday morning supplementing the official practice periods scheduled for Friday and Saturday afternoon. Wietzes' task was made much more difficult on the Thursday when Clark, who was out in 49/R1, had an accident which badly damaged the car. At first it seemed that he had simply slid gently off the track at the right-hand corner immediately after the pits and backed up a grass bank, with relatively minor damage. However, closer examination revealed that both right rear radius rods had been kinked and a rose joint broken off by the wheel. More seriously, the impact had pushed the radius arm bracket forward into the tub and rippled the side skin. The result was that a serious re-building session was (once again!) on the agenda for the overworked Lotus mechanics: Wietzes would have to wait until this was completed before he could have his first taste of DFV power and Chapman engineering.

Pre-race publicity had focused on the fantastic speed of the new breed of Grand Prix cars, so the obvious target was to try and improve on the existing lap record, which was held by Dan Gurney at the wheel of a Can-Am Lola T70, at 1.23.1. In the first official session, Clark managed to get under this benchmark, this being achieved on Goodyear tyres rather than those of the team's official supplier, Firestone. The reason for this change was that the latter was awaiting a shipment of tyres, which were overdue, and therefore did not have the right combination of tread and profile available. Hill also tried the Goodyears in the Saturday morning session, and was quite enthusiastic about them. All this must have been rather galling for Firestone but it was all resolved satisfactorily, since both Lotuses appeared on Firestones, as contracted, for the final official session.

Repairs to Wietzes' car were completed in time for him to have a few exploratory laps on the Saturday morning, but it was no surprise that he was well off the pace, as he recalls. 'All I did was a couple of laps. Jimmy didn't want to drive the car again, after he'd crashed it. I said, "Well, why don't you do a couple of laps just to

The Canadian Grand Prix started and finished in the wet. Clark excelled in the dry spell in between the rains but eventually the water got in his electrics and he was forced to retire. Here the field splashes away from the start, with Clark joined by (left to right) Hill, Gurney and eventual winner Brabham. (The Dave Friedman Photo Collection)

make sure the car feels the way it should be because I don't know?" Anyway, to me, it never did feel right. It wasn't a "happy camper" in the back end. After all that, I thought probably the bump steer was out somewhat because the tub got bent a little bit.'

Although Hill had been below par (he was suffering from the 'flu) for the earlier part of the meeting, he was right on the pace when it counted, and Lotus-Fords were first and second on the grid when the final times were announced. They were very evenly matched and the only cars to beat the 1.23 barrier, with Clark recording 1.22.4 and Hill only three-tenths adrift. Wietzes scraped in on the back row, which wasn't a bad effort given his lack of track time and distinctly limited experience. After practice it was found that both the leading Lotus cars had a petrol leak, fortunately at an early enough stage to allow the problems to be rectified in time for the race.

After his brief foray onto the track on Saturday, Wietzes had to wait until the race for his next taste of Formula 1 and, even then, only after a fuel scare like that of the other two Team Lotus cars. 'The next time I got in the car was on Sunday morning. I got in and all of a sudden the gas was coming in the cockpit, they had to take it away and change a fuel bag. So I just made it on to the grid – good thing I was on the back!'

Race day dawned drizzly. A crowd of 55,000 turned up, which, at the time, was claimed to be the biggest ever to see a sports event in Canada. From the start, Clark jumped into the lead, ahead of Hulme, Hill and Brabham. Brabham team had made a last-minute decision to start on Goodyear intermediates, whereas the two Lotuses were on Firestone dry tyres. This decision was to be crucial for, as the rain got harder, the Repco-Brabhams moved up, with Hulme taking the lead and Brabham taking Hill for third place. By lap twelve, Hill had been caught and passed by a flying Bruce McLaren, who was having his first outing in his V12 BRM-engined eponymous chassis and by lap 18, Clark had been demoted to third by the Kiwi.

The tide turned briefly after around 20 of the 90 laps when the rain stopped and the track began to dry. Clark started to fly, catching and passing McLaren and starting to reduce the gap on Hulme, which stood at 26 seconds by lap 27. He closed relentlessly until, on lap 58, he swept majestically into the lead, the tyre tables turned. However, his glory was shortlived for, just as he regained the lead, a fine drizzle began to fall once more, and this turned into heavy rain. Six laps later, Clark suddenly disappeared, while Hulme pitted for some new goggles, leaving Jack Brabham in an unexpected lead which he maintained to the finish. The problem with Clark's car was electrical, with the rain shorting the ignition. The resourceful Scot leapt out of his car and dried the offending components off, but it was to no avail and, although he got the car running again, it soon succumbed to the same problem.

At about this time, Wietzes also stopped by the pit wall with water in the electrics, leaving Hill in the only surviving Lotus. He had been far from comfortable and after one spin in the wet had stalled the engine and been forced to get out and push start the car. Nonetheless, it was a race of severe attrition, with the result that, at the finish, he had moved up to fourth place, merely as a result of keeping going. Incredibly, it was his first finish of the season in a Lotus 49 from six starts, which goes to show what rotten luck he had experienced to that point. Even though he finished, Lotus records show that he did so with no clutch!

The problem with water in the electrics effectively spelled the end of Clark's title hopes, for Championship leader Hulme had finished second to add further to his points total. Duckworth attributes the problems experienced by Team Lotus at Mosport to nothing more than a failure to waterproof components as effectively as they

Beefing-up The Weak Spots

should have done: 'Most of our problems at that stage were the coil lead. You had to have a lot of silicone MS-2 around the plug caps and the distributor leads and the coil leads. You definitely had to smother them with waterproof silicone and I don't think we'd done it when we should have!'

Amazingly, the Grand Prix circus returned to Europe for one race, the Italian Grand Prix, before completing the final two North American races in the United States and Mexico. The grandly titled the 'Grand Prix of Europe' was, as usual, to be held at Monza. As it turned out, it was an appropriate title, for it was without doubt the most exciting and entertaining European round, and probably the best race in the entire World Championship series of 1967.

For this race, the 49s were little changed, the main difference being that solid torsion shafts, which had a degree of 'give' in them, had replaced the original BRD tubular halfshafts which did not flex at all. Clark was in his regular car 49/R2, Hill was in 49/R3, while a third driver, Giancarlo Baghetti, had bought a drive in the spare chassis, 49/R1. Baghetti's career was on the wane at this stage, having begun promisingly enough with Ferrari when he became the first ever person to win on their World Championship Grand Prix debut at Reims in the 1961 French Grand Prix. Subsequent spells with the disastrous ATS venture, which he left Ferrari for along with Phil Hill for the 1963 season, and then the Scuderia Centro Sud in outdated BRMs, did nothing to enhance his reputation and he had only a handful of drives in the 1965 and 1966 seasons. In fact, this appearance for Team Lotus was to be his only Formula 1 drive of 1967 and his last ever in a World Championship race.

Practice was arranged in the form of two three-hour sessions, one on the Friday and the other on Saturday. This meant that it was quite an unhurried affair which gave all the drivers a chance to experiment with settings as well as different aerodynamic aids. Surprisingly, it was the Brabham team, not Chapman's Lotus outfit, who were the innovators, trying wrap-over bubble canopies over the cockpit and streamlined engine cowlings. However, this was abandoned after Brabham found the canopy stiflingly hot and fumey; he also had trouble seeing where he was going through the perspex.

Clark was out early on in the Friday session in 49/R2 and decided he wanted a change of ratios, which necessitated a change of gearbox. For a while he circulated in Baghetti's car, although it was not set up for him and therefore rather uncomfortable to drive. When he got 'his' car back, he went straight out and set a 1.28.5, which was some way inside Mike Parkes' 1966 pole time of 1.31.3 and this remained the fastest time throughout practice on both Friday and Saturday. Hill was toying around with the set-up of his car as normal, but still managed a 1.29.7, while Baghetti did not get to practice on this first day as a result of Clark using his car.

On the second day a change to softer springs did not benefit Clark and he could not do better than 1.29.7, while Hill also went half a second slower than he had managed the previous day. However, the main reason for the lack of improvements was that rain began falling after only half an hour of the session: the track did not dry out again, even though practice was extended by a further half an hour to 7pm. Baghetti struggled to come to terms with the power of the DFV and his final practice time put him on the back row of the grid, almost seven seconds slower than Clark. Towards the end, Clark went out again and did some comparitive tests between Firestone and Goodyear tyres once more, feeling that the Firestone tyres were not allowing him to slide the car as much as he would like to round the Monza curves, while Baghetti ran out of fuel, curtailing his time in the car.

Try as he might, 'Black Jack' Brabham could not better the Scot's time and had to settle for second on the grid, with the surprise package being Bruce McLaren, the McLaren-BRM V12 having only its second race and outqualifying the works BRM cars with their complex H16 units.

The day of the race was hot and hazy and a mood of expectation hung in the air. The grid had been given an odd look to it by the events of the previous day and it was given an even odder look when the starter dropped his flag to indicate that the cars were to move forward from the dummy grid to their proper starting positions and Brabham, McLaren, Gurney and Hill took this to be the start proper and streaked off at full chat towards the Curva Grande. Clark's hard work in winning pole counted for nought and he set off in hot pursuit. However, at the end of the opening lap he was already up to fourth behind team-mate Hill, by the next lap he was second and, as the field completed lap 3, he swept past Gurney's Eagle into the lead.

Clark then began to pull out a bit of a lead, but already the car was beginning to handle strangely. Amazingly, he had picked up another puncture, but it was not immediately obvious because the centrifugal forces generated at the speeds the cars were achieving down the straights meant that the tyre was holding its shape. It was only round the corners that the loss of pressure became more apparent. As Clark grappled with his car's changing handling characteristics, his pursuers caught up with him and it was only a brave manoeuvre from Brabham, up the inside of the Scot into the Parabolica on lap 12, where he pointed to Clark to indicate that he had a visible tyre problem, which finally alerted Clark. That lap, he came past the pits looking over his shoulder and he could see

Practice at Monza was a relaxed affair, apart from the armed policemen milling around. Here one of their number keeps a watchful eye on Hill's R3 while Dougie Bridge fits the nose cone. Also in shot are Dick Scammell (standing nearest right front wheel) and Gordon Huckle, while Hill is leaning against the pit wall chatting to Colin Chapman. (Aldo Zana)

for himself that he had a puncture, so he toured round and pulled into the pits for a wheel change. Although the mechanics worked quickly, he rejoined just as the leaders shot past the pits, putting him a lap down, in 15th place.

So began a drive of truly epic proportions. Whereas it had been fairly obvious that Clark had been driving well within himself so far that season, hardly needing to extend the Lotus 49 to its full capabilities, now he really put pedal to metal. It took him around ten laps to catch the leading group of three, which consisted of the Brabhams of Hulme and Brabham and Hill's Lotus, to unlap himself. However, it proved more difficult to break the tow and get away from them. By lap 30 he was up to ninth place and had broken away from Hill and Hulme, with Brabham dropping back slightly, having over-revved his Repco engine. Then, on lap 30, Hulme's Brabham slowed suddenly and pulled into the pits with a blown head gasket, leaving Hill comfortably in the lead by some 10 seconds. Clark was still lapping faster than anyone else on the track, having set a new lap record of 1.28.5 on the 26th tour. By lap 33, he passed Baghetti in the number three 49 to move up to seventh place and, for a brief time, the Lotuses were running in 1-2-3 formation on the road. The demise of McLaren's BRM engine and a pit-stop for Amon's fourth-placed Ferrari moved Clark and Baghetti up to fifth and sixth places respectively, so that all three Team Lotus entries looked to be in line for a points finish.

It was not to be, for, on lap 51, Baghetti coasted into the pits with a broken camshaft, the first sign that all might not be well. By this stage, Hill had pulled out a healthy lead of 55 seconds over the ailing Brabham, who was gradually being reeled in by John Surtees in the brand new Honda, christened the 'Hondola' because it had been built up around an unused Lola Indy chassis. By lap 54 of 68, Clark had swept past Rindt's Maserati to take fourth place and Hill followed him through, leaving only his team-mate, Surtees and Brabham on the same lap. The Scot urged his Lotus on as the Honda and Brabham were now within his sights. In contrast, Hill looked set for a comfortable victory, which would go a long way towards compensating him for the miserable season he had endured to this point. Sadly, on lap 59, his engine blew spectacularly on the approach to the Parabolica, the crankshaft having broken.

On the same lap, Clark was able to slingshot past Surtees to grab second place and, as they went round the Curva Grande for the 61st time, he passed Brabham to regain the lead - an incredible recovery considering he had made up an entire lap on his rivals. The stage seemed set for a heroic victory, although initially the Scot had trouble pulling away as the canny Brabham had jumped into his slipstream, with Surtees in hot pursuit. However, after 65 laps - and with three remaining - he had eked out a three second advantage, while Surtees had managed to force his way past Brabham. As they crossed the line to start their last lap the commentator's voice had risen to fever pitch, for the gap was visibly smaller and Clark seemed to be in trouble. His engine coughed as he rounded the Curva Grande and Surtees and Brabham jinked round him in unison. The final time into the Parabolica, Brabham dived down the inside of the Honda but got out of shape on the cement dust thrown down to absorb the oil from Hill's blow-up and Surtees slipped inside him and held on to the chequered flag, to win by a car's length. Clark coasted over the line a disconsolate third, apparently out of fuel. The crowd went absolutely wild, because they had witnessed an incredible battle for victory and a remarkable drive on the part of Clark.

Back at the Lotus pit, the inquest began as to what had gone wrong. Although the official Lotus explanation at the time was that a faulty fuel pump had failed to pick up the final three gallons of fuel, Keith Duckworth remembers the day's events differently, and has a simple explanation for the Scot's problems on the final lap: 'He ran out of fuel … ran out of fuel! Chapman would never have more than the odd extra ounce of fuel in and Clark's absolutely masterful effort – he was obviously going "Harry Flatters" everywhere – meant that he used more fuel, in catching up this lap and a bit, than he would have done normally. An absolutely brilliant performance.'

This version of events is confirmed by Clark's mechanic Allan McCall, who tells a familiar tale of Chapman's obsession with the amount of fuel put in his cars: 'Chapman did the fuel calculation and decided that we only needed 31 gallons or something crazy like that! Everybody thought '**** off!'. So Dick Scammell took me to one side and muttered "Put 33 in it" and I'd done my own calculations and I put 36 in it! I think we were about five or six gallons more than what Chapman had calculated for that particular weekend. People say we had a fuel pump pick-up failure - that's bullshit! We simply ran out of fuel because Chapman had screwed up chronically with his calculations …

'After the race, he was shouting "Who put the fuel in, how much fuel did you put in the car?" I think Scammell spoke up and said "Actually Mr Chapman, you said 31 gallons and I told Allan to put 33 in it …" and I said "… and I put 36 in it!" and I got this enormous lecture about not doing what I was ****ing supposed to do. He was saying "Don't you EVER do that again. If I tell you to put 31 gallons in it … blah, blah, blah." But that was the wonder of Chapman: the man was brilliant but, now and again, he had also to realise that some of us had brains. We could calculate – we went to school ourselves you know!'

The other problem that contributed to Clark running out of fuel was Chapman's use of foam in the bag tanks. As well as stopping the fuel from surging too much inside the tanks, it was later found to hold up the fuel inside it, preventing the last few gallons from being delivered to the seat tank and from there on into the catch tank, and it seems this may have occurred at Monza.

Clark's drive was, and still is, hailed as one of his best ever. However, Jabby Crombac was surprised to find out from the man himself that he did not consider this to be the case: 'Well, it looked like that to us, to external people, but I specifically discussed it with Jimmy and he said "No, no, no! My biggest drive was at the Nürburgring in 1962", that was when he switched off the fuel pump and he caught up and finished fourth. He said "This is where I put everything into it." You have to bear in mind that Monza was not a very difficult circuit but, of course, it took a lot of guts because, at the time, you had some very fast corners. But, you see, Jimmy was a strange guy

Beefing-up The Weak Spots

It is lap 63 and Clark has retaken the lead, having made up a deficit of more than a lap after a pit stop to change a punctured tyre. He leads Brabham by 2.8 seconds and Surtees by 3.5 seconds with only five laps standing between him and a fairy-tale victory. It was not to be... (Phipps Photographic)

because he was surprised that the others were so slow, he said. "Why aren't they quicker here? It's a piece of cake, flat out, no problem, why aren't they quicker?" - he just couldn't understand it!'

The result of this race meant that only Brabham or Hulme could now win the World Championship. Clark - who had needed to win - had lost any last mathematical chance of taking the title with the final lap cough and splutter of his engine.

Chapter 9
A Winning Streak At Last

The teams returned across the Atlantic for a final showdown in North America, starting with the US Grand Prix at the Watkins Glen circuit near New York. The main change to the cars for this race was the replacement of the original small round access hatch on the top of the scuttle above the drivers' knees on Clark's 49/R2 with a slightly larger oval. However, unlike 49/R1 and 49/R3, the aperture was not enlarged to include the fuel filler cap, these two items remaining separate on Clark's car, giving the tub a unique characteristic that would enable it to be easily identified in subsequent years. At the same time as this modification, the D-shaped aluminium sheet strengthening baffle was fitted towards the front of the tub in place of the original strengthening brace and the front rocker arm fairings were also extended to spread their loads over a wider area, thus bringing Clark's car into line with the other two 49s.

The rapid pace of tyre development during 1967 is also illustrated by the fact that, for this race, Lotus were using 10-inch wide front wheels for the first time, to cope with the latest Firestone covers.

The spare car, 49/R1, was again rented out, this time to the Mexican driver Moises Solana, who was renewing a relationship with Team Lotus which dated back to 1964. Solana had made a habit of appearing once a year in a hired car at his home Grand Prix, but this year was also taking in the US round as he had done in 1965 when he last drove for Lotus.

The race's large prize fund (the last-placed finisher was guaranteed as much as the winner of the British Grand Prix that year!) had attracted a large complement of competitors. The accomplishments of Dan Gurney during the year, as well as the presence of a Ford engine in the field, ensured that a large crowd would attend. Two four-hour practice sessions were laid on, one on the Friday and the other on Saturday. There was plenty of incentive to set a quick time, with US$1000 on offer for the pole position winner.

Heavy mist and rain descended on the circuit for the Friday session. As had become his custom, Clark swapped around between his usual chassis and the one intended for Solana, finding that the latter was better on a drying track, but eventually setting his best time in his regular car. This was a stunning 1.06.80, which was well inside Jack Brabham's previous record of 1.08.42. Hill in the sister car was the only other car inside 1.08, with 1.07.09. On the Saturday, the weather was better from the start, and Clark improved his times further, first with Solana's car, getting down to 1.06.58 and then back in

The 'Tech Shed' at Watkins Glen was a cramped facility but at least it was dry. Here Allan McCall works on Clark's car nearest the camera, with Solana's R1 (car 18) and Hill's R3 (car 6) in the background. (Ford Motor Company)

A Winning Streak At Last

his own, getting down to 1.06.07, which seemed likely to net him the US$1000. However, his team-mate had other ideas. Concentrating his efforts on the one car appeared to benefit Hill as he first equalled Clark's time and then went more than half a second faster, with a 1.05.48, to secure the 'loot'.

It was quite clear that the two Lotus-Fords were the class of the field on this circuit, for no-one else was able to get close to them. Dan Gurney upheld local honour by putting his car on the second row, with Chris Amon in the 48-valve Ferrari alongside. The two Brabhams were on the next row, with the 'boss' getting the better of his Championship rival Hulme. Solana was a superb seventh, his performance all the more creditable given that he completed only nine laps of practice. Denis Jenkinson was moved to remark at the time that this 'either says a lot for his improvement since last year, or that the Lotus is an easier car to drive than it looks.' One suspects that it was the latter but it was, nonetheless, a good performance.

On Ford's home ground, Walter Hayes was acutely aware that a win would help enhance the manufacturers' reputation and that of the DFV project in general among the company's top brass. However, he was so concerned about the dominance of the two Team Lotus cars that he feared they would end up racing each other for the win, with the possibility that they would overdo things and neither would finish. To avoid such a scenario, he called a meeting in his hotel room the night before the race to propose a novel arrangement: 'I wanted very much to win there because it was where 'Uncle' Henry [Ford] lived. Even more remarkable was the dominance of the engine because I did something at that race that I don't think anybody had been able to do before. In practice for the race, Jimmy and Graham were enjoying blowing each other off. They were both trying to do the faster lap and I said to Chapman, "I'm not going to have this, bring them to my room at the end of practice." So they were both paraded in and I said "We are going to toss up and decide who is going to win. Whoever wins the toss will win the race." And Graham won the toss.' At the same time, it was agreed that the loser, in this case Clark, would be allowed to win the Mexican Grand Prix as a reward for sticking to the arrangement.

Starter Tex Hopkins brought the cars forward with his back towards them, turned and did one of his trademark leaps in the air as he swished the flag down to get the race underway. Hill jumped straight into the lead, hotly pursued by Gurney, who soon split the Lotus pair. The first 'Lotus incident' occurred on the second lap, when Solana spun and stalled his car. Although the mechanics went out to his car and got him going again, he was subsequently disqualified for a push-start and so his race came to a premature end.

On the eighth lap, Clark slipped past the Eagle to make it a Lotus 1-2, and he dutifully stayed in the number 2 position. However, Hill was beginning to experience

Tucked away in the corner of the 'Tech Shed', and on a makeshift seat, Hill undertakes a bit of essential goggle maintenance. A far cry from today's motor home and marble-floored garage culture... (Ford Motor Company)

Flag-man Tex Hopkins leaps in the air to start the US Grand Prix at Watkins Glen. Pole-sitter Hill leads off with Clark alongside. They are followed by Gurney (Eagle-Weslake) and Amon (Ferrari), while the third Team Lotus entry of Moises Solana is just visible behind Gurney. (Phipps Photographic)

Lotus 49

In the early laps at Watkins Glen, local hero Dan Gurney split the mighty Lotus 49s until engine maladies intervened once more. Here he chases Hill through the Esses while Clark maintains a watching brief, in accordance with the agreement made the night before with Walter Hayes... (Richard Spelberg Collection)

Mexican Moises Solana surprised everyone at Team Lotus with his pace at Watkins Glen, in a car he had not driven before the meeting began. (Ford Motor Company)

problems with his car's clutch, and had slowed his pace, which allowed both Clark and Amon to close up onto his tailpipes. On the 41st lap, Hayes and Chapman decided that it would be better for Clark to take the initiative, given Hill's problems, as the Ford man describes. 'I was a bit nervous, so I waved Jim Clark by and told him to get on with it.' After 65 of the 108 laps, Amon finally found a way by the Lotus as Hill struggled to find a gear out of the final corner, but the intermittent nature of the Englishman's car's maladies was illustrated by the fact that he put in a new record lap four laps later, to retake the position and on the 81st lap broke it again, to leave it at 1.06.0, thus netting him a further US$2000 to add to his practice bounty.

Three laps later, the Kiwi was back in front, a position which he held on to for a further 14 laps before his engine expired. By this time, Hill's clutch problems had got worse and he also had to contend with fluctuating oil pressure and - it was discovered after the race - a broken flywheel. Consequently, he eased off and settled for a second place finish, given that he had almost a one-lap advantage over his nearest challenger, Hulme. Clark seemed to be heading for a comfortable victory but, halfway round the 106th lap, with two and a half laps remaining, the top link of the right rear suspension broke at the weld and the wheel on that side fell inwards in a manner reminiscent of Hill's problem with his left rear at Silverstone in July. The Lotus pit had their hearts in their mouths as he continued, at a much-reduced speed. Although Hill closed right up, he was still 6.3 seconds behind Clark at the chequered flag.

Once again, circumstances had played to the advantage of the Scot rather than the Englishman and the win that should have gone Hill's way went to his team-mate. Clark was very aware that things hadn't quite worked out as agreed, as Hayes recalls. 'He was a gentleman. Immediately the race was over, he leapt out of his car and raced back to Graham and said "I didn't pass you, they told me to".' Despite this apology, Hill still felt slightly wronged, for he felt that, had he known of his team-mate's difficulties, he probably could have upped his pace slightly and caught and passed Clark to take the win as planned, as Allan McCall recounts. 'Graham was a bit p****d off because Jimmy won the race and he felt that he should have been notified that Jimmy had a problem on the car.'

Nonetheless, it was mission accomplished for Lotus and Ford, with newspaper headlines proclaiming 'The Ultimate Race Car - It Breaks At The Finish Line!'

A Winning Streak At Last

With the right rear wheel of his car leaning inwards at a rakish angle, Clark gives the thumbs up to a leaping Tex Hopkins after nursing his car home. In front of Ford top brass and on the Blue Oval's home territory, this was an extremely important win. (Phipps Photographic)

On the winners' rostrum at Watkins Glen. Clark is about to give Ford's Walter Hayes an impromptu shower of Champagne as, from left, Gordon Huckle, Colin Chapman, Graham Hill, Keith Duckworth (far right) and Allan McCall (squatting) look on. (Ford Motor Company)

Lotus 49

Pictures of a smiling Clark kissing Miss US Grand Prix surrounded by Duckworth, Hayes, Chapman, Hill and the mechanics all decked out in their Ford jackets were beamed around the world. From a mechanic's point of view, yet another failure in the suspension department was a big disappointment, as McCall says: 'That was just an embarrassment running around with the back wheel falling off!' Once again, Clark's incredible mechanical sympathy had been displayed for all to see - only the previous year at the same circuit he had coaxed the recalcitrant H16 BRM engine to its one and only victory against all the odds.

Although there was no Lotus interest remaining in the outcome of the Championship, the Team headed for Mexico determined to trounce the opposition without any last-minute dramas intervening. The specification of the 49s was basically the same as at Watkins Glen three weeks earlier, save for the addition of more substantial subframes holding the rear suspension in the light of the failures experienced in previous races.

There was one small change to Clark's regular chassis, 49/R2. This was the addition of a bracket between the two top link mounting points in an effort to try and spread the loads more evenly and prevent any further failures in this area. As Allan McCall recalls, this was something he had done at his own initiative and was not well received by Colin Chapman: 'Quite honestly I got so p****d off with the car and the fragile part and what happened at Watkins Glen with the rear wheel falling over. What we called the pine trees on the back, which held up the rear suspension, they used to break on a very regular basis. When we got to Mexico I spent my own money and my own time in Mexico City and I found the bits and pieces and made up a little support bracket that went between the two links. I even found a place and had it nickel-plated. This was before anybody turned up. I put it on the car and Chapman came along and went apes**t!'

McCall's move reflected frustration among some of the mechanics about the number of breakages suffered by the 49s during 1967, which they felt had compromised their title challenge unnecessarily: 'There were a lot of things that we, as mechanics, would have fixed if we had been allowed to, that would have possibly let us win that Championship. I might be speaking a bit high here, but that's the way we felt about it in those days.'

With the possibility of a car breakage increasingly preying on Jim Clark's mind (and not surprisingly, given the number of failures experienced during that season), McCall felt his move was justified, even though it was not authorised by the Lotus boss: 'I was getting too big for my boots I suppose ... but, by the same token, I know that stuff like that mattered to Jimmy. Contrary to what people think, Jimmy was very conscious of the fact that

Once again, Moises Solana made seasoned observers sit up and take notice with his pace in the third Team Lotus car at the Mexican Grand Prix meeting. Here he is seen entering the hairpin at the Magdalena Mixhuca circuit, the scene of many a drama in races gone by. (The Dave Friedman Photo Collection)

mechanics could hurt him. He figured I always had his health in mind, that I was taking care of him. That's the reason I got his car because in the Tasman series, the previous one where we came down to NZ with the 2-litre V8 Climax, we got on quite well and it never missed a beat, never frightened him, nothing ever fell off, so, I suddenly ended up with his 49, on his request.'

Although the concept and execution of McCall's extra part was actually very sound, Chapman's reaction was probably more attributable to his NIH (Not Invented Here) attitude. The thought that one of his mechanics could have come up with a solution which alleviated the problem was probably too much to swallow! Nevertheless, the new part stayed on the car for the rest of the weekend, so perhaps he thought there was something in it after all ...

On the first day of practice, the mechanics found bearing metal in the oil after warming up Solana's car, 49/R1, so a decision was taken to install a fresh engine. As this happened just at the start of the session, it was decided to let Solana have a run in Clark's car. Again, he was impressive, working down to a respectable time after only a few laps. Clark then reclaimed his car and ended the day quickest, with a time of 1.48.97 round the long sweeping Magdalena Mixhuca circuit. Hill ended the day fourth fastest, on 1.50.63, with Solana down in 12th as a result of his limited track time. The next day Clark and Solana played musical chairs, with the Scot commandeering 49/R1 and the Mexican taking the wheel of Clark's regular mount. The main reason, as McCall remembers, is that 49/R1 had the fresh engine in: 'It was one of those things where the motor in the car [49/R2] had done a lot of mileage, and R1 had a fresh engine, so Moises, who was renting a car, got to drive the car with the worn-out engine, so to speak.' As a result, he also got to drive the car with McCall's newly fabricated bracket.

On Saturday, Hill's 49/R3 began by smoking badly, a problem traced to a fractured oil pipe. Clark was clearly driving well within himself, getting down to 1.47.56, but capable of getting into the 1.46s if required. Behind him, the grid was tightly bunched, with Hill ending up fourth fastest on 1.48.74. Solana improved his time by more than two seconds, although this only moved him up three places on the grid to ninth.

Overnight, yet another crack had been discovered in the tub of Clark's 49/R1 in the area of the mounting bracket for the lower right rear radius arm. This was the part which had been patched up in Canada after Clark's Thursday shunt and which was already supposed to have been 'strengthened' many times since those failures on the first day of testing at Snetterton back in May.

An estimated 125,000 people crammed into the circuit on race day, which was hot and dry. Crowd control was negligible - as it always was at this circuit - and it was not long before spectators pushed through the fence and took up station on the grass and the grass banks at the side of the track, which afforded a much more spectacular view of the racing!

As the flag dropped, Clark got bogged down, with the result that the fast-starting Gurney shunted him from behind. One of the Lotus tailpipes went through the top of the Eagle's radiator and the American's race was over virtually before it had begun. Hill jumped past Amon into the lead and Clark quickly regained position, boosted by the 'push-start' he had received from Gurney and was into third position by the time they reached the hairpin on the far side of the circuit for the first time. Solana had made a storming getaway and was up into fifth place so that, as they crossed the line for the first time, Lotus were running 1-3-5. On the second lap, Clark displaced Amon and the next time round he was ahead of Hill, a position he would not relinquish for the rest of the race.

The order settled down, although Hulme managed to slip past Solana on lap 12. On the following lap, the Mexican disappeared, his car having suffered front suspension failure. The pin holding the lower left link to the upright had broken under braking, putting the entire load on the top of the upright, which also then broke. Solana held onto the stricken car well and was able to bring it to a halt unharmed. The fortuitous nature of the decision to switch Clark into the car with the fresh engine was immediately obvious, for if the Scot had been in this, his regular car, his race would have been run. It also confirmed McCall's worst fears about the propensity of the 49 to break, and he left the team soon after this race to join McLaren.

Five laps later Hill was also out, his car's left-side driveshaft having broken at the wheel end. This caused the shaft to flail around and removed the spring/damper unit, giving the Englishman a nasty moment. Like Solana, he was able to bring the car under control and coasted in towards the pits and retirement. As it turned out, even if the driveshaft hadn't broken, he would have gone out fairly soon after, because when the car was drained for airfreighting back to the UK after the race, it was found that it had virtually run out of water.

Not for the first time, Jim Clark was left as the sole Team Lotus representative in a Grand Prix. He did not falter, and took the chequered flag a comfortable one minute and twenty-five seconds ahead of his closest challenger, Jack Brabham. In finishing third, Denny Hulme did enough to clinch the World Championship title, the second successive year that a Repco-Brabham driver had carried off the spoils. The team's victory was a testament to the reliability and ruggedness of the car and engine package, a complete contrast to the record of the much faster Lotus 49, which broke too often to give its drivers a realistic shot at the title.

The final Formula 1 race of 1967 was the non-Championship Spanish Grand Prix held at the Jarama circuit in Spain, just north of Madrid, three weeks after the Mexico race. This was intended as a 'dress-rehearsal' for the circuit and the organisers, the Real Automovil Club d'Espana (RACE), who were to host a full World Championship event the following May. The tight, twisty circuit had been designed by the Dutchman John Hugenholtz, owner of the Zandvoort circuit, and bore more than a passing resemblance to it. If anything it was twistier: Jim Clark reckoned that at least 23 gear changes a lap were required, which was more than at Monaco! Although it had been opened in July with a Formula 2 meeting, this was the first time that Formula 1 cars had run in anger on the circuit.

Team Lotus arrived hot-foot from Mexico, having

literally thrown their cars and the crates of spares into a transporter and driven straight from the airport down to Jarama. Clark was driving the same chassis, 49/R1, that he had taken to victory in Mexico, while Hill had 49/R3, which was still sporting damage to the rear as a result of the errant driveshaft in Mexico. As a result of this, he practised the first day in the spare car, which was a Formula 2 Lotus 48. As only four Formula 1 cars were entered, a number of Formula 2 machines were invited, and proved to be quite a competitive proposition on the tight track.

Clark took his by now customary pole position with a lap of 1.28.2, while Hill was a little further back on 1.29.7, a time which was equalled by Jackie Stewart in the Matra MS7 Formula 2 car. As this was fitted with an FVA engine, it was an all-Ford Cosworth front row! The race itself was something of a procession, with Clark and Hill not really troubled, the Scot finishing fifteen seconds clear of his team-mate, who had been forced to speed up at one stage when he came under attack from Jo Siffert's BMW-engined Lola T100. Once again, car problems afflicted the luckless Hill, and he finished the race clutchless.

The end of the season allowed Lotus, Ford and Cosworth to take stock and assess their achievements in this, their first year of Grand Prix racing together. Undoubtedly, it was a year of missed opportunities but there had been many prominent successes too. It seemed that, despite Duckworth's success in developing the FVA engine, rival engine designers did not take him seriously and were caught napping by the sophistication and sheer power of the DFV. There was only one other engine which seemed competitively powerful during the 1967 season - the V12 Weslake - and this was also designed and developed in England by another specialist engine tuning firm, Harry Weslake's small concern in Rye, Sussex.

The vital difference between the two engines was that, while the DFV was built to a standard design with the intention of being able to be put into small-scale production, the Weslake engines were individual masterpieces, hand-crafted but each subtly different, which made interchangeability of parts from one unit to the next virtually impossible. Looking back today, Duckworth cites this engine as the one that he thought most likely to provide a serious challenge to the DFV: 'We were fairly fortunate that the Gurney Weslake just wasn't built well enough. It looked the best one. They did win a race at Spa. Whether they would have done if we hadn't had our two failures ...'

Of course, much of the credit for the success of the Lotus-Ford combination in 1967 had to go to Jim Clark because his affinity with the cars he drove enabled him to bring them home to the finish in an almost undriveable state, when other drivers might have given up the unequal struggle. In a way, it seemed rather unfair that, despite his heroic performances, and the fact that he scored four wins - twice that of the man who won the World Championship - he only finished third in the final points standings. However, that merely served to underline the 'win or bust' reputation of Lotus and the fact that consistency is the key to winning Championship titles.

Despite not managing to put up a serious challenge in the end for the World Championship, morale in Team Lotus was still high. As Jabby Crombac explains: 'Each time they did not win was a lost opportunity. But you see, on the Sunday night when we hadn't won, the air in the team was 'S**t, we missed one, but it's OK, we'll win the next one'. There was so much confidence that it was sort of healing the wounds. The confidence for the next outing was the best way to forget about the disappointment. The disappointment comes if you don't win a race that is the only chance in your life to win. At Lotus we didn't have that, we knew that, OK we missed it today, but the next time it will be ours.' By the end of the season, the 49 had developed into quite a reliable racing car (in the hands of Clark, at least) and had won the last three races it had contested. Therefore, the prospects for 1968, with a number of enhancements to the basic package planned, looked very good.

Walter Hayes' decision to insist on two top-line drivers in Clark and Hill had been vindicated: a Lotus 49 had been on pole position for every Grand Prix race since its debut at Zandvoort, with Clark taking six of these and Hill three; a Lotus 49 had led every race it had contested; and a Lotus 49 had won every race it had completed without problem. The project had easily accomplished one half of Hayes' predictions of its likely achievements and the stage was set for the second part of this to be fulfilled in 1968.

Incredibly, the first World Championship round of 1968, in South Africa, was only seven weeks away when Jim Clark took the chequered flag at Jarama that Sunday afternoon in November, while the first round of the Tasman series was but a week later. Ahead lay a busy schedule for Team Lotus, for it was committed to fielding cars in both and this period also included the Christmas holidays!

The original pair of chassis, 49/R1 and 49/R2, were prepared for the Tasman series for Hill and Clark respectively to drive. At the same time, two new cars - 49/R4 and 49/R5 - were built towards the end of 1967, to the same basic design as 49/R3 but incorporating the modifications introduced during the latter half of the season. The third and fourth chassis, 49/R3 and 49/R4 (which first turned a wheel at Snetterton in late November, with Jackie Oliver at the wheel) were prepared for the South African Grand Prix, which called for the cars and spares to be packed away in crates and sent off by sea via Southampton to Durban, some three weeks before the event.

The fifth chassis, 49/R5, remained at Hethel, for it was to form the basis for the new improved version of the 49, the 49B. At around this time, it was also announced that a car was to be sold to an old friend of Lotus, Rob Walker, for Jo Siffert to drive, but that he would not take delivery until after the South African race. The team had no cars available since, for obvious reasons, they did not want to sell 49/R5. However, when the two Tasman cars returned, the situation would be eased and they could afford to sell a car or two.

The latter part of 1967 was also a frenetic period for Cosworth because they had not only agreed with Lotus to develop a de-stroked 2.5 litre version of the DFV suitable for use in the Tasman series but, following Hayes'

Colour Gallery 1

Guvnor's perks: Colin Chapman takes his latest creation for a quick run up the runway at Hethel in May 1967. (Ford Motor Company)

The simple, sleek lines of the tub allied with the neat installation of the DFV engine and its unrivalled power set the Lotus 49 apart from other Formula 1 cars of 1967. (Goddard Picture Library)

This view from the main grandstand at Zandvoort during the Dutch Grand Prix in June 1967 captures the moment when Hill's race came to an end. Mechanic Dougie Bridge helps Hill push his car the last few yards to the pits and into retirement. (Tony Walsh)

An historic moment: Jim Clark settles into a Lotus 49 for the first time at Zandvoort in June 1967 in front of a handful of onlookers, assisted by (from left) mechanic Allan McCall, Firestone's Brian Hayward and mechanic Gordon Huckle. (Goddard Picture Library)

Colour Gallery 1

Above – The tremendous torque advantage of the Lotus 49 was not such a factor at the high-speed Spa-Francorchamps circuit, venue for the 1967 Belgian Grand Prix. Even so, Clark still ran away with the race until spark plug problems intervened.
(Phipps Photographic)

A grand entrance: the first appearance of a 3-litre Formula 1 engine with "Ford" on the cam covers, at the 1967 Dutch Grand Prix, saw Graham Hill take pole and lead early on before dropping out and leaving team-mate Clark to take victory. (Richard Spelberg Collection)

Home win: Clark drove a well-judged race to come home first in the 1967 British Grand Prix. (Ford Motor Company)

Clark in full flight: the Nürburgring circuit was punishing on cars, particularly their suspension, and this shot – taken during the 1967 German Grand Prix – shows why. (Goddard Picture Library)

Colour Gallery 1

Out of luck: Hill seemed to have more than his fair share of problems during 1967. Even when he won the toss for the US Grand Prix, his car let him down and he had to allow Clark past into the lead. Here he leads the Scot through the Esses at Watkins Glen. (Phipps Photographic)

Maestro in Mexico: Clark drove away from the rest of the field in the 1967 Mexican Grand Prix and was never challenged. (Phipps Photographic)

Lotus 49

Colour Gallery 1

Opposite – Clark's last: the Scot started 1968 as he finished the previous season - with a dominant victory in the South African Grand Prix, his 25th and last. Here he leads Stewart's Matra, the Brabhams of Rindt and Brabham and the Honda of Surtees. (Phipps Photographic)

A congratulatory pat on the shoulder from a clearly delighted Clark for Chief Mechanic Bob Dance after the Scot's victory in the opening Grand Prix of 1968 at Kyalami in South Africa. (Ford Motor Company)

Lotus 49

It is January 1968 and a tanned Jim Clark sits proudly at the wheel of R2 in a Christchurch garage where it had been resprayed in the colours of the cigarette brand, Gold Leaf, after Players' sponsorship deal with Team Lotus had been announced. (Ford Motor Company

Fine weather, huge crowds and top Formula 1 stars pitted against the best local talent gave the Tasman series a unique appeal. Here Clark dives into Creek Corner at Warwick Farm in February 1968. (Paul Cross)

Colour Gallery 1

For the European stars, the Tasman races offered the chance to get away from the winter weather, hone their racing skills and earn a bit of money without the pressure that World Championship racing inevitably brings. Here Clark and Hill relax in the pits at Warwick Farm, February 1968. (Paul Cross)

Graham Hill joined the 1968 Tasman series for the Australian rounds in R1, but never really got the best out of his car. (Paul Cross)

Lotus 49

First time out: the virtually untested 49B was produced in around six weeks and took victory in its first race, at Monaco in May 1968. Here Hill rounds Station Hairpin. Note the wedge profile adopted with this car. (Phipps Photographic)

Below – A superb cutaway drawing, which appeared in the Italian magazine *Quattroruote*, details the second evolution of the 49B, with the high wings, as raced from the French Grand Prix onwards. (Richard Spelberg Collection)

1968 - Lotus Ford "49 B mk II"

La Lotus «49 B» monoscocca con motore Ford 8V - 3 litri, nella sua ultima versione. Il grosso alettone deportante ha i supporti fissati sui portamozzi posteriori.

Colour Gallery 1

After the problems with turbulence at Rouen, the works cars ran much higher-mounted rear wings for the 1968 British Grand Prix. This is Hill on the approach to Druids hairpin. (Ford Motor Company)

Wing mountings had to be hurriedly added to R2 when it was returned to the works from its period of extended loan to the Rob Walker team in time for Oliver to drive the car at the 1968 British Grand Prix. Note also the early DFV with the patched-up cam covers where the tin boxes had previously been and the strengthening strut between the two top links, similar to the one made up originally by Allan McCall back in 1967 and scorned by Chapman ... (Douglas Booker)

Still flying: despite the addition of a rear wing and nose fins providing considerable downforce, the cars were taking off when they got to the Nürburgring in 1968, just as they had the year before. (Goddard Picture Library)

Jo Siffert's victory in the 1968 British Grand Prix was richly deserved, coming after he saw off a race-long challenge from Chris Amon's Ferrari. (Doug Mockett)

This bustling scene of pit activity at the 1968 Italian Grand Prix meeting is one of those photographs where, the more you look, the more you see. From left, Bette Hill, mechanic Bob Sparshott using special wheel tool, Autolite man Brian Melia, Jackie Oliver chatting to a journalist, Colin Chapman dreaming up next year's car, Graham Hill in car, mechanic Eddie Dennis, Racing Manager Dick Scammell, mechanic Trevor Seaman on right rear wheel and Chief Mechanic Bob Dance on left rear wheel. (Doug Mockett)

The Rob Walker team took delivery of a brand new 49B, chassis R7, just in time for the 1968 British Grand Prix. It came complete with a 'customer-specification' height rear wing, but this did not stop Siffert running with the works cars in the race. He would go on to embarrass them on several subsequent occasions during the season. (Ford Motor Company)

Lotus 49

He might have worked his mechanics to the point of exhaustion, but Colin Chapman could also be generous when recognising success. This is Dave Sims' reward for being part of the 1968 Championship-winning team. The only problem was, they never had enough time to take a holiday ...! Bob Dance recalls that most took the cash instead - £200 is equivalent to around £2000 at late 90s values. (Dave Sims Collection)

With crowds lining almost every conceivable vantage point, local hero Bill Brack swings into the first corner at Mont Tremblant during the 1968 Canadian Grand Prix. (Bill Brack Collection)

Colour Gallery 1

...to the wire, literally! Hill's cable-operated pivoting wing made a crucial difference - actual and psychological - in the battle for the 1968 ...pionship during the final round in Mexico. Here he leads the flying Jo Siffert and title rival Jackie Stewart. Note the flapping bungee ... right strut, for which Bob Dance received considerable grief, and the rings of black adhesive tape on the wing to smooth airflow over ...vet heads ... (Phipps Photographic)

The 49s sent to the 1969 Tasman series were hybrids, being basically original spec 49s with 49B-style lower radius rod mountings and a high wing. Note the clearly visible control cable at the rear of the wing strut and the bungee attached to the leading edge of the wing. (Paul Cross)

Lotus 49

Left – Conditions for the Warwick Farm round of the 1969 Tasman Series could not have been more different to the year before. Here Hill splashes round before water got in his car's electrics. (Paul Cross)

Rindt was in a class of his own at Warwick Farm in 1969 with his spectacular, tail-out style of driving. (Paul Cross)

The spirit of the sodden 1969 Warwick Farm crowd was lifted by Rindt's superb demonstration of wet-weather driving. (Paul Cross)

A Winning Streak At Last

(... continued from page 64) discussion with Chapman in August at the Nürburgring, they had also undertaken to expand the supply of engines. DFVs were to be provided to a further two teams - Ken Tyrrell's Matra International which had signed up Jackie Stewart from BRM and Bruce McLaren's ambitious team, which was to have World Champion Denny Hulme on its driving strength. In addition, a new specification of DFV, designated the 8-series (the eight standing for 1968) was to be introduced for the South African race.

The conversion of the DFVs for Tasman use was a relatively straightforward undertaking, as Duckworth recalls: 'They [Lotus] wanted a 2.5 litre engine for the Tasman series. So I think some small deal was done ... and it must have been sorted out relatively early because it did tend to take a few months to make cranks. So we must have had forewarning of a fair time to get a few cranks made and some longer con-rods.' Had it not been such a simple exercise to convert the existing engines, it is unlikely that the project would have gone ahead. As it was, when the Tasman series was over: 'We just took the cranks and rods out, stuffed them on the shelf and put the standard rods, cranks and pistons back.'

- First Person -
Keith Duckworth on
Jim Clark and Graham Hill

"You could actually tell the difference between a Graham Hill engine and a Clark engine by the fact that Clark would have apologised for having over-revved it on two or three occasions and the valve gear would show no signs of having been over-revved, whereas Graham's had never been over-revved and the valve gear was quite often tatty! I think Clark just changed gear gently, didn't he? There was never any hurry about anything, he had bags of time because he was incredibly good.

Graham was really an exceedingly courageous driver, a very brave driver, because I think that he was running at a higher percentage of his 'tenths', a higher number of 'tenths' than Jim ever did. I think Jim had prodigious natural ability, whereas Graham was working hard at it."

Clark: a natural (Ford Motor Company)

Hill: courageous (Brier Thomas)

Chapter 10
From Triumph To Tragedy

Jim Clark's dominant displays in the final 1967 Championship round in Mexico and the non-Championship race at Jarama had suggested that another Clark/Lotus walkover was in prospect for 1968. As it turned out, the year proved turbulent in more ways than one.

Changes within the Formula 1 team saw Dick Scammell move across to the Indy section and Bob Dance appointed Chief Mechanic. Bob was another Team Lotus old hand, having joined in August 1960. He was trained to build the 'Queerbox' sequential change gearbox (another Lotus concept some decades ahead of its time), after which he had become involved in the development of the Lotus-Ford twin-cam engine. Between 1963 and 1967 he worked on the fabled works Lotus Cortina team, with which he enjoyed considerable success. His meticulous approach and attention to detail had not escaped Chapman's attention, with the result that he joined the Formula 1 team for the final non-Championship race of 1967, and was elevated to the top mechanic's job for the following season. Jim Endruweit, carried on in his role as Racing Manager for the whole of Team Lotus, with responsibility for the technical side of the business, while Andrew Ferguson also continued as Competitions Manager, looking after the administrative side of things. The laconic Endruweit, a former Air Articifer with the Fleet Air Arm division of the Royal Navy, had worked his way up through the ranks from when he joined as a gearbox mechanic in 1958 (this seemed to be a favourite starting position at Team Lotus!) to the position of Chief Mechanic in the early-to-mid 1960s, and thence to Racing Manager.

Because of the rush to prepare for the South African Grand Prix, the Tasman series and building a further two 49s, little testing, development work or design changes occurred in the latter stages of 1967, and the cars that appeared at Kyalami were to the same specification as the ones which had finished the Jarama race.

Practice for the South African Grand Prix got underway on Thursday, 28th December 1967. This made for a rather hectic festive season for Team personnel, many of whom arrived late on the Wednesday, having flown from England on Boxing Day evening. Due to delayed flights, some drivers even missed the first session, but that did not really matter as there were three full days of practice in which to get accustomed to their cars and the track.

Clark was one of the first drivers onto the track and, with a well-sorted design (albeit with a car that had barely turned a wheel) he was quickly down to a time almost two seconds under the qualifying record. However, this performance was not just attributable to the superiority of the Lotus 49/Clark combination, for the entire 2.5 miles of the Kyalami track had been resurfaced since the teams last visited. The Scot covered more laps than any other driver and steadily worked his way down to a time of 1.23.9 by the end of the day, an impressive six seconds below the record.

Team-mate Hill arrived in time for the Friday practice session but, in contrast to the untroubled progress of Clark, he experienced a frustrating session, with a fuel pump cutting out and causing him to stop out on the circuit, clutch problems and then the link from the rear anti-roll-bar to the left rear upright worked loose - not once but twice. Despite these tribulations, he ended up second quickest, although some 1.6 seconds slower than his partner.

Clark's ability to adapt to the changing characteristics of a racing car was amply illustrated by the sequence of events that took place during the Friday and Saturday practice sessions. Intrigued by the return to Grand Prix racing of Dunlop, who were supplying the tyres for Jackie Stewart's new Matra, Clark asked to try them. Considering that Lotus at the time were contracted to Firestone, it seems amazing given today's cut-throat environment that he could get his wish, but he did. Having further reduced his lap time on his regular Firestones to 1.23.0, he went straight out on the Dunlops and knocked a tenth off this, while at the same time complaining of understeer! A bit more pressure in the tyres saw him go faster still, finishing the day at 1.22.4.

On Saturday he continued with the Dunlops. By now the car had been better set up to accommodate the tyres, with wider front rims, and he consistently circulated in the low 1.22s, well ahead of all his rivals. However, eventually he switched back to his Firestones for his final runs (although he maintained the Dunlop suspension settings!) to record a stunning pole position lap of 1.21.6. This was a second faster than his nearest rival Hill, who was fractionally quicker than Stewart, experiencing Cosworth power for the first time at a Grand Prix meeting, and who had acquitted himself very well in a car so new that it hadn't even been painted in Matra's familiar blue.

This was to be the last Grand Prix meeting in which the Lotus 49 would compete in the traditional Team Lotus colours of British Racing Green with a yellow stripe. Already, there were signs of the growing influence of the commercial sponsor or partner, with more prominence than in the past afforded to stickers on the cars advertising Firestone tyres, Shell fuels and oils and STP fuel additives. Even by this race the cars had begun to lose the clean, unfettered look they had displayed so beautifully during the 1967 season. By the next Grand Prix it would be gone altogether ... and so would Clark - truly the end of an era in more ways than one.

Shell had taken over as Team Lotus supplier of fuel and oil from Esso, which had pulled out of motor racing at the end of the 1967 season. Esso's withdrawal was quite a significant blow, for the company had not only been the Team's long-time supplier but its principal source of financial support since 1953. Of the other backers, the link with STP had been established in 1966 through the Indy programme - although this was the first time that the company had been identified on the Formula 1 cars - while Firestone was merely continuing a long-standing relationship with the team.

The Sunday following practice - New Year's Eve - was a day off for the drivers, even if the mechanics still had a lot of work to do preparing the cars. The race, to be held over 80 laps, started at 3pm on the Monday and was run in beautiful, if extremely hot, sunny conditions and clear skies. With a track temperature estimated at 130 degrees F (54 degrees C), cooling was clearly going to be a major challenge for teams, since this was considerably hotter than practice conditions and many cars had experienced severe problems then. After the morning warm-up, an oil leak was found on Clark's car in the fresh engine that had been installed on Sunday, which had itself been fitted because of a suspected cracked liner in the unit he'd used in practice. For once, all was well with Hill's car, which sat isolated in the pits while Team mechanics swarmed over the Clark car in an effort to solve the problem before the race. However, the work was completed in good time and Clark took his rightful place on the grid.

At the start, Stewart got the drop on Clark, beat him to the first corner and led him over the finish line at the end of the lap. His glory was shortlived, for Clark dived past on the second lap and began to open up a commanding lead, increasing the gap at around a second a lap. Meanwhile, Hill had got bogged down at the start and, from second on the grid, ended the first lap in seventh place. He quickly began to regain lost ground and, by lap 13, he was up to third, having passed Amon, Brabham, Surtees and Rindt.

By quarter distance, Clark was a comfortable 15.7 seconds ahead of Stewart, with Hill only 4.2 seconds behind the Matra. Seven laps later, Hill took second place from his former BRM team-mate and Clark stretched his lead to 17 seconds. Clark continued to steadily increase his advantage until it reached around 25 seconds, a margin that he had little difficulty in maintaining until the end of the race. The main challenge to Lotus supremacy, Stewart's Matra, was removed just after half distance when the Scot's Cosworth put a con-rod through the side of the block. After this Hill was able to consolidate on his second place although, in the closing laps Rindt, in his first race for Brabham, began to close on him and finished only 5.1 seconds behind.

Thus, the first race of the season concluded in the

South Africa 1968 marked the last appearance of R3 in Team Lotus colours. Multiple South African Drivers Champion John Love struck a deal with Colin Chapman to buy the car at Kyalami, although it was taken back to England and shipped out later after being prepared for sale. Hill drove to a steady second place behind teammate Clark. (Ford Motor Company)

Lotus 49

A garlanded Clark looks very content with his day's work as he drives into the paddock at Kyalami having completed his lap of honour. (Ford Motor Company)

The South African Grand Prix was the only race start that R4 made before it was burnt out in a fire shortly after having been sold to the privateer, Rob Walker. Here Clark guides his car through the crowd at the back of the pits after the race. (Ford Motor Company)

best possible way for Team Lotus, with a resounding 1-2 result and a new record race lap for Clark, seven laps from home, in 1.23.7.

The following weekend, on the Saturday, was the first Tasman race of the year, so Clark and Dale Porteous set off for New Zealand to join up with fellow Kiwi and chief mechanic Leo Wybrott and Dale's brother Roger, who didn't work for Lotus but was helping out on the trip down under.

The two cars sent to the Tasman series were designated 49T, the 'T' standing for Tasman. They featured a short-stroke version of the Cosworth DFV christened the DFW. This was necessary in order to conform to the regulations of the Tasman series, which called for a 2.5 litre unit instead of the 3 litres being used in Formula 1 at the time. This gave the DFW a bore of 85.67mm, a stroke of 54.10mm and an estimated power output of 350bhp at 9200rpm. Despite the problems associated with such a method of reducing cubic capacity, at the time this was still reckoned by Duckworth to be more than sufficient to give the Lotus cars an advantage over the opposition.

The only other changes made to the car were cosmetic ones to the bodywork, specifically the side of the nosecone, where flares were added just before the outlets on either side for the expended air from the water/oil radiator. This was an effort to increase the efficiency of the exits and, as a result, stop the drivers' feet being roasted. These flares were not added to the cars that went to South Africa, although the outlets had been enlarged for the same reasons.

Chassis numbers 49/R1 and 49/R2 were readied for the series, the former having been driven to victory by Clark in the Madrid GP and the latter having been the car raced with such promise by Solana in Mexico and by Clark before then. The plan was for Clark to drive 49/R2 in the first four races in New Zealand, after which Hill would join him in the other car for the Australian half of the series, having declined to take part in the early races due to a dispute with organiser Ron Frost.

The first race was held at the 1.75 mile long Pukekohe circuit, which took just under a minute to cover, giving an average lap speed of 106mph. Clark seemed to have little difficulty in adjusting to the lower power output of his DFW-engined car, although he did have to get used to something else that was new - Dunlop tyres. The latest Firestones had not turned up in time and so the Scot experimented with Dunlops, as well as the Firestones that were available, during practice. Since he set his fastest times on Dunlops, these were the covers he ran in the race, much to the embarrassment of Firestone, the team's official supplier. After putting his car on pole (for which he won 50 bottles of Champagne!) he was holding a steady lead over Chris Amon's 2.4 litre Formula 2 Ferrari when, on the 45th of the scheduled 58 laps, the engine's valve gear broke and it seized, leaving Amon to cruise home a comfortable victor and to bag fastest lap.

A week later, the series moved to the tiny 1.2-mile circuit at Levin on the North Island of New Zealand near Wellington, the nearest thing to Amon's home race. Clark - by now back on Firestones - again took pole position and led comfortably, only to make an uncharacteristic mistake and leave the track, allowing Amon through into the lead. In trying to regain lost ground, Clark slid his car into a rather solid kerbstone, which bent the rear suspension and forced him into retirement. This left the way clear for Amon to take a very popular second consecutive victory, and another fastest lap.

From Triumph To Tragedy

Clark was back in his regular R2 for the Tasman series. This bustling pit-lane shot is taken at Pukekohe in New Zealand, one of only two races the team contested in the green and yellow colours of Team Lotus before the Players deal was finalised. (Ford Motor Company)

The race at Levin was to be the last ever appearance of the Lotus 49 in British Racing Green, for the following week, Colin Chapman announced the link-up of Team Lotus with the John Player tobacco firm, which would sponsor the team at all levels with its Gold Leaf brand of cigarettes. Henceforth, the team was renamed Gold Leaf Team Lotus and all cars were to race in the brand's red, white and gold colours. Consequently, Clark's car had to be hastily repainted in the garage of a Ford dealer closest to the venue for the next race, in Christchurch.

The third race of the Tasman Cup, the Lady Wigram Trophy, proved to be a fairytale debut for Gold Leaf Team Lotus. The 2.3-mile aerodrome circuit was ideally suited to the 49 and the Scot romped home to a comfortable victory ahead of Amon and also claimed joint fastest lap with the New Zealander. He also bagged the NZ$1000 prize for being the first person to lap at 100mph in the race, making sure of it by doing so on only the second lap of the race!

Seven days later, the cars assembled again for the Teretonga Trophy at the Invercargill circuit. During practice, in an apparent effort to combat the aerodynamic lift being caused by the new generation of wider tyres, Clark and his mechanics tried an experiment which involved the fitting of a wing section on the back of his car above the gearbox, with the objective of providing download. Jabby Crombac recalls that Clark's interest in aerodynamic devices had been sparked by a race (one of only two single seater races that he ever did in which he didn't drive a Lotus) in a USAC car at Riverside in November 1967: 'It was one of those peculiar small

The wing: after driving an Indy car with a vestigial wing, Clark was keen to try one out on his Lotus and so his mechanics made this, from a sawn-off section of helicopter rotor. Angle of attack was determined by hanging it out of the window on the way back from the airfield! When Colin Chapman found out about it, he ordered it to be removed forthwith and it was never raced ...
(Roger Leworthy)

Indycars [a Vollstedt] and this car had wings on it, vestigial wings. Anyway, Jimmy was very impressed by that and from California, he flew to the Tasman series [via South Africa].' Leo Wybrott takes up the story: 'By the time we reached Christchurch down in the South island we'd obviously worked on the idea and kept thinking about it. And so, what with Roger [Porteous] being a helicopter pilot, we came up with the idea of using a piece of helicopter blade.

'We figured that the best place to go was the airport. At that time, deer-stalking with helicopters was very big in New Zealand in the South Island. They used to fly in to collect the carcases and smashed the tips of the rotors all the time. So there were plenty of helicopter rotor arms about. And that's what we acquired, for free, from the helicopter place at Christchurch Airport. Coming back, we decided that we would try to see what sort of effect it had. We'd taken quite a long section, a lot longer than we needed. So we had this out of the window of the car and we were working out how much force you needed to hold it up as you cranked up the angle. By doing this, we came up with a very, very rough guide as to the dimension down from the trailing edge to the edge of the window. Those were the original ideas of how much angle we gave it when we very first set it up on the car! It was supported at each side at the front off the top of the gearbox ears, and there was a stay that was adjustable with two rose joints down from under the middle of the wing towards the back, down onto the centre of the gearbox.'

Clark never actually drove the car with the device and it was discarded for the race because, according to Crombac, when Chapman heard about it 'he went bonkers! He said "You will not do anything to my car without me being there and designing the change" and he made them remove it. But it showed that Jimmy was very keen on it, very keen.' The experiment was not repeated again during the series, driver and team having no understanding of the significance of their endeavours. However, a young Ferrari engineer, Giovanni Marelli, who was working on Chris Amon's car at the time, took a number of photos of the device. Wings would not be seen on a Grand Prix car until the Belgian Grand Prix in June that year, despite the fact that they had been used successfully during the 1967 season by the Chaparral 2F sports car. It seems too much of a coincidence that Ferrari was, along with Brabham, the first to appear with such a device, although it was mounted in a different position to the Lotus experiment, being positioned just behind the roll-over bar, over the engine.

Back at Invercargill, rain played havoc with the cars in the heats, with the result that Amon's car did not manage to start. Appearing to favour their local hero, officials scratched the normal practice of putting the winner of the preliminary heat - in this case, Clark - on pole position and, much to Clark's disgust, reverted to original practice times to decide the grid for the final. This saw Amon on pole, with Clark alongside.

Despite the torrential rain, the Scot jumped into his customary lead and looked a certain winner until, with only seven of the 69 laps remaining, he spun off the track at 140mph after hitting a bump on the concrete main start/finish straight. The car veered onto the in-

These two shots, taken at the Surfers' Paradise circuit in Australia, illustrate the difference in stature of Hill and Clark. While Hill positively filled the cockpit, his head well up in the airstream, Clark sat further forward (courtesy of his aluminium shoulder support) and lower down. The shots also illustrate the flared nostrils on the side of the nose cone introduced for the Tasman series to try and improve the exit of air from the radiator, as well as the different fuel pump arrangements on the two cars, visible at the rear of the tub where the lower rear radius arm is mounted. (Brier Thomas)

From Triumph To Tragedy

This is the assembly area at Surfers' Paradise, with Hill togged up and ready to go, waiting for Bob Sparshott (squatting), Roger Porteous and Leo Wybrott (check shirt) to finish what they are doing on his car. The number of fingers in ears suggest that a raucous DFW has been fired up! Meanwhile Clark is more relaxed, as is his mechanic Dale Porteous (standing at back of car). On the far right, Firestone representative Brian Hayward looks on. Ford Australia on the nose was in deference to the financial support put up by the local subsidiary of the Blue Oval. (Brier Thomas)

field, cleared a safety ditch, flattened a wire-stranded fence and knocked down a heavy post but otherwise did not hit anything solid. However, the nose cone was knocked askew and so, after a quick inspection of the damage, Clark drove round into the pits to have it removed. The determined Scot then stormed back into the race and, despite the denuded look of the car, finished second, only 10 seconds behind winner Bruce McLaren in a BRM. The Kiwi had inherited the lead after the spins not only of Clark, but Amon and Frank Gardner too. Clark's only consolation on this occasion was the fact that he had set fastest race lap.

A fortnight after the finish of the New Zealand leg of the Tasman Cup, the series reconvened at the two mile Surfers Paradise track, near Brisbane, Australia. Graham Hill, with mechanic Bob Sparshott, had now joined Clark in chassis 49/R1 (which had been sent out by boat to Brisbane) to bolster the Lotus ranks for the remaining four Australian races. For the second half of the series, Amon was armed with a brand new four-valve V6 Ferrari engine, developing an extra 20bhp, so it seemed likely that the Kiwi would be providing even stiffer opposition than he had done to this point.

A petty dispute about advertising on cars had seen the BRM and Lotus teams sit out the preliminary heat (Clark had the paddock gate slammed in front of him by race officials to prevent him going out!) but, after a clarification of the rules, they were all back on the grid for the final and Clark raced to a comfortable victory. His team-mate had a more testing time trying to get rid of the Formula 2 McLaren of Piers Courage from his mirrors, something he never really accomplished, although he just managed to stay ahead to provide the team with its first 1-2 of the series.

This finishing order was repeated a week later at the

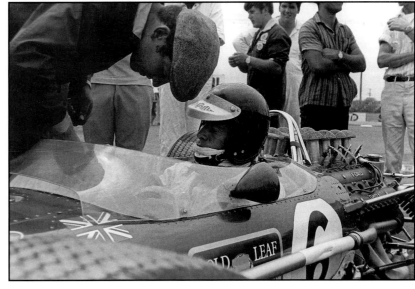

Warwick Farm circuit near Sydney, with Clark cruising to an easy win while Hill worked hard to keep Courage at bay, only crossing the finishing line four seconds ahead. The next weekend, the cars moved West to Melbourne and the Sandown Park circuit. With an average lap speed in excess of 100mph, its characteristics better suited the Ferrari than the tighter confines of Warwick Farm. Amon and Clark were therefore very closely matched and throughout more than 100 miles of racing were rarely more than a car's length apart. At the finish, the Lotus just had the advantage, but they were virtually side-by-side and the official margin of victory was given as a tenth of a second. Unbeknown to Clark,

Final preparations: Dale Porteous adjusts Clark's mirrors prior to practice. Note the sticker over Players sailor man on Gold Leaf logo and the dent in the side of the water pipe... (Brier Thomas)

The unmistakeable style of Jim Clark: head to one side, the Scot sweeps to a comfortable victory at Surfers' Paradise. (Brier Thomas)

Into Creek Corner on the first lap at Warwick Farm and, before a packed house basking in the sunshine, Clark leads Hill, Amon (Ferrari), Courage (McLaren-FVA) and Denny Hulme (Brabham-FVA). (Paul Cross)

Hill was less happy than Clark with the handling of his 49 in the Tasman series and often found himself fending off lesser cars. Here he is pursued by Leo Geoghegan's Repco-engined ex-Clark Lotus 39 and Piers Courage's McLaren-FVA Formula 2 car. (Brier Thomas)

Clark contemplates: the Scot used to get very keyed up before races and this manifested itself in his habit of biting his nails down to the quick. (Roger Leworthy)

Thumbs up for victory: A delighted Clark takes the chequered flag at Warwick Farm, having led from start to finish (Paul Cross)

Legends together: three of the greatest drivers of their era. From left Jack Brabham and Stirling Moss share Jim Clark's joy at winning at Warwick Farm. (Paul Cross)

Lotus 49

The Australian Grand Prix at Sandown Park was undoubtedly the most exciting Tasman race in the series. Jack Brabham upset the form book in his 2.5-litre Brabham-Repco BT23E and took pole. Here he takes off from the start with fellow front-row men Chris Amon (Ferrari) and Clark, with Geoghegan and Hill (just visible at left) on row two. (Chevron Publishing Group)

In the early stages of the race at Sandown Park, the top three ran in close company. After Brabham had made a shaky start, he closed up rapidly and looked likely to challenge for honours until his engine tightened up and he was forced to retire. (Ford Motor Company)

Close-run thing: Clark looks across at Amon's rapidly-closing Ferrari as they cross the finish line at Sandown Park. This was his final victory: five weeks later the unthinkable had happened and Clark was dead ... (Chevron Publishing Group)

From Triumph To Tragedy

and those who witnessed it, this close-fought win was to be his last in a Formula 1 car.

The concluding round of the series was held at the Longford circuit in Tasmania, the only genuine road circuit in the series. The roughly triangular shaped track, 4.5 miles in length, required drivers to go under a railway viaduct, cross two wooden bridges and a railway level crossing and ran through the outskirts of the town of Longford. Although tricky in places, on the main straights the fastest cars were reaching 150mph.

Normally run at a very fast pace, this race was ruined by heavy rain, with the start delayed and the race distance cut almost in half, from 28 to 15 laps. The difficult conditions were ideally suited to a more nimble Formula 2 car and Piers Courage duly obliged, fulfilling all the promise he had shown in previous races to take a well-deserved victory in his McLaren, powered by a 1.6 litre Cosworth-FVA (effectively one half of a DFV). Both Clark and Amon struggled round, the former struggling to put the power of his DFW down on the rainwashed track and finishing in a lowly (for him) fifth place. However, his Kiwi rival was back in seventh, just behind the second Lotus of Hill, and this was enough to secure the Tasman Cup title for Clark, who finished with a points tally of 44 to Amon's 36 and the 34 of the impressive Courage.

Although it had succeeded in taking Clark to the series title, the on-track performance of the DFW didn't truly fulfil Keith Duckworth's expectations: 'I think we were disappointed that it didn't go better. We lost more than we thought.' Duckworth's disappointment illustrated his constant quest for perfection and the fact that the engine had not been designed as a 2.5 litre unit from the start and, therefore, in his opinion, was a compromise. In contrast, the 2.4 litre engine used in Amon's Ferrari was a smaller capacity block which had been bored out and this had no difficulty keeping up with the DFW-engined Lotus because it was a more compact, lighter unit altogether.

Back home in the UK, a fifth Mark I specification 49 car, chassis 49/R5, had its first runs in late December and early January, shortly before the announcement of the link-up with John Player. When the deal was announced it was repainted and displayed to the press with a metal frame taking the place of an engine. The reason for this was that the team had no engines spare at the time! When 49/R3 and 49/R4 returned from South Africa, the 8 series engine from Hill's car was removed and fitted to 49/R5 for three days of testing at Snetterton in the hands of Jackie Oliver, regular test driver Graham Hill being otherwise engaged in the Tasman series.

While the two Tasman cars made their way back to the UK, work was beginning in earnest on new developments aimed at taking on the basic Mark I 49 concept a stage further. In particular, Chapman and his designer Maurice Phillippe sought to bolster the weak areas of the car that had become apparent during the 1967 season. The main concerns were the propensity of the lower rear radius arm mounting point to crack, the tendency for the existing arrangement to cause rear wheel toe-steer, the general weakness of the rear top links and the strains placed on the cylinder block by the existing mounting arrangements for the rear suspension assembly using the triangular tubular structure incorporated in the original design. Other important objectives of the revamp were to improve the car's weight distribution and its aerodynamics.

The problem of weight distribution was tackled through moving the oil tank to the rear of the car, 'saddling' it over the gearbox, with a small 'Coventry' alloy oil cooler mounted on top of the tank. In addition, various alternative positions would be tried for the battery, which was normally located under the driver's legs. One option involved the battery being recessed in the oil tank, but this was discarded as it resulted in too much weight being placed to the rear. While these experiments were being carried out, the prototype oil tank was held onto the car by a number of bungee cords! At the front, the team experimented with a revised interpretation of the 'beard' spoiler they had first tried back at Spa the previous year in an effort to compensate for the removal of weight from the front.

In February that year, the South African driver John Love had finalised arrangements to buy chassis 49/R3 from Team Lotus. The car was prepared for sale by the works on February 12th and the engine (DFV number 705, which had been first used by Hill at Mosport) was installed on March 8th. For the princely sum of £13,000 (about £134,000 at late 90s values) he acquired a machine which would enable him to dominate the South African Formula 1/Formula 5000 series for two full seasons until a technically superior rival (ironically a later specification 49) came along to eclipse it.

In early March, Rob Walker also took delivery of a 49 chassis, to replace his venerable old Cooper T81-Maserati that Jo Siffert had campaigned for the previous two seasons. The car sold to Walker was 49/R4, which Clark had driven to victory on its first ever outing in South Africa. Records show that the car was prepared for sale at the same time as 49/R3 and that DFV number 707 (the engine Clark had used in practice in South Africa) was fitted on January 31st.

Walker's ownership of the car was to be shortlived. In unofficial practice on the Friday before the Race of Champions meeting, Siffert crashed heavily, badly damaging the car, although luckily without injury to himself. The story goes that he hit a damp patch on the track, but it could also be that he overdid things in his enthusiasm at getting his hands on a machine capable of challenging for outright victories.

The bitterly disappointed team packed up and returned home to Walker's garage at Beare Green, near Dorking, Surrey, with no hope of repairing the car for the following day's meeting. The objective was to get it ready in time to go testing before the International Trophy at the end of April. However, during the process of stripping it down, the car caught fire, with catastrophic consequences, as mechanic Tony Cleverley recalls. 'It was a spark from a drill that caused it. It was something that had always been used and no one even thought about it. I suppose some vapour got out somewhere and just popped it off nicely. It wasn't too clever but it happens.' The fire quickly spread through the whole of the garage and what remained of the 49 (as well as several other

Lotus 49

For the Race of Champions at Brands Hatch an interim 49 appeared, with an oil tank and cooler mounted in 'saddle' fashion over the ZF gearbox and still held on by…bungee straps! This was due to a desire on Chapman and Phillippe's part to place more weight at the rear of the car and thus improve rear wheel traction. (Ford Motor Company)

A beard spoiler, similar to the one tried at Spa the previous year, was fitted to Hill's car in practice for the Race of Champions but was discarded for the race. Here the car is being pushed away from the old scrutineering bay at Brands Hatch. (Steve Wyatt/Quilter House)

Hill grapples with the appalling handling of his 49 while trying to fend off the attentions of Amon (Ferrari) and Hulme (McLaren) during the Race of Champions. According to Chapman, the car's poor performance meant 'we had to go back and scrap the whole lot and start again'. Six weeks later, the first 49B emerged from the race shop. (Classic Team Lotus Collection)

valuable cars including a Delage and a Lotus 18) was completely destroyed.

It was a devastating setback for the small, privately run (and financed) team, for it could not afford to replace the car and, anyway, another 49 was not available for purchase from Team Lotus. Walker faced the prospect of having to close down the team but his brother-in-law, Sir Val Duncan, intervened and lent him the money required to stay in business. An order was placed for a replacement 49 chassis, although it turned out that the team would have to wait some considerable time for it, an experience that would be repeated two years later when it ordered a Lotus 72.

At the Race of Champions, only one works 49 - the interim 'development' car, 49/R5, was entered for Hill, since tax-exile Clark did not want to waste his precious days in the UK competing in a non-Championship Formula 1 race. In the various guises it was tried in, the car's handling proved rather unpredictable. In the morning the arrangement with the battery sunk into the oil tank was tried, but for the afternoon it had returned to its normal place. Wider, deep conical rims were also being used for the first time to make the most of new Firestone rubber. The final difference was that the water pipe that had always run along the side of the tub was now relocated to the place it was supposedly intended to run - along the channel cut into the side of the tub at the bottom.

The combination of all these new elements, plus the unwanted distraction of a row over the displaying of the Players' Sailor Boy logo on the car - which saw him black-flagged until it was covered up with masking tape - meant Hill struggled in vain to find a balance. In the end, he qualified the car in a lowly (for a 49) sixth place. On race day, he made a lightning start to jump up to fourth, after which he proceeded to hold up a queue of cars consisting of Amon's Ferrari, Hulme's McLaren and Ickx's Ferrari. It was almost a relief - both for Hill and his pursuers - when a driveshaft broke and punctured the oil tank, causing a moment which tested his reflexes to the full as he sought to bring the car to a safe halt without hitting anything.

To keep them racing, Team Lotus lent chassis 49/R2, just back from the Tasman series, to Rob Walker/Jack Durlacher Racing. This was essentially in 1967 trim apart from the small modifications that had been made to the ducting on the side of the nose. This extended loan lasted until midway through the season, when the private team finally took delivery of another 49.

Following the Race of Champions, 49/R5 was dismantled and rebuilt to become the first definitive 49B. While the battery was retained in its original position, the 'A-arm' top rocker arms, which had previously been at a 90 degree angle to the monocoque, were swept forward, lengthening the car's wheelbase by some three inches and having the effect of moving weight distribution slightly more towards the rear. The whole of the rear suspension was heavily revised, with the introduction of a new, more robust fabricated steel crossmember, to which the suspension, oil tank/cooler unit, less sharply angled spring/damper units and a revised rear upright were attached.

The new crossmember had the effect of reducing

the loads on the cylinder heads of the DFV, because it was mounted on the gearbox casing, thereby solving a problem which had worried both Team Lotus and Cosworth for some time. A revised arrangement for the lower rear radius arms was also adopted which saw the abandoning of the mounting points on the side of the monocoque which had proved so troublesome during the 1967 season. They were replaced with lower, longer rods mounted in cut-outs on the underside of the tub. The water pipe was also positioned back on the side of the tub, due to the fact that it had been found to interfere with access to the fuel pump and the lower rear radius rod.

Extensive consideration was also given to improving the transmission. While the reliability worries concerning the ZF box seemed to have been largely solved by the modifications that took place as the 1967 season progressed, there remained one crucial problem - the ratios were fixed and only the final drive could be changed, which meant that it was an immensely time-consuming process to change gear ratios. This resulted in Team Lotus taking three gearboxes, with differing ratios, for each car to all the meetings, the consequence being that if none of them was quite right the drivers were stuck with these ratios for the duration of the meeting. With the sudden surge in power of the DFV, it was also particularly important to get the gearing just right for each corner at a circuit. In addition, all the strengthening which had taken place during the 1967 season had added to the weight of the ZF unit, thereby negating some of its original advantages.

The obvious alternative to the ZF box was an off-the-shelf unit from the British Hewland Engineering concern. The Lotus boss had never really rated Hewland, and they had never seen eye-to-eye, maintaining a mutual antipathy towards one another which never really changed, as company founder Mike Hewland recounts. 'Chapman didn't like me, and I didn't like him, but that's neither here nor there! He said I was a 'bloody blacksmith', or something. Well, that was allright. I didn't care, I rather admire blacksmiths!' Nonetheless, Chapman was forced to swallow his pride and admit that the Hewland FG400 box, which would allow relatively easy ratio changes to be made at the track, was a more sensible option for the 49B.

This decision came as no surprise to Hewland, who had long predicted that the ZF box wasn't up to the job. 'Chapman said that the business of having a gearbox with the ability to change ratios was nonsense, it wasn't required, and what you did was you changed the gearbox. I went up [to Hethel] and I saw the size of the final drive gears and knowing what the torque and that was on the engine I thought 'Well, he'll be lucky for that to work'. In fact, what happened was, the side plates weren't strong enough and the diff came out of the side of the casing. Then they went into a stupid programme of making stiffer side plates and this, that and the other. I think they wound up with cast iron sideplates or something! By the time they'd finished it was very heavy and it didn't work, because the final drive gears weren't big enough.'

The FG400 gearbox that was chosen was a hybrid unit, as Hewland explains. 'It consisted of the FT200 gearpack put onto a DG300 final drive section.' The naming of Hewland's gearboxes is a story in itself, as the company's founder recalls. 'The FG stood for half of one and half of the other. It was the FT and DG boxes put together, as simple as that! FT200 stood for what it says - Formula Two, 200 horsepower. That was what I was told in 1965, when I did the original design work. LG stood for Large Gearbox because it was a large gearbox! At that time, the use of American V8 engines was getting fairly common in McLarens and Lolas, and so on, and it was used in Can-Am a great deal.

'DG stood for Different Gearbox. I can tell you exactly what happened there. I was drawing away at this thing and I said "Oh Christ, what are we going to call this gearbox?" And the woman from accounts came in and said, "Oh, is it a different gearbox?" and I said "Right, that's it - DG!" The DG box came into being because there was a mooted four and a half litre sports car formula that died a death. We made about 12 of them and they sat on the shelf for a year or more. Then they went into the back of Jack's Repco-Brabham and very conveniently won the World Championship twice. On the back of that engine it was bullet-proof.' Later in the 1968 season, the more robust DG300 would also be used by Team Lotus in an effort to ensure maximum reliability and, subsequently, was retained for the rest of the 49's active Formula 1 career. Other improvements in the driveline area included the adoption of constant velocity (CV) type driveshafts.

Several aerodynamic improvements were also introduced. Moving the oil tank and cooler out of the nose and to the rear of the car meant that the only remaining function of the radiator in the nose was water cooling. For this reason, it could be smaller than before, and this allowed the nosecone to take on a more downwards sloping attitude than before, lowering the front of the nose. Small adjustable fins were added to the side of the nosecone to provide downforce at the front. This was trimmed out at the rear by a gently sloping engine cover, the objective of which was to provide download on the rear of the car. It incorporated a small NACA duct to permit a clear flow of air into the rear oil cooler, with an outlet at the back to facilitate extraction.

One might imagine that hours of extensive windtunnel research had gone into determining the optimum shape and angle of attack for this engine cover, but this was not so, as Bob Dance explains: 'We had to help fabricate things and make bits and pieces. I got the job of making that streamlined tail in clay, which I'd never ever had anything to do with in my life! Chapman said "What I want is this" and gave me a sort of thumbnail sketch and I thought, "Christ, where do I start?" So I was lobbing this clay all round the back end of the DFV and gearbox, which were covered up. Then I started scraping away and the Old Man would come along every now and then and look and say "No, no ... that's not what I meant at all!" until it was how he wanted it.'

Deeper, cone-shaped split cast magnesium wheels were introduced to accommodate ever-widening Firestone tyres. At the front these were 12 inches across (a 50% increase over the car's original Zandvoort specification), while at the rear they had grown to a width of 15 inches, compared to the 13 inches seen in 1967. Both front and rear rims remained 15 inches in diameter.

Lotus 49

Close-knit: the team which went out to the 1968 Tasman series was very small and they spent a lot of time together. Chief Mechanic/Team Manager Leo Wybrott, right, was devastated by Clark's death and left Lotus shortly afterwards, although he was to return later that year. (Ford Motor Company)

Other 'under-the-skin' changes made were a switch to Cam Gears rack and pinion steering, and the introduction of Girling AR-type brake calipers front and rear. Due to these changes, the wheelbase of the car had grown from 95 to 98 inches, with the front track having widened to 62.6 inches from 60 inches. The weight of the car also increased by 78lb, to 1180lb 'dry' and 1640lb 'ready to race'.

In the midst of all this work, Team Lotus was completely devastated by the news of the death of the seemingly immortal Clark in a Formula 2 race at Hockenheim in Germany, on April 7th. Clark had been competing in the first round of the European Championship when he left the track at a flat out right-hand curve and hit a tree side-on. The impact killed him instantly. Although the precise cause of the accident was never determined, the most probable explanation is considered to be a punctured right rear tyre.

In *Jim Clark - The Legend Lives On* (Patrick Stephens Ltd, 1989), Graham Gauld suggests the likely sequence of events: 'Jimmy probably picked up something in the tyre on the fourth lap. (In practice the previous day, Walter Habbegger had broken the crankshaft of his car, and a small fragment of metal may have been lying in the road). It didn't need much loss of pressure for the effect to be felt and, when Clark started to go through the twisting piece of track in front of the grandstands, the rear-tyre pressure was possibly down from 15psi to 10psi. When he reached one sharp left-hand bend a fellow driver saw him slide and then quickly correct it. As the next corner was a sharp right-hander, the weight of the car would be thrown on to the left-hand side and away from the right rear tyre, and it is assumed that Jimmy put his trouble at the previous bend down to the greasy surface of the track. Even if the pressure had dropped to 8psi, the tyre would have stayed on the rim as it was a slow right-hand corner. However, once he opened out on the back straight, the steadily rising speed and the steadily deflating tyre brought the power of centrifugal force into play.

'In tests following the accident it was found that, under these conditions, the tyre would begin to grow and at a crucial point the sidewalls would be drawn in, and there would be an explosive depressurisation as the air rushed out. From that moment an accident was inevitable.' Not even Clark's legendary car control could have saved him in those circumstances. The only eyewitness of the accident reported how Clark's car had slid one way and he corrected it, then the car slid the other way and speared off into the trees at virtually unabated speed.

Ironically, it was probably Clark's uncanny ability to instantly compensate for the changing handling characteristics of his car that contributed to his demise. Jack Brabham had several firsthand experiences of this, including the Italian Grand Prix at Monza and a Formula 2 race at the Rouen-les-Essarts road circuit. 'He didn't have a very good feel for the car which, I am absolutely certain, was what got him in the end. I raced against him at Rouen once. There were four of us in the race for the lead, always passing one another, and Jim had a tyre going down and everybody knew about Jim's tyre going down except Jim! We could not believe that he didn't know his tyre was going down, and when he went off down the hill, on his last time down there, I just chickened out. There was no way Jim was going to get to the bottom of the hill with that tyre - he just couldn't do it.

'Sure enough, on the first right hand bend on the way down, he lost it. I had backed off to give them all plenty of room to have the accident. Unfortunately, the other two boys went past while he was in the bank, he bounced back and I took the whole radiator off the front of his car. Anyway, what I'm getting around to is that Jim just did not know that his tyre was going down - I couldn't believe it. And at Hockenheim, I am sure that is what happened to him. The tyre went down and he didn't realise it.'

Whatever the cause of his accident, the fall-out from Clark's death was immense and it altered the attitude of many drivers, mechanics and followers of the sport forever. For the drivers, it shattered their self-confidence, their belief that they could cope with almost all eventualities. If it could happen to someone as supremely gifted and talented as Clark, what hope did the rest of the 'mere mortals' have of surviving? For the mechanics and fans it seemed as if some of the magic had gone out of the sport. It was truly the end of an era. Even twenty five years on, conversation between the mechanics who were working for the team at the time becomes hushed and slightly awkward if the day of Clark's death comes up in conversation. In the same way as many people can remember where they were and what they were doing when President Kennedy was assassinated, so most people with an interest or involvement in motorsport can remember where they were when they first heard that Jim Clark was dead.

It seemed like the heart had been ripped out of Team Lotus, leaving many people, from Chapman downwards, wondering whether what they did was all worthwhile. Certainly, for some time after the accident, there was a question mark over whether Team Lotus would pull out of racing altogether, such was the devastating effect it had on the Lotus boss. It seemed that without his great friend, racing just did not hold the same appeal any more and that when Clark died, something inside Chapman was extinguished too. He had also lost a vital competitive edge, in the form of one of the fastest and most consistent drivers on the Formula 1 grid, as Jabby Crombac points out: 'He had such a special relationship with Jimmy that his loss was not only the loss of somebody he was fond of, but also he could see that his main trump card had been taken away from his game.' While he would establish good working relationships with several more of his drivers in subsequent years - notably Mario Andretti - a different Colin Chapman emerged post-Clark, vowing never to get too attached to his drivers again.

Crombac is in no doubt that, had he lived, Clark would have gone on to clinch a third World Championship that year, given the improved reliability of the 49, the technical superiority of the Cosworth engine and the high wings developed by the team: 'There is no doubt in my mind that Jimmy would have been the World Champion in 1968. The fact is that Graham won and Jimmy was quicker than Graham, period. Therefore Jimmy was bound to be the World Champion.' Such an outcome would have been justifiable reward for the frustrations he had endured during 1966 and 1967 but, sadly, it was not to be.

Despite what was by all accounts a terrible atmosphere in the racing shop - not helped by the presence of the wreckage of Clark's car under a sheet - work continued apace on the preparation of the first 49B. The intention had always been to have the car ready to race in the second round of the World Championship at Jarama on May 12th. As a result, little priority or importance was attached to the team's entry for the BRDC's International Trophy meeting at Silverstone at the end of April. Consequently, a singleton entry, in the form of chassis 49/1, still in 1967 guise, was made. Following the death of Clark, the event was approached with even less enthusiasm. Amidst all the chaos and upset, preparation for the Indy 500, with the Lotus 56 turbine-powered cars, was also reaching fever pitch towards the end of April, with Hill being heavily involved in the track-testing. Only a few days before the Silverstone event, he and Mike Spence had driven the turbines at the same track, prior to them being despatched in readiness for the beginning of the 'month of May' that constitutes qualifying and the race.

Hill qualified for the International Trophy in a lacklustre sixth position, unable even to match the pole time set by Clark for the previous year's British Grand Prix. This may have been something to do with it being less than three weeks since the death of his team-mate, but probably also reflected the fact that, armed with Cosworth-DFV engines too, other teams were catching up (and, indeed, showing signs of overtaking) Team Lotus in terms of speed and sophistication of design. Hulme's McLaren - which emerged fastest - was able to knock a second off Clark's record. The new, improved 49 clearly could not come soon enough.

After a moving silence in memory of Clark had been observed before the race - held on a Saturday - Hill was

running a steady third until he went out on the 14th of the scheduled 52 laps with a split fuel line. This brought an end to a depressing weekend for the team, although not for the ever-active Hill, who was entered in a Formula 2 race at Jarama the next day. Jo Siffert was also having his first outing in 49/R2 at the Silverstone meeting. Qualifying carefully in eighth spot on the grid (the consequences of another accident having been made very clear to him), he ran a steady sixth in the race until exactly half distance, when clutch failure forced him to retire.

On Tuesday May 7th, a month to the day after the death of Clark, Team Lotus was rocked by a second fatality: that of Mike Spence at Indy. He had taken over the seat in the 56 that Jackie Stewart was to have occupied. Stewart was to have taken the place of Clark but couldn't drive because he had injured himself in practice for the Formula 2 race at Jarama. Spence was the victim of a freak set of circumstances when the right front wheel swung back on impact with the retaining wall and hit him on the head; he subsequently died from his injuries. Coming so close after Jim Clark's accident, this was all too much for Chapman who returned to Hethel, instructing his Competitions Manager, Andrew Ferguson, to take over at Indy, saying he wanted nothing more to do with Indianapolis and racing in general. He had also decided not to go to the Spanish Grand Prix at Jarama the following weekend. As a consequence, although the new 49B was taken to the meeting, it did not turn a wheel and was kept under wraps in the garage, as Chapman wanted to personally supervise its first outing.

Hill and Racing Manager Jim Endruweit were left to hold the team together during this time, a job they accomplished very well. Hill's wife Bette recalls the weekend very well. 'Everyone in the team was distraught after Jimmy's death, but Graham pulled them together and we just knuckled down, the whole lot of us. They were brilliant, absolutely brilliant.' In an effort to lift spirits in the team, Hill took everyone out on the night before the race, as Bob Dance recounts. 'Graham and Bette took us all out to the Villa Rosa flamenco restaurant in the middle of Madrid and it ended up being quite a late night.'

With the 49B not available, Hill drove the same car as at Silverstone, the faithful old prototype chassis, 49/R1. This car was still in original 49 specification and it showed in his grid position of sixth, back on row three. Siffert, in the Walker car, was a further four places back. At the start, Hill lost a place to Surtees' Honda but within ten laps had passed both Surtees and Bruce McLaren to

In a bid to lift team morale in the wake of Clark's death, Graham Hill and his wife Bette paid for the entire team to have a night out at the Villa Rosa flamenco restaurant in Madrid on the eve of the Spanish Grand Prix. From left are Hill, Bob Sparshott (with his back to the camera), Sid Carr, Bette Hill and Bob Dance. (Bob Dance Collection)

For much of the Spanish race, Hill and Hulme ran this close, and it was Hill's gritty determination to win in an older car that kept the McLaren at bay. (Ford Motor Company)

move up to fifth place in a group running nose-to-tail behind the BRM of surprise leader Pedro Rodriguez. Frenchman Jean-Pierre Beltoise, in a Matra, eventually got past the little Mexican but only four laps later he was in the pits (his car had been smoking badly). In the smoke shroud following Beltoise, pole-sitter Amon (Ferrari) had managed to nip past Rodriguez and Hill had taken Hulme's McLaren, so that when the smoke cleared the order was Amon-Rodriguez-Hill-Hulme.

Approaching one-third distance, Rodriguez slid straight on under braking while trying to pressurise Amon and was out. This left Amon with a comfortable lead, which at half-distance in the 90 lap race was 16 seconds. Having seemed almost certain to break his 'duck', Amon's car ground to a halt with fuel pump failure at two-thirds distance. This suddenly galvanised Hill and Hulme into action; they had been circulating together, but not in real combat. Hulme pressurised Hill for the next twenty laps but, try as he might, could not find a way by as the Briton defended his position superbly. Eventually, Hulme eased off with around 10 laps to go, settling for a secure second place, and Hill crossed the line to take a well-deserved, popular and very timely win for Lotus. A race of attrition due to the searing heat in which the event took place, saw Brian Redman (Cooper-BRM) join Hill and Hulme on the podium, despite having finished a lap down. Although only 13 cars started in the first place - one of the poorest fields seen at a World Championship round for years - the rate of attrition was such that Beltoise, who lost nine laps in the pits, finished last, but in fifth place!

Hill could not have chosen a better occasion to score his first Grand Prix win for Lotus. It was an achievement that many felt was long overdue, since he

Siffert had his first race in the Rob Walker team's borrowed R2 at Jarama and ran well before his retirement. (Ford Motor Company)

had led so many times only to be forced out by mechanical trouble. It also provided Team Lotus with just the lift it needed at a time when morale could not possibly have been at a lower ebb, as Bette Hill remembers: 'Winning was the best thing that ever could have happened. Not only for Graham but also for the team and for Colin. It was wonderful. We were introduced to King Juan Carlos before the race and then he was able to present Graham with the trophy afterwards. It was magic - what was needed for everyone.'

Siffert's race in the Rob Walker car had been dogged by petrol leaking onto the pedals, which made them slippery, and then a transmission noise and vibration. Unable to discover quickly what the problem was, the team withdrew the car before anything broke.

And so the curtain fell on the front-line Grand Prix career of the Lotus 49 in its original form, for the next race at Monaco would see the debut of the long awaited 49B. Although the 49 would continue to be campaigned by Team Lotus for one more race and by Rob Walker's team until mid-season, team-leader Hill was about to step into a new era of wings and downforce.

Vital victory: a relieved Hill takes the plaudits of the crowd on the victory rostrum in Spain, flanked by (left) third-placed finisher Brian Redman (who looks thoroughly underwhelmed by his first 'podium' finish) and second-placed man Denny Hulme. This win really boosted morale at Gold Leaf Team Lotus and was instrumental in encouraging Colin Chapman to continue with Grand Prix motor racing.

- First Person -
Bette Hill on Graham's relationship with Jim Clark and Colin Chapman

He and Jimmy both had the same aim and they both knew that Colin was a hard task master, be it there were favourites. They wanted to beat each other, there was no question about that, and Colin wanted Jimmy to beat Graham, there was no question about that, either! Yet they had tremendous respect and admiration for each other - all three of them. It was difficult for Graham. He had come into an established relationship. It was like being the lover on the outside. Jimmy was, without a doubt, Colin's favourite and in many cases both Graham and I felt that Graham was not getting fair do's.

Colin was a tough character from the word go. I always used to say I wished I'd liked him all the time, or hated him all the time! He used to floor me by his toughness and then he would do something quite gracious and I'd say "He's not that bad." He was a hard taskmaster and he didn't ask anything of anyone that he wouldn't do himself. I never heard Graham say anything against Colin except: "He's tough and that's why he's successful. He wants the best out of everything that he's put into it." But there were times when I just had to walk away!

Hill at Team Lotus: 'like being the lover on the outside.' (Ford Motor Company)

Chapter 11
Turbulent Times

The problem of who was to take the seat left vacant by Jim Clark was resolved early in May when, on the recommendation of Endruweit, Chapman selected the young, up and coming Essex driver, Jackie Oliver, who remembers it well: 'I tested a car with a view to taking over Jimmy's place. It was really down to Jim Endruweit, who'd seen me driving in all the other formulae in testing and racing.' Oliver, who had been a works or works-supported Lotus driver since 1965, driving Lotus-Cortinas and sports cars as well as Formula 2 and 3, found out that he had got the nod, after a test to evaluate him at Snetterton, only a matter of days before the Monaco Grand Prix: 'It was conclusive enough with regard to the performance of the car. They had records of what a 49 could do round there in the hands of Graham. And that performance persuaded Colin to give me that Formula One opportunity.'

Hill was to debut the reworked chassis 49B/R5, while Oliver was entrusted with the Spanish Grand Prix-winning car, 49/R1. Not only was it the first event for the 49B, but this race also marked the first appearance of the Cosworth DFV on the streets of Monte Carlo. Hill was into his stride straight away in the new car, setting fastest time in the first session on Thursday afternoon. Oliver played himself in gently, emerging joint sixth in a time that would ultimately be his quickest of the three practice sessions, while Siffert suffered transmission woes with a broken crownwheel and pinion and universal joint.

Following the deaths of Clark and Spence, Chapman had vanished without trace for several weeks. On the Thursday evening, he suddenly reappeared in the Lotus garage as if he'd never been away. For the mechanics, their work almost completed, this was bad news since he began to draw up one of his legendary job-lists, as Bob Sparshott recalls: 'He breezed in, about nine o'clock at night. We were working away and he played it just as if he'd been in the day before, despite the fact that no-one had really seen him since the tragedy. He wanted to look round the 49B because he hadn't actually seen it! Bob [Dance] followed the "Old Man" round as he filled up his pad with things that needed doing until there were at least two pages of jobs. I threw a wobbly and he took me out to the back where there was a little balcony. He put his arm round my shoulders and asked "Now, what's the matter?" I said, "Well, it's ridiculous. We've done all this work, carted this car around. Now you've breezed in and we've got all these things to do before the morning. It's impossible. We can't do it, it's not right". Anyway, in about ten minutes he'd sweet-talked me, and I was good for another 24 hours!'

The early morning practice session on Friday saw the young French driver Johnny Servoz-Gavin throw down the gauntlet to his rivals, dipping below Hill's Thursday time by a tenth of a second. Substituting for the still-injured Jackie Stewart in the Matra-Cosworth, it was his first appearance in a Formula 1 car, although he had driven a Formula 2 car in the same event the previous year. Demonstrating that he was now coming to grips with his Lotus 49, Siffert equalled the Frenchman's time, but suffered a recurrence of the previous day's trouble, preventing him from going for a faster time. However, just towards the end of the session, Hill popped in a quick one to pip his younger counterparts to the post by over half a second: a time that would subsequently be good enough for pole on Sunday. More importantly, the new car ran like clockwork and experienced only a few minor problems with the oil overflow, which were soon solved. In contrast, Oliver's lot was not a happy one, for he had similar problems to Siffert and returned to the pits on foot mid-session with a broken piece of driveshaft in his hand. As a result, it was decided to put the cast-iron side plates - which had first been used in 1967 - back onto the ZF gearbox of his car and that of Siffert, since it appeared that the box was flexing under power.

By the time the cars went out for the final session on Saturday afternoon, two Formula 3 heats had already laid

Turbulent Times

rubber and oil down on the track and a steady drizzle further dampened drivers' hopes of improvement. Although the track dried towards the end of the session, it was clear that it was generally slower than the previous day and only a handful of people - none within the top six on the grid - improved. This left Hill on pole, with the upstart Servoz-Gavin alongside him on the front row. A delighted Siffert was third, his highest ever grid position. Although 'Seppi' set the same time as Servoz-Gavin, the Frenchman took the front-row spot by virtue of having done the time first.

Despite the problems in final practice, which had prevented him from improving on his Friday time, Oliver acquitted himself reasonably in his first appearance for Gold Leaf Team Lotus, and he ended up on the seventh row in 13th place. Even today, there are very few drivers who would choose to make their Formula 1 debut on the streets of the tiny Principality, since even for the most established and experienced drivers it was (and still is) an exhausting, demanding and nerve-wracking circuit. With admirable honesty, Oliver admits that he was struggling: 'Not having started the season and driving a Formula 1 car for the first time at Monte Carlo – I'd never been there before – it was too difficult for me.'

To make matters worse, he had already incurred the wrath of Chapman, as Endruweit recounts: 'Oliver was in bad odour. One of the problems was that he'd taken his girlfriend to Monaco with him and this got right up the Old Man's nose. Here he was giving this young driver an opportunity and he had the temerity to take his girlfriend! Instead of sitting in the garage pondering over the car at night he was out swanning around ... so he was in trouble long before the race!' Endruweit had driven down to Monaco via Germany, where he had picked up some driveshafts for the team. When he arrived, he received the full stinging force of Chapman's anger: 'I got there, walked in and got a b******ing. Nothing about "Oh good, nice to see you, have you got the driveshafts?!" He was ranting on about Jackie Oliver and it was all my fault!! So I had to go and give Oliver a real b******ing.'

After Bandini's fiery and, ultimately, fatal accident the previous year, the race distance had been shortened from the traditional 100 laps to 80 to prevent drivers tiring too much, and the circuit featured considerably more armco barrier than had been seen before. Certainly straw bales were conspicuous by their absence. Race day thankfully dawned dry, warm and sunny. At the start, Servoz-Gavin leapt away into the lead and pulled out a good advantage in the opening laps, pursued by Hill, Surtees in the Honda and Siffert.

However, Oliver's race was shortlived: on the very first lap, shortly after coming out of the tunnel, he hit the back of the Bruce McLaren's car, which had been in an incident with Scarfiotti's Cooper-BRM. He takes up the story: 'Well, I didn't qualify very well and was stuck at the back of the grid. The tunnel was the old-style tunnel with light-bulbs hanging down the centre so hitting daylight was always a visual problem and, on the opening lap, I ran into somebody else's accident. Maybe someone a bit more experienced could have avoided it. Scarfiotti and Bruce had tangled and I collected Bruce as I came out of the tunnel and took all the wheels off one side.'

The accident pretty well wrote the car off, so it was with a heavy heart that Oliver set off back to the Lotus pit to explain to Chapman and the rest of the team what had happened. The situation was all the more painful for him, since the last thing that the Lotus boss had said to him on the grid was that he should stay out of trouble on the early laps, and drive to finish. Today he smiles ruefully as he recalls: 'He'd told me to stay out of trouble, and those words were still ringing in my ears. It was one of the first of many times I got fired!'

Bob Dance remembers the occasion with great clarity: 'Before the race the Old Man briefed Oliver and said "Whatever you do, you've got to finish, right? If you finish you're in the points, I guarantee you are in the points! So, all you have to do, first lap, you take it sensibly, don't shunt it, don't get involved in anybody else's accident because, at the end of the race, you'll get a point because there can only be five or six cars finish."

The result of Chapman's rethink of the 49 was the 49B, which was completed in May in time for the Spanish Grand Prix but not raced due to Chapman's non-attendance. These shots, taken on the runway at Hethel, show the new nose fins, fatter tyres, wedge engine cowl with NACA duct for the oil cooler, swept forward front rocker arms, and new method of mounting the lower rear radius arms in cut-outs at the bottom of the tub. (Classic Team Lotus Collection)

Lotus 49

The 49B's wedge engine cowl was neatly waisted at the rear, with an exit duct in the back for the air from the oil cooler. Also visible are the Hewland FG400 gearbox, which replaced the ZF, and the CV joint driveshafts. (Aldo Zana)

Jackie Oliver made his debut at the Monaco Grand Prix, taking the place left vacant by the death of Jim Clark. It was a difficult start to his time with the team, his race ending in a first-lap shunt... (Ford Motor Company)

'So everybody all lines up and we're off. First lap - no Oliver! Finally he appeared on foot and he got a huge blasting: "What did I tell you before the start of the race? Why did you go and shunt the bloody car!" And he couldn't get away. The pits had the race track passing on both sides and he was hemmed in for the rest of the afternoon!'

Oliver also felt he was a convenient target for Chapman's anger at the time, in that the Lotus boss was still trying to come to terms with the loss of Clark. 'He was fine up until Jimmy died, but afterwards he became quite difficult in terms of what he demanded - expecting me, and other people, to step into Jimmy's shoes and, of course, no-one ever could. That frustrated him and that frustration was vented on everybody else.' Looking back, Bob Dance tends to agree: 'The Old Man was extremely difficult to work with at that point in time.'

Chapman's prophecy proved absolutely right for, incredibly, after only 15 laps, the field was reduced to just seven cars! Servoz-Gavin's lead was shortlived for he went out on the fourth lap with a broken driveshaft. This left Hill in command, able to dictate the pace from the front, ahead of Siffert and Surtees. Siffert soon went out with a broken crownwheel and pinion - the third time that weekend - and after 17 laps Surtees limped into the pits with a broken gearbox. This brought the total number of retirements to 11 out of 16 starters. Amazingly, there were no more retirements in the remaining 63 laps of the race, which become something of a procession with Richard Attwood's BRM the only other car on the same lap as Hill at the chequered flag.

For the third consecutive race, a Lotus 49 had taken victory, with Hill scoring his second win in a row, going some way to make up for his run of bad luck in 1967. More surprising to the mechanics, the 49B had lasted the distance despite a minimum of testing and all the concerns about reliability that come with the first race for a new car.

After the race, the mechanics set about retrieving Oliver's damaged car and were looking forward to following this up with a celebratory beer. However, this was not to be so for all of them, as Bob Sparshott recounts. 'We took the truck round and, after we'd got Hill's car in at the top, Oliver's car was manhandled into the bottom. We closed the doors and went over to the pension house to have a something to eat and a wash and in the middle of dinner we got a phone call from Chapman's hotel. "I want the driveshafts taken off the winning car, stripped, measured, marked up, bagged up and delivered to the hotel, ready for the morning because I'm running the same sized joints on the Indy cars and I want to see how they've done in the race". And I think it was Dick [Scammell] who had to come out and say to us "I want some volunteers". He knew he was not going to be very well received because we were all wanting to go out and

Hill sweeps round Station Hairpin on his way to a comfortable victory first time out in the 49B (Ford Motor Company)

Turbulent Times

Loneliness of the long-distance driver: It is 4 o'clock in the afternoon as Jackie Oliver sweeps down into Eau Rouge at the start of another 8.76 mile lap of the Spa-Francorchamps circuit. He finished 5th to score his first World Championship points. (Ford Motor Company)

have a beer and it wasn't straightforward. If it had been on the ground and someone had said "You've got to do this job" you'd have said "OK, we'll do it in an hour, no problem", but because it was in the truck, you couldn't get the wheels off, you had to get the car out. I don't know who did it, but it took them hours and hours to do it and a lot of ingenuity …'

The fact that Oliver had badly damaged the only remaining 49 that Team Lotus possessed added a sense of urgency to the task of finishing the second 49B, chassis R6, in time for the next race. This was the Belgian GP at the Spa-Francorchamps circuit in the Ardennes. In the end, it was a close-run thing, with the car being completed only the day before official practice began. The fact that it was finished at all was due solely to a mammoth 48-hour effort on the part of the mechanics, Bob Dance, Eddie Dennis, Sid Carr and Dale Porteous. It was during this gruelling stint that an incident took place that illustrated the tremendous pressure under which the team's mechanics operated. Carr suddenly decided that he'd had enough, said as much to his colleagues, downed tools and walked out, never to return! As Porteous recalls: 'We were on our second all-nighter when he walked out.' As soon as the car was finished, it was loaded onto a trailer which was hitched up to a van (the team transporter with Hill's car had long since departed) and Dance, Porteous and Dennis set off for Dover, one mechanic short before the meeting had even started. Dance was desperate for some sleep, so while Porteous took the first stint of driving, he went into the back of the van and slept on a camp bed during the trip to Dover, trying to recover!

By the time Oliver's car arrived at Spa, the circuit had already been shut for Friday afternoon's practice and it was impossible - no matter who you were - to gain access to the centre of the circuit where the pits and paddock were located. So the mechanics and car were forced to sit it out until the end of practice and Oliver didn't get to drive at all. For the 49B that was inside the circuit, Hill's chassis 49B/R5, the session was little better. Gearbox gremlins meant that he had to sit in the pits while the mechanics Bob Sparshott and Trevor Seaman took it apart. Eventually the problem was traced to a broken pinion thrust bearing. When he did get out, he was slowest of all those who set a time, as he was constantly in and out of the pits trying to find the optimum set-up and was struggling to maintain a steady oil pressure. In addition, he was trying to make up his mind about whether or not the car was better with or without the wedge engine cover and nose fins. Eventually, after trying both configurations he decided to run without any aerodynamic aids for the race.

Unfortunately for both Hill and Oliver, Saturday dawned wet and miserable, precluding any fast laps. This left Hill in a lowly 14th, on the sixth row of the grid. Meanwhile, Oliver, having his first outing in a single seater at Spa (he had driven a Ford GT40 there the previous year) and his first ever drive in a 49B, was having a baptism of fire in the wet. While the Monaco street circuit was daunting because of the close proximity of kerbs, walls and barriers, the 8.76 mile Spa circuit was equally challenging because of the high speeds attained and the close proximity of ditches, houses, cafés and

Lotus 49

Water in the electrics and a sticking throttle thwarted Oliver's attempts to repeat his Spa points finish at Zandvoort. (Ford Motor Company)

woodland ... Despite this, he settled in well and ended the session third quickest behind Surtees' Honda and the BRM of Pedro Rodriguez, although his time was only good enough for 15th, alongside team-mate Hill. This gave him the confidence he needed at that early stage in his Formula 1 career: 'I'd been there before, driving sports cars. I was quite quick in the wet, so the performance took a big uplift at that circuit. The 49 was a fantastic car round there.'

Over at the Rob Walker team, Siffert hoped that this would be his last race in 49/R2 as the team's new car, chassis R7 was supposedly nearing completion at Hethel. By dint of the time he set in the dry session on Friday, 'Seppi' lined up two rows ahead of the works 49Bs.

It seemed that Chapman had slightly lost the initiative that he traditionally always held in terms of pushing back the frontiers in Formula 1 development. Both Ferrari and Brabham appeared with small wings mounted above their engines to give added downforce, trimmed out at the front by nose tabs in the case of the Ferraris and proper fins like the Lotus on the Brabham. It seemed to be working well for the Ferraris, at least, because they were first and third, both cars on the front row of the grid, with Amon and Ickx respectively. However, due to

Both Ferrari and Brabham appeared at Spa with chassis-mounted wings. Ferrari's version was mounted behind the driver's head. This shot shows Jacky Ickx at Zandvoort. (Ford Motor Company)

various problems in practice the Brabhams started at completely the opposite end of the grid, last and last-but-one.

The day of the race was dull but dry. Hill's car was looking naked without its wedge engine cover or nose fins, while Oliver had opted for the full aerodynamic package, including an aluminium strip added at the trailing edge of the wedge like a modern day Gurney flap to give extra downforce. Within five laps of the start, Hill was out with a broken driveshaft UJ joint. This was to be the first of many such failures through the year as the stresses on the car increased through greater aerodynamic download, although of course in this case, Hill was running without aerodynamic devices.

Meanwhile Oliver drove a steady, sensible race and eventually found himself in the top five. However, all was not well with his car, for he was experiencing the beginnings of the same problem that put Hill out: 'I pitted near the end because the UJs were giving up, although I didn't know that at the time. I heard them going and I though perhaps it was something dangerous that was vibrating. Colin looked at it and - we didn't have radios in those days, you used to have to shout into the driver's helmet - he shouted "It's OK but be careful of the UJs", but I only heard the "be careful" bit! So I promptly put my foot down as I left the pits which probably didn't do them much good ...' Contemporary reports describe how the Lotus left the pits with a lot of smoke coming from the outer UJ on the right-hand driveshaft. Sure enough, it didn't last another lap, but Oliver was far enough ahead of his pursuers for that not to matter, and he secured his first Championship points of the season, for fifth place.

It was a race of attrition, won by Bruce McLaren, after leader Jackie Stewart had been forced to pit to take on more fuel with one lap to go, he rejoined to take fourth place, behind second place finisher Rodriguez and Ickx, third in his home race. Siffert was one of the last retirements, going out on his 26th lap at Stavelot, with a blown engine caused by a lack of oil while running some way down the field.

For the next Grand Prix, the Dutch at Zandvoort, the same cars - 49/R5 for Hill and R6 for Oliver - appeared in unchanged specification, except for the fact that Hill was now committed to the full aerodynamic package of wedge-shaped engine cover (complete with aluminium flip-up as run by Oliver at Spa) and nose fins. He was determined not to get caught out in the way that he had been in Belgium and set his fastest practice time in the first session, second only to Chris Amon. His time was around eight tenths of a second faster than his 1967 pole mark, when the 49 had made its stunning debut, showing the progress that the car had made in 12 months.

However, the fact that there were cars faster than him on the grid also served to illustrate how the likes of Ferrari and Repco-Brabham had clawed back the performance gap that was all too evident the previous year. Only Rindt in the Brabham went fast enough to threaten Amon's pole on Saturday and most of the grid went slower. The majority of the session was run off in wet conditions and it was only towards the end that the rain abated and the strong winds that were a constant feature of Zandvoort dried the track.

Oliver was around one and a half seconds slower than Hill and ended up on the fourth row alongside John Surtees in the Honda, while Siffert was three places back in the Walker car, complete with new engine after the Spa blow-up. Not having access to the latest 'works-spec' wedge engine cowling, the Walker mechanics had fashioned their own interpretation, which consisted of two paddle-like 'flip-ups' either side of the engine, trimmed out by a beard spoiler similar to that tried by the works team in practice for the Race of Champions earlier in the year. Sadly, it didn't seem to do much for his pace in what, after all, was a car that was little different to how it had been 12 months earlier.

The race started in torrential rain. Although Rindt made the best start from the middle of the front row, it was Hill who led the field across the line on the opening lap, and the stage looked set for a situation where he could dictate the pace of the race from the front as he had done in Monaco. However, looming in his wheeltracks was Stewart, his Matra shod on Dunlops which were generally to prove superior in the wet throughout the 1968 season compared to Firestone's offerings. Going into the fifth lap, Stewart out-braked Hill into the Tarzan hairpin and pulled away quickly thereafter.

The revelation of the race was Beltoise in the V12 Matra, who had started virtually at the back of the grid in 16th place, but by the end of the first lap was up to an astonishing eighth. He continued his meteoric progress through the field so that by the end of the fifth lap he was in third and closing on Hill, who seemed unable to maintain the same sort of pace as the French cars. When Beltoise spun off at Tarzan, it seemed like Hill might be able to secure his hold on second place but, despite slipping down to seventh as a result of his 'off', Beltoise got going again and began scything his way through the field once more, passing Hill for the second time in the race on the 50th of the scheduled 90 laps.

The Englishman was increasingly suffering from a problem common to the early DFVs, and one magnified by the sand which was everywhere at Zandvoort, as the similarly afflicted Oliver recounts: 'All the Cosworths used to suffer from jamming of the throttle slides. At Zandvoort it was an absolute nightmare because of the sand that used to blow about. I went off on the first lap in the middle of the pack and the throttle didn't close at the second hairpin.' At two-thirds distance, and shortly after having suffered the ignominy of being lapped by leader Stewart, Hill, his throttle problem worsening, spun off at Tarzan, knocking off his car's left-side nose fin against a wooden post and damaging the steering. He managed to rejoin but, twenty laps later, the slides jammed again and he spun in exactly the same place, except that this time he removed the right-side nose fin and broke the front suspension, putting him out of the race.

Meanwhile, Oliver was fighting the dual problem of sticking throttle slides and a waterlogged engine and, although running at the finish, had lost so many laps in the pits that he was unclassified. At least he finished. Siffert had given up an unequal battle against the elements at approaching two-thirds distance, experiencing gear selection and vision problems, which were enough to convince him that an early bath was by far the more

Colin Chapman's answer to the wings of Ferrari and Brabham was, as ever, bigger, better and more outrageous than anything anyone else had come up with, on the Grand Prix grids at least. The wing struts were mounted on the rear uprights. This car was termed the 49B Mark II. (Classic Team Lotus Collection)

preferable option to finishing the race.

By this stage, it was clear that wings were here to stay, and Lotus were in danger of being left behind if, before the next Championship round in France, Chapman did not come up with something to match the aerodynamic appendages that were sprouting on other leading cars. In his own inimitable style, the solution he arrived at was bigger, and in his opinion, significantly better than the opposition's. Whereas Ferrari and Brabham had opted to mount their wings above, and supported by, the engine, Chapman chose to attach them directly to the suspension uprights. He reasoned quite rightly that there was little point in loading something that did not need to have any extra forces exerted on it and that the maximum benefit could be had by such a direct transmission of aerodynamic download to the wheels and tyres. What emerged from the Hethel workshops was a car with a high-mounted rear wing almost the width of the car - not dissimilar to that seen on the Chaparral sports car the previous year - mounted on two struts.

Lotus 49

ecco la verità sull' incidente di OLIVER

CHAPMAN si tranquillizza

Right - photos from an article in the German magazine *Auto Motor und Sport*, which focused on Oliver's accident. After having given an impromptu TV interview, a shaken Oliver walked back to the pits, with Chapman putting a consoling arm on his shoulder. (Motor Presse/Auto Motor und Sport)

Die häufigen Defekte der Lotus haben dem Leichtbau-Fanatiker Colin Chapman schon viele Vorwürfe gebracht. Obwohl sich der geniale Konstrukteur oft an der Grenze des Vertretbaren bewegt, sind nicht alle Lotus-Unfälle auf die Filigranarbeit zurückzuführen. Paul Frère nimmt aus Anlaß des Oliver-Unfalls in Rouen zum Thema Leichtbau Stellung.

Miraculous escape: Oliver's massive accident at Rouen had a big impact on drivers, teams and journalists. The page from *Autosprint* shows that the car began spinning as can be seen from the tyre marks and hit a brick gatepost (point of impact circled). This broke the car in two, with the rear wheels and gearbox and the rest of the car ending up some way apart. As the accident happened in view of the pits, Chapman, Chief Mechanic Bob Dance, and mechanics Dale Porteous and Eddie Dennis were quickly on the scene to try and ascertain the cause of the crash. (Autosprint)

Now that they had the high wing, the 49Bs no longer required the wedge engine cover and this was discarded. Other small changes included a new steering set-up featuring a rack and pinion derived from the Ford Escort of the time! It used helical gears and was cited by the team as being superior because of its smoother action and slightly quicker ratio compared to its predecessor. One other change was the switch to more robust Hookes-type halfshaft joints (the normal type of UJ), from the CV-type versions that had been used since Monaco.

The team had almost no time to test the new high wings prior to going to the French Grand Prix, which was to be held on the ultra-fast road circuit at Rouen-les-Essarts. Bob Dance's records show that Hill drove chassis 49B/R5 at Snetterton on the Tuesday preceding the race, but that Oliver's car, R6, was sent to France without having been tested! The first time he drove it was late on the Thursday afternoon in first practice, when he pulled out of the pits and set off through the fast downhill sweeps towards the Nouveau Monde hairpin! There's nothing like being thrown in at the deep end ...

Interestingly, the two Team cars for Oliver and Hill differed quite significantly, with the former's having a much wider rear wing, the downforce from which was trimmed out by significantly larger nose fins. In contrast, Hill retained nose fins the same size as at Zandvoort, and had a markedly narrower wing than his team-mate, perhaps reflecting his distinct lack of enthusiasm for such appendages at this early stage in their development, something which the following day's events would show was not entirely without foundation. After the first session, the two Gold Leaf Team Lotus drivers were ninth and thirteenth on the grid, both having failed to crack the magic two-minute barrier round the dauntingly fast 4.065 mile circuit: hardly a ringing endorsement of the effectiveness of the new wings! Siffert was even further back, complaining of instability under braking in his now quite ancient (in racing terms, at least) chassis 49/R2.

On the Friday, both Hill and Oliver were quick to improve but still some way off the pace, despite their apparently superior aerodynamics. In reality, the big wings were costing them revs on the long straights, which was impacting on top-line speed, essential on a course like Rouen. None of the corners were really quite suitable to allow the cars to show off their superiority when it came to high speed cornering. Even the fast sweeps after the start/finish line were not sharp enough to make much of a difference.

While Hill suffered from gear selection woes, his

A lone marshal was left to stand guard over the wreck of Oliver's car after its shunt. This shot shows very clearly just how well the tub stood up to the impact, undoubtedly saving Oliver's life. The Rouen pits are just visible in the background. (Ted Walker/Ferret Fotographics)

team-mate had a much more serious incident, which saw him totally lose control of his car and become a passenger on the flat out section just before the pits. He takes up the story: 'Dickie Attwood was in front of me in the BRM and he shifted from one side of the road to the other when I was coming up behind him and I flicked the other way and as I did it just went buh-buh-buh-buh-buh [he makes rapid circular motions]. I can't think that the flick caused me to lose control.'

'The only thing I can think of is that the wing was on very tall - for the first time - flexible pillars. They were probably under a fair amount of load because you are doing 190mph there and I think the turbulent air at speed and the sudden change of direction wobbled the wing. I think one of the supports gave way and, as a consequence, it fell over backwards. As it fell, it picked the rear wheels off the ground. All I know is that I lost control in that manner and of course it nearly wrote me off and certainly wrote the car off! I should have been dead, because I hit the wall at such a speed and came out of the car.'

On impact, the car hit a brick gatepost and broke up, scattering wreckage all around but the tub did its job and, miraculously, Oliver emerged shaken up but with little more than a few cuts and bruises. He probably survived due to the fact that the rear suspension and gearbox took the bulk of the impact, this entire assembly being ripped off and ending up some distance down the track from the tub. Ashen-faced, Oliver reported back to Chapman, to tell him that he'd written off a second 49 chassis in four races: 'Colin was never sympathetic. I mean he used to have cars fail all the time, that was part of his stock-in-trade. He used to push things right to the limit and, as a result, he had very competitive cars. Now we've got engineers that can analyse things, so we reduce the risk. But then, the analysis was done with the driver in the car to see whether it would break. When things went wrong, being the type of person he was, he would normally be suspicious of the driver and he'd also be suspicious that the mechanics might not have bolted it together properly. Then, if those possibilities could be eliminated, he would consider if it could have been a design failure.'

Initially Chapman was worried from Oliver's description of the accident that a gearbox bellhousing had broken, and began to warn other teams using the same component that they should pull their cars in and check them. Although over the years this act has been portrayed as a bit of Chapman gamesmanship and opportunism, Bob Dance maintains that, at the time, Chapman was genuinely concerned about preventing any repetition of the incident and that it was only later, when he had the chance to inspect the wreckage, that it became apparent that the whole rear end had been detached as a result of the impact rather than being the cause of it. The monocoque stood up very well to an impact that ten or even five years before would probably have proved fatal for the driver. However, Oliver also thinks the fact that it was the rear of the car that hit the gatepost - and not the

Lotus 49

Foot perched on the front wheel of his car, Siffert takes a swig of drink before embarking on one of the greatest drives of his career. (Ted Walker/Ferret Fotographics)

A youthful-looking Jackie Oliver, seen here at the British Grand Prix, faced a difficult task stepping into Jim Clark's shoes at Lotus for the balance of the 1968 season. (Ted Walker/Ferret Fotographics)

tub - which saved his life: 'All the energy was taken sideways by the rear wheel and gearbox assembly, it just tore the thing off the back. The actual monocoque didn't take the impact, that went spinning on down the road. So it was how I hit the wall that saved my life. Pure luck!'

All this left Oliver without a car for the race, since there was no spare at the time, so Hill's was the only works car left on the grid. After his problems on Thursday, Siffert knocked nearly three seconds off his practice time, despite the fact that his mechanics had been unable on stripping the car down to find any explanation for its strange behaviour under braking the previous day.

The forecast for the day of the race was not good and, as the day wore on, the skies got darker and darker. The race was not scheduled to start until 4pm, by which time a steady drizzle had begun to fall. This quickly turned to heavy rain, which continued unabated until about two-thirds distance. Ickx took an early lead in the Ferrari which he surrendered for less than a lap to Pedro Rodriguez during a particularly heavy squall. As his challengers fell by the wayside, the young Belgian drove off serenely into the distance to take his first Grand Prix victory. The event was marred by a dreadful fiery accident on lap three, which claimed the life of the French driver Jo Schlesser, who had been driving a new, virtually untested air-cooled Honda car. For Team Lotus, the race came to an early conclusion, with Hill's car breaking a driveshaft accelerating out of Nouveau Monde. By that stage, he had moved up to fourth, having caught and passed Stewart's Matra which, having been so dominant in the wet two weeks earlier, was struggling to keep up on the French road circuit.

This left the private Walker car as the only Lotus representation in the race. It was already way behind, having lost a minute at the start as the mechanics tried in vain to get the clutch to work. Somehow they got the car running again but, by the time Hill had gone out, the Swiss driver had already been lapped by Ickx and was suffering from dirty, misted-up goggles. On seeing Hill standing by his car, he stopped and asked if he could borrow the Englishman's visor, which Hill promptly helped to clip on, and Siffert went on his way... He struggled round to finish a lowly eleventh, six laps behind winner Ickx. However, he was content in the knowledge that this was to be his last race in the outdated 1967 car, and that he would be getting his hands on the latest 49B for the British Grand Prix in two weeks time.

At this halfway stage in the season, after six rounds of the World Championship, Hill still held a commanding lead in the points standings, by virtue of his two wins in Spain and Monaco and his second in South Africa at the start of the year. His tally of 24 points was some eight points ahead of his nearest rivals, Jackie Stewart and Jacky Ickx, with Denny Hulme a further four points adrift. Hill's team-mate Oliver was down in 17th place, with the two points he had secured at Spa all he had to show for his so-far eventful debut season, while Siffert's woeful lack of reliability in the first half of the year was evidenced by the fact that he had not garnered any points by this stage.

The Grand Prix field moved to Brands Hatch in Kent for the next round of the Championship and the crowds flocked to see the winged wonders (or monstrosities, depending on your viewpoint at the time!) streaking up and down the undulations that made the circuit both stimulating to the driver and easy on the eye for the spectator.

From a Team Lotus point of view, the biggest challenge had been trying to find the time to build the 49B chassis R7 ordered by the Rob Walker team to replace the car that Siffert had written off at the same circuit in March. Mechanic Tony Cleverley remembers that he spent some considerable time up at the Lotus works helping to build the car. 'We built that from scratch up at Hethel over three or four months. They were going to do it themselves and deliver it, but there was nothing forthcoming, so we had to go up there ourselves and build it up ...' It was touch and go whether or not it would be ready in time, as Cleverley explains: 'We were finishing it off going down there in the lorry actually!' The car was finally completed in the paddock on the morning of first practice. This 'customer' specification

- First Person -
Jackie Oliver
on the Lotus 49

"Well it was the modern era, the start of what we've now become used to, which was the use of the engine as a stressed member. The whole compartment of the car was modern by comparison to the other F1 cars, because of the stressed structure and it was probably lighter than any other car because of its construction.

The DFV engine was very easy to use and had a beautiful power range to it and the high wing just transformed the car completely. You had aerodynamic grip on the car that you couldn't imagine, so it was the best Formula 1 car I've ever driven."

Turbulent Times

It is three o'clock and the starter has just dropped the flag to get the 1968 British Grand Prix underway at Brands Hatch. Oliver (centre) has made the best getaway, pursued by Hill (left) and Amon (right). Eventual winner Siffert is just behind Oliver, while Dan Gurney's Eagle can be seen pulling off to retire. (Ford Motor Company)

The two 49s fielded by the works team at the British Grand Prix were quite different, not only to the rest of the field but to each other. After his Rouen accident, Oliver was in R2, which still had the original specification lower rear radius arm mountings and the oil tank and cooler in the nose. The only concession to modernity was the addition of a high wing. In contrast, Hill's car has the deep cut-outs for the radius rods and the oil tank/cooler over the gearbox. The high wings on the Lotus entries made the efforts of Ferrari (Amon, car 5) and Brabham (Rindt, car 4) look positively puny by comparison. (Ted Walker/Ferret Fotographics)

car differed only marginally from the works 49B, in that it had a lower height of rear wing and longer exhaust pipes. For the first time, Siffert had a truly competitive Grand Prix car with which to demonstrate his undeniable talent, and it was an opportunity he viewed with relish.

Having written off his 49B at Rouen, the only car left for Oliver to drive was the old 49 that had been on extended loan to the Walker team. This was now back at the factory, so it was hastily repainted and a 49B-style nosecone with fins and high rear wing were added. Otherwise, the car was original 49, with the old-style 'fir-tree' tubular assembly on which the rear suspension uprights were mounted, ZF gearbox, oil radiator and tank in the nose, and original rear radius rod mounting points. The result of this was that the mounting points for the rear wing were closer together than for Hill's car, making it look even more spindly and precarious. Hill's car had adopted the wider nose fins seen on Oliver's car at Rouen, while the overall height of the rear wings on both cars had been raised in an effort to avoid the sort of turbulence that appeared to have been largely to blame for the latter's high speed crash.

On Thursday, both Gold Leaf cars were on the pace right away, finishing practice in first and third for Hill and Oliver respectively, split only by Amon's Ferrari. Oliver was feeling considerably more at home on a track he was familiar with. The high wings really came into their own on the undulating Kent track and it was gratifying for Lotus, in its 'home' race, to be back at the top of the time sheets for the first time since Monaco. The increase in lap speeds attributable to the development of wings was clear for all to see, for the first nine cars were all inside Bruce McLaren's lap record set at the March Race of Champions meeting! Meanwhile, Siffert was playing himself in gently, finishing up 13th at the end of the first practice session, more than three seconds a lap slower than the works cars.

In the second session on the Thursday afternoon, Amon topped the timesheets, with Hill and Oliver second and third, while Siffert improved by a second and a half to finish up tenth. Final practice took place on Friday morning, and this saw the Gold Leaf duo really pull all the stops out. Hill took the coveted pole spot with a lap fully 2.7 seconds inside the record, while team-mate Oliver was next up, half a second slower. However, the biggest improvement came from Siffert, who by now had taken to the 49B like a duck to water, and he catapulted himself up the grid order with a near-two second improvement on his Thursday lap time to put him fourth, on the second row alongside Jochen Rindt's Brabham.

107

Lotus 49

Jackie Oliver really made his mark at Brands Hatch. On a circuit he was extremely familiar with, he challenged his far more experienced team-mate for pole and led the race for many laps before his engine seized. (Classic Team Lotus Collection)

The front row was completed by Amon, whose Ferrari had consistently proved to be the closest challenger to the 49Bs during practice.

After two days of dry sunny conditions, Saturday was overcast, but conditions remained dry. An examination of the cars on the dummy grid revealed that Stewart's Matra was the only one of the leading runners without a rear wing, looking naked among a sea of struts and aerofoils. Of the other works teams, only BRM did not have a wing and therefore it was no surprise to find their cars towards the back of the grid.

As the flag dropped, Oliver got a superb start from the middle of the front row and shot into the lead ahead of Siffert, who had made an even better start, getting the drop on Rindt, Amon and Hill. As the field streamed under the bridge and out onto the 'country loop' of the Kent circuit, Lotus 49s were running 1-2-3, with Amon in hot pursuit, trying to find a way past Hill. The Englishman wasted little time in getting past Siffert on the first lap, but he held station behind Oliver until he passed him on the third lap, by which time the older 49 was smoking visibly, the result of an oil breather being too close to an exhaust pipe. This order was maintained, with Amon hanging on grimly, until the 26th of the scheduled 80 laps, when Hill's car broke an outside UJ joint on the left-hand driveshaft, damaging the suspension beyond repair in the process as it flailed around.

To say that Hill and the team (and a large proportion of the crowd) were desperately disappointed, would be a major understatement. The 49s were in dominant form and he had missed a perfect opportunity to open up a huge lead in the title chase. However, a Gold Leaf car was still in the lead and, although it was still smoking, it was less obvious than it had been before. What was more, the Walker 49 was waiting in the wings, should the other works car fail ...

Approaching half distance, on lap 37, Amon tired of following Siffert and slipped past. However, he made little impression on Oliver and the Swiss clung tenaciously to his exhaust pipes, maintaining the pressure all the time. Within six laps, Amon had made a mistake and Siffert was past him. However, greater drama was to unfold on the next lap as race leader Oliver coasted to a halt with a seized engine caused by oil loss which had manifested itself in the form of the smoke which had been coming from the car: 'The old Cosworths used to breath quite heavily and the only way they could keep the oil level up was to put the oil back into them. A mechanic had put the breather on the wrong side of the exhaust system so it burnt through the pipe and it was breathing oil out into the atmosphere instead of into the collector tank and back to the engine. Eventually it ran out of oil and seized.'

Suddenly, Siffert and Amon were fighting for the lead! They ran virtually nose-to-tail for the remainder of the race, the Ferrari only falling back slightly in the final few circuits as it was suffering from excessive tyre wear. After 80 hard fought laps, during which these two had lapped the entire field, the popular Swiss took the chequered flag to score his first ever Grand Prix victory and the Walker team's first since Moss' victory in the German Grand Prix in 1961. What was more, a Lotus 49 - chassis R7 - had again taken victory in its first race! Siffert finished 4.4 seconds ahead of a frustrated Amon - the

Chris Amon gets well crossed up as he fights to keep up with the flying Swiss, Jo Siffert, during their race-long battle in the 1968 British Grand Prix. (Autocar/Ted Walker/Ferret Fotographics)

Jo Siffert was a tremendously popular winner, driving R7 to victory in its first race. (Ford Motor Company)

bridesmaid yet again - with the Kiwi's team-mate Ickx finishing third to complete a good day for Ferrari.

Everybody in the Rob Walker team was over the moon. It was the culmination of a long relationship between Walker and Siffert and the aftermath of the race was a moment to savour, for it would be the private entrant's (and, indeed, any private entrant's) last ever Grand Prix win. Today, Cleverley looks back on that race with fond memories. 'It was a bloomin' good effort. Especially as a private entrant too'. Although the team was a private one, it enjoyed a close relationship with the works team by virtue of the links established in the early 1960s between Walker and Chapman – for it was a Lotus run by Walker which scored the marque's first Grand Prix victory at the 1960 Monaco Grand Prix, as Cleverley recalls. 'Wherever we went we were always alongside the team or with them, or something like that. They were very helpful. Whenever we wanted anything, we had it. New things came out we had them the next week or something like that. In those days everyone mucked in.'

Siffert's exuberance on the track led to plenty of excursions off of it, which provided plenty of work for the mechanics, according to Cleverley. 'He was easy to work with, but at every race we went to he knocked a corner off it somewhere. We used to have suspension built up ready to go on each corner because each time he went out he knocked one off!' For all this, he remembers the 49 as an easy car to work on. 'Compared to what we had been looking after - the Cooper-Maserati - it was a doddle.' He also holds Rob Walker in high esteem as someone to work for. 'A gentleman. One of the nicest people you could ever have worked for.'

For all the happy, smiling faces in the Walker pit after the British race, there was an air of frustration among the Gold Leaf personnel, for Hill had failed to score any points for the fourth race in succession and three of these retirements were directly attributable to driveshaft failures. Clearly, reliability had to be improved if Hill was to stand any chance of maintaining a challenge for the Championship, for his closest rivals, Stewart, Ickx and Hulme, had all scored points at Brands Hatch and the Belgian was only four points adrift of Hill.

While Chapman and Phillippe continued to disagree over driveshaft sizes, the most pressing task facing the mechanics was to convert Oliver's 49, chassis 49/2, into a pukka 49B before the next race, the German Grand Prix, to be held at the Nürburgring in two weeks time. This entailed the removal of the whole tubular rear suspension mounting assembly, and its replacement with the B crossmember, the swapping of the ZF box for a Hewland FG400, the modification of the lower rear radius arm mounting points (which involved considerable cutting away of the monocoque) and the relocating of the oil tank and cooler from their original position in the nose to over the gearbox. In scenes reminiscent of the rush to get R6 readied in time for Spa in June, the car arrived on a trailer on Friday morning and was finished in the paddock, although not in time to get Oliver out for first practice that afternoon.

The solution arrived at for Hill's car was a revised driveshaft set-up, utilising Mercedes and BRD components, with the aim of making the whole arrangement more robust. Unfortunately, he had little opportunity to test this out in dry conditions, since the circuit was

Siffert did not manage to reproduce his stupendous British Grand Prix form at the Nürburgring, qualifying well down the grid and going out with ignition trouble. Here one can see the concentration on his face as he tackles the Karussel corner at the Nürburgring - in the wet. (Goddard Picture Library)

After an 'off' on the morning of the race, Jackie Oliver was under strict instructions to bring his car home to the finish in the German Grand Prix. This race marked the first appearance of R2 after it had been rebuilt to 49B specification. Here he has just climbed up the hill and is about to drop down into steeply-banked Karussel. (Goddard Picture Library)

shrouded in mist and a steady drizzle was falling. He ended the session fourth after several cautious laps, with Rainmaster (and, arguably, Ringmaster) Ickx comfortably fastest, ahead of team-mate Amon and Jochen Rindt's Brabham.

Conditions were even worse for final practice on Saturday afternoon, evidenced by the fact that Ickx went a minute slower than he had done the previous day. However, a new contender emerged, in the form of the Dunlop-shod Matra of Jackie Stewart, which lapped four seconds faster than Ickx. Oliver did get out for this session, but was a disappointing 13th fastest, finding the combination of a new car, tortuous circuit and appalling conditions a bit overwhelming.

Since the two scheduled official practice sessions had been wet, restricting drivers to a handful of laps of the 14 mile track around the Eifel Mountains, a further session was arranged on the morning of the race. Although the mist had lifted, it was raining steadily and the track was awash from overnight rain. Despite these conditions, Stewart went round some ten seconds faster than the previous day, while Oliver improved by half a minute. However, on the fast run of sweeping downhill bends before Adenau Bridge, he overcooked it and ended up in the hedge, damaging the left-side suspension.

With the race due to start in just over two hours, the Lotus mechanics faced a desperate race against time to retrieve the car and try and repair it, as Bob Dance recalls: 'The car came back with the wheels pointing this way and that. Not seriously crashed but bent. We didn't have time to set the car up with parallel bars or anything. The Old Man stood in front of the car and eyed up the wheel alignment and when he decided the wheels were pointing in the right direction he said "Right that's it, lock it up there." Because there was no time - you had to get the car, the pit kit and the tools up the slope to the track and we just got out by the skin of our teeth.' The car made it onto the dummy grid with minutes to spare and in very 'patched up' form. Consequently, Oliver was under strict instructions to take things easy and make it to the finish.

The weather got steadily worse, with a return of thick mist and the rain continuing to pour down. When the race finally got underway after having been delayed for almost an hour, Hill splashed through from the second row to lead into the South Curve, ahead of Amon and Stewart. The Scot was revelling in the conditions and by Adenau had forged ahead of the Ferrari and was closing in on the Lotus. Two-thirds of the way round the lap, he swept past and was gone, a class apart on his Dunlop tyres. By the end of the first lap he led by eight seconds and he continued to stretch this gap relentlessly until flag-fall. Behind him, Hill and Amon engaged in an almost race-long dog-fight for second place until, on the 12th of the 14 laps, the luckless Ferrari driver spun off due to problems with his car's limited slip differential.

Unaware of his rival's fate (by now the mist was so thick and the wall of spray from each car so dense that it was like driving in a cocoon) Hill pressed on and spun at the Esses after Hohe Acht, stalling the car in the middle of the track. Expecting Amon to come tearing out of the gloom at any moment, he frantically tried the starter but to no avail. Instead, he was forced to resort to climbing out of the car and turning it so that it was pointing downhill, whereupon he jumped back in and, once the car had gathered a bit of momentum, bump started it by engaging a gear and hitting the starter at the same time. This whole episode cost him around a minute yet, amazingly, nobody passed him, illustrating the extent to which the Briton and Amon had pulled away from the pursuing pack. However, Rindt had almost caught him and closed to within a few seconds of Hill at the end of that lap. The Austrian was a constant threat thereafter, but was unable to find a way past the wall of spray thrown up by the Lotus and finished a bedraggled third, six seconds adrift.

Meanwhile, Stewart had sailed serenely on, his progress only rudely interupted by a near miss with Oliver's Lotus, which loomed suddenly out of the mist as he came up to lap it. His margin of victory - four minutes and three seconds - was stupendous even though Hill's spin and subsequent problems in restarting had exaggerated it. It was clear that here was a significant threat to the Championship aspirations of Hill and Team Lotus and the nine points resulting from his victory brought the Scot to within four points of Hill in the title standings. Of the other contenders, Ickx finished a distant fourth but the points enabled him to stay in touch, while Hulme finished just outside the points in seventh, denting his title defence. Oliver had followed his instructions and stayed out of trouble, bringing his rather ill-handling 49B across the finishing line a whole lap down on Stewart, in 13th place: 'The car was cobbled together. It had a straightened track rod or something like that, so it wasn't tracking very well. Anyway, Colin said "The car's not very good, I want you to finish the race", so I did just that.'

The wet conditions at the Nürburgring had flattered to deceive Lotus that their transmission woes had been solved. At the next event, the non-Championship Gold Cup at Oulton Park, Hill had put his 49B on pole, with Oliver back in the middle of the grid but with only seven laps of the race gone, Hill's crownwheel and pinion broke and he coasted into the pits to retire. Having strengthened the drive train in one area, all that had happened was that the stresses had been transferred to another area. At the time Hill had been chasing Jackie Stewart, who had made a better start from alongside him in the middle of the front row. Stewart went on to win the race at record speed, chased home by Amon's Ferrari, while Oliver benefited from other people's unreliability - having passed Jack Brabham early on in the race - to finish a good third. That Oliver's car ran without problem to the finish must have been a source of frustration to Hill, while the reliability and speed of Stewart's Matra meant that it was fast becoming the car to beat in Formula 1. This was in stark contrast to the 49B, which was failing to fulfil its early promise due to a lack of reliability. The Scot was also emerging as Hill's strongest challenger for World Championship honours.

Hill was renowned for his fastidious approach to setting a car up which, as Bob Sparshott recounts, could be slightly frustrating for his mechanics: 'He had a brown book that Bette used to keep. All through his early racing days, he got her to write down what the car settings were, so he could tell you what BRM used five years before - roll bars, springs, the whole lot, it was all in there. That used to work against us because he would turn up at the track with it. We would painstakingly have set the car up at the factory on a nice flat bed and it would take ages to get the bump-steer right, the toe-in, the corner weights down to the last bit. We'd get to some godforsaken paddock somewhere and it would be sitting there all lovely.

'He always started off the same way: "Morning lads. Now Robert ..." and he used to twiddle the old moustache, "... what have we got on the settings?" and I'd get my folder out and say "Well, we've got these springs on the front and so and so on the back." And he'd say "Have we got a chart?" He knew we had all the spring charts and he'd look at the graphs and spring frequencies and rates. Then he'd say "That's a bit stiff, you know. I think we're going to be a bit stiff here. Have we got something a little bit softer?" And that's how it would start.

'As soon as you change the springs, you've lost your corner weights and we couldn't find anywhere level. It was a nightmare. He'd go right through the car and there'd be anti-roll bars, gear ratios, he'd have all these charts showing everything. And we hadn't even turned a wheel! Then the Old Man would arrive and say "What are you doing?" and we'd tell him. His response was "Never mind what Graham wants, I pay your wages and I will tell you what to do ..." and then he'd have an argument with Graham and they'd have a real set-to and time would be ticking away, and we'd be waiting for instructions ...'

One infamous occasion was on the grid of the Gold Cup at Oulton Park. Sparshott takes up the story: 'He [Hill] wanted a bump rubber cut off because he wanted less bump rubber in the damper. Those heavy old Armstrong ones. The only way you could do it *in situ* was with a little knife and you had to cut round it to the shaft and then pull it out through the spring with a pair of pliers. We were in the middle of doing this when the Old Man arrived. He was horrified! He said "What the hell

Lotus 49

All the works 49Bs appeared at the Italian GP with new top ducting for air exiting the radiator and the sides of the nose filled in. This provided significant additional downforce, enabling the nose fins to be run at a lower angle of incidence. The nose cone was also split into half, with the bottom half and the fins remaining in place while the top half could be removed for access as required. This is Oliver's R2 in the pits at Monza. (Aldo Zana)

The 49B Mark III, as the cars introduced at Monza were termed, featured the now standard high wing but the driveshafts were considerably more chunky than in earlier shots, reflecting the increased drive-train loads created by the wings and the ever-wider tyres. (Classic Team Lotus Collection)

would always win, because ultimately he's got the sanction of 'I can't do it, I can't drive the car like this', so he would get his own way. But Graham would very often go full circle and you'd end up with the car back where you'd started. It was irritating, I'm afraid! But he was a good guy. He treated us well and he always gave a load of money to Bob [Dance] to share out saying "Make sure the boys have something to eat and a drink on me".'

The day after the Gold Cup Siffert, who had elected not to make the long trip north to Oulton Park, thrilled his home crowds by taking part in the St Ursanne-Les Rangiers hillclimb event in the Walker 49B. Clearly he wasn't there just to make up the numbers, for he scorched up the three-mile climb in a record time to take victory, demonstrating the sheer versatility of both car and driver.

With three weeks to go until the Italian GP, the team set about trying to improve the reliability of their cars, while at the same time working on new developments that would give their performance a boost. It was clear from recent races that Ken Tyrrell's Equipe Matra International team and Dunlop had made great strides forward in performance terms and Lotus needed to respond. The new development introduced was in the form of two ducted air exits on the top of the nose, replacing the exits on either side of the nose which had been used since the 49's debut. This arrangement, which had first been seen on the McLaren M7A at the Race of Champions in March, provided quite significant additional downforce and enabled the front nose fins to be reduced in size.

Hill had just enough time to give the car with its new nose a quick run at the National Cortina Day at Mallory Park, setting a new joint lap record with Jackie Stewart's Matra, before it was despatched to Monza. This trek was not popular with the mechanics because it meant the extra mileage of going up to Mallory Park and then back down south to Dover. However, it was important in publicity/PR terms for both Team Lotus and Ford, since Hill was a major draw for the organisers, the BRSCC.

Another development introduced in time for Italy was a revised exhaust layout on the Cosworth DFV, which was reckoned to be good for an extra 15bhp. Since Monza was all about power and straight line speed, this would provide a welcome boost. Instead of having the four pipes from the engine divide into two lots of two pipes and thence into the tail pipe, the new layout saw all four pipes run straight into one. The only 49B runner who was not able to use this system was Siffert. The 'works' contracted teams of Lotus, Matra and McLaren were given preference when it came to having the new systems installed and Cosworth simply ran out of time.

For the first time that season, Lotus was to field three cars at Monza. For some time, Chapman had been making overtures towards the talented young Italian-American driver Mario Andretti, with a view to him joining the team and he had finally agreed to drive in Italy. Chapman was fulfilling a promise made to him some three years earlier, at Indianapolis, as Andretti recounts: 'I first met him in 1965, when he was at Indianapolis with Jim Clark. I was a rookie there. That's when I first expressed my desire to do Formula 1 eventually. It was quite simple. He said "Whenever you feel

are you doing?" So I told him and he said "What the bloody hell does he know about this? Where is he?" They had a fearsome argument while we were waiting there, with a knife cut halfway through this thing, waiting to know whether or not it's going to stay in or come out!' After the race, Bob Dance recalls the whole team, drivers included, being rebuked over this episode by Chapman: 'The Old Man had us all together in the crew cabin of the transporter and he gave us a rocket - Graham as well! He said "They're my cars and I'll set my cars up the way I want them. So listen Bob [Dance] and Bob [Sparshott], you don't take orders from the drivers, if the car has got to have a change in its set-up, you take the orders from me. Is that clear?" The answer just had to be "yes!"'

Although what he regarded as drivers fiddling unnecessarily with their cars always infuriated Chapman, they would usually get their way because, as Sparshott pointed out, it was they who had to drive the cars on the track and risk their lives doing so. 'In the end the driver

This overhead shot shows the sculpted top ducts in the nose cone (the radiator can just be seen in the gloom of the ducts), as well as the fact that by this stage Jackie Oliver - probably in the wake of his Rouen crash - was wearing seat-belts. Also visible are the separate fuel filler and large round access hatch, marks which identify this as the chassis used in 1967 by Jim Clark. (Classic Team Lotus Collection)

you're ready, just call me." And that's what I did in 1968. I called him and he said "See you in one week in Monza, we will test." I had so much reverence for Colin, I had confidence that he was the one that I wanted to be with, and he was so accommodating besides. He said "You want to race, we'll make a spot for you, we'll add another entry." And he was true to his word with no drama, no complications just "Go and race." So that was great, the best way I could possibly do it.'

Andretti had secured special dispensation from the organisers to start the Italian race, because he and fellow American Grand Prix debutant Bobby Unser (who had taken Dickie Attwood's place at BRM) were due to compete in a USAC race at the Indianapolis road circuit on the Saturday. In so doing, they would infringe the regulation which said that drivers could not contest two international races within 24 hours. The plan was for the two drivers to practice on the Friday at Monza, hop on a transatlantic flight back to the US, do the USAC race and then fly back to Monza for the race on Sunday!

Andretti's inclusion in the team created a further problem in that there was no car for him to drive! Since Oliver's accident which wrote off chassis R6 at Rouen, the car that was to have been retained as a spare, chassis 49B/R2, had been pressed into service as a race car and there had been no time to build up a replacement for the chassis he destroyed. As a result, a new chassis, the eighth in the sequence, had been built up but, instead of being given the correct chassis number of R8, it was assigned the number R6. This was mainly because it was more convenient since customs documents already existed for that chassis number, but also because some of the parts that could be salvaged from the wreckage of the original R6 had been used to complete this eighth chassis, as indeed they had been used to help finish R7. All this extra work, as well as the added complication of running a third car at the event, did not sit too well with the mechanics since, as Bob Dance points out, they had the lead of the World Championship to defend: 'Not only were we trying to win the Championship but we had a third car as well. It was all extra work when we were trying to win GPs for Graham.'

Hill switched from his regular mount, chassis 49B/R5, to the new car for Italy. However, Oliver did not get Hill's old car to replace his venerable 49B/R2. Instead, this was entrusted to Andretti, leaving the young Essex driver with his regular car and rather disgruntled at being usurped by some American upstart. Chapman went out of his way to help Andretti become accustomed to a Formula 1 car, sending the team down early to give him some unofficial practice. Mario was immediately on the pace and, if he made a mistake at all, it was probably that he went far too fast for his own good, for he reportedly lapped in 1m 27 seconds, which was one and a half seconds under Jim Clark's 1967 lap record and a second quicker than Graham Hill on the day. More importantly, it was quicker than the Ferraris ...

No longer the expected midfield runner, he seemed capable of running with the best and this did not sit well with the organisers. By the eve of the race, their 'dispensation' had somehow evaporated, something Andretti attributed to some subtle pressure from the 'home' team: 'I think that is what incensed the Ferrari people because in retrospect - I've never been able to prove it - I think that the protest about me driving, even though we had a deal, came from Ferrari. And, of course, being Monza, they prevailed.'

When the first official practice session began at 3pm on the Friday, Andretti and Unser had gone straight out to do their compulsory minimum number of laps and

Lotus 49

The entire rear-end of the 49B, including gearbox, oil tank/cooler and rear suspension could be detached as a single unit, to facilitate easy engine changes. This shot shows the fabricated steel cross-member to which the coil spring/damper units were attached, which replaced the tubular steel subframes of 1967. (Ford Motor Company)

by 3.15pm they were off in a helicopter on the first leg of their epic trip. Mario had recorded a very respectable 1.27.2 which, although not as fast as he had gone the previous day, still ended up being the eighth quickest time of the session. Unser, on the other hand, struggled and his time would not have been fast enough to qualify if he and Andretti had been permitted to start.

Having invested all their energies into getting Mario out early in the session, the Gold Leaf Team Lotus mechanics then faced a marathon session of gearbox installation. Hill's had failed the previous day in unofficial practice, while Oliver's car had arrived at the circuit minus its gearbox and then, when it did arrive, it was without any output shafts, so Team Lotus had to borrow some from Matra! Both cars got out right at the end of the session, with Hill uncharacteristically getting straight

onto the pace to finish up second quickest behind Surtees' Honda, while Oliver struggled round some three seconds slower.

For the second session they were better prepared but, despite having three and a half hours in which to find a good balance, Hill actually went almost one and a half seconds slower... Oliver, on the other hand, did improve and set a time that moved him up to 11th on the grid, although still 0.2 seconds slower than the time Andretti had managed in his brief lappery the previous day. Towards the end of the session, his DFV blew spectacularly, while as the flag fell to end it, the fuel-metering unit drive on Hill's engine sheared, leaving the mechanics with a busy night in prospect.

Meanwhile, Siffert had been experimenting with a variety of different set-ups. While the works cars had retained their nose fins and high rear wings throughout the two days of practice, the Swiss had been assessing whether there was an advantage to be had from running without them. The answer was that there was a straight line speed gain, but that the car was more of a handful round the corners. The gain in speed was quite important to Siffert, and worth trading for less cornering stability, as his engine had not had the power-boosting modifications of the other Cosworth runners and he felt that he might otherwise struggle to keep up. His time from Friday was good enough to put him on the fourth row alongside Rindt's Brabham and just in front of Oliver.

With Amon, Ickx and Bruce McLaren improving in the second session, Hill found himself bumped back to the second row of the grid for the race. However, Ickx aside, his other main rivals for the Championship - Stewart and Hulme - were behind him on the grid, which was a start, at least.

Overnight, a decision was taken to put Oliver in Andretti's car since, although efforts were still being made to try and re-instate the American after his departure, they were deemed unlikely to be successful. It seemed more practical to swap cars rather than have to switch Mario's engine into Oliver's car, since there was also an engine in Hill's car that needed changing, so all Jackie's settings were transferred to chassis 49B/5 and 49B/2 was put in the transporter.

Race day was hot and dry and a huge crowd had gathered to see whether their beloved Ferraris could vanquish the competition. Surtees was on pole in his V12 Honda, having been fastest in both practice sessions. Alongside him was Bruce McLaren, his McLaren M7A powered by a Cosworth DFV, and, on the outside of the front row was Chris Amon in his V12 Ferrari. As the flag fell, McLaren made the best start and he led a snaking field of 20 cars across the line at the end of the lap. By lap seven, a leading bunch had emerged, comprising Surtees, McLaren, Amon, Siffert, Hill, Stewart, Hulme and Oliver (who had made an exceptionally good start from the middle of the grid). These eight had opened up a gap on their pursuers, led by Ickx, Servoz-Gavin (getting another chance in the second Matra-DFV) and Rindt.

Then, within two laps, two incidents took place which effectively ended the Gold Leaf Team Lotus challenge. On lap nine, Amon's Ferrari crashed in the Lesmo bends on oil from the oil pump which operated the

moveable wing on his car and in the ensuing mêlée, Surtees also crashed and Oliver left the track while trying to miss them: 'I had to avoid them, went on the grass and spun. I collected it up again but right at the back of the pack and, of course, if you lost the slipstream at that time, you were out'. As he gathered the car up, Oliver had been passed by Ickx, Servoz-Gavin, Rindt and Hobbs. Although he managed to re-pass Hobbs (driving the second Honda) he was unable to rejoin the first or second groups and eventually succumbed to a by now familiar Lotus malaise: 'I rode around on my own, out of the slipstream of the leading pack and then the transmission broke.'

Two laps later, Hill, who had been running consistently in the top five, suffered a rear wheel failure which saw the wheel detach itself near the hub and he slithered to a halt, fortunately without hitting anything. The sole remaining Lotus representative was the private Walker entry for Siffert and he was now one of a four-strong leading group after all the events of the preceding laps, along with McLaren, Stewart and Hulme. However, by lap 30, McLaren had begun to lose touch with the leaders due to a problem with his engine and, on lap 34, exactly half distance, he peeled off into the pits, his challenge effectively finished.

At this stage of the race, in terms of Championship points, the situation was starting to look dismal for Hill, since his closest challengers were all running in the points. Hulme was first, Stewart second and, with all the retirements, Ickx was up to fourth. However, nine laps later, the Matra's Cosworth engine unexpectedly blew up and his chance to go top of the points standings was lost. This left Hulme and Siffert running up front, but Siffert could not mount a challenge to the McLaren, because he was only just managing to keep up due to his horsepower disadvantage. Cruelly, with only nine laps to go, a mounting eye on a rear spring/damper unit broke and, as there was no short term solution to this problem, he had to retire. All this left Hulme to cruise to a comfortable victory, while Ickx was just pipped on the line for second by Servoz-Gavin due to fuel starvation problems as he accelerated out of the final corner and towards the chequered flag.

So what could have been a pretty disastrous day for Lotus turned out to be not so bad for, amazingly, Hill came away still in the lead of the Championship with 30 points. However, only six points separated the top four drivers, with Ickx on 27, Stewart on 26 and Hulme having jumped up to 24. It was now clear that, barring miracles, one of these four drivers would be World Champion at the end of the year.

The Italian American driver Mario Andretti joined the works team for the Italian Grand Prix and showed a good turn of speed before officialdom and bureaucracy stepped in to prevent him racing. Here he is just exiting Parabolica corner, with the fabled Monza banking in the background. (Ford Motor Company)

Chapter 12
Challenge For The Championship

To the thinly disguised incredulity of Graham Hill, Canadian Lotus importer Bill Brack joined the works team for his home Grand Prix at Mont Tremblant. The driving technique required by the high wings made this a particularly difficult debut. (Bill Brack Collection)

For the Championship finale, the Grand Prix circus moved to the American continent, for two races in North America - Canada and the US - with a concluding round in Mexico. Being so far away brought its own problems - trucks were hired to transport the team and cars, cars had to be hired for the drivers and mechanics to get around, and the scope for repair work was pretty limited as it all had to be done on site. Effectively, what could be packed into crates was what could be taken on these 'away' trips. Attention had to be paid to the weight of spares because of the high cost of air freight and weight limits on the chartered freighter.

The Canadian round was held at the Mont Tremblant circuit at St. Jovite in Quebec, near Montreal.

Once again, Team Lotus were running a third car, this time for a local pay driver with strong Lotus connections, Bill Brack: 'I was the Lotus importer for Canada at the time and we were pretty successful, we'd bought a lot of cars from Colin Chapman. I had just won the Eastern Canada Formula Championship in a Lotus 41 Formula B car and I thought I was pretty good. So I called Colin Chapman up and said "What do I have to do to get a ride in a Formula 1 car?" He said "Well, we'll see if we can arrange that for you Bill". And he wanted C$6000. So I paid the C$6000 and got my ride!'

The presence of the Lotus Grand Prix team in Canada was also seen as a big sales opportunity by Brack: 'We decided to have a press announcement, a bit of a show of the cars that were around at the time. I think it was the Europa, the Elan and the Elan 2+2, and we asked if Graham Hill could come to Toronto on his way to Quebec for a press day here with the Lotus cars.' That Brack would be driving the third car had obviously not been explained to Hill prior to his arrival in Canada, as he now recalls with a laugh: 'Graham flew into Toronto and I picked him up at the airport - in a Lotus of course!

'I drove him downtown and on the way I said "Oh, Graham, I'm really excited because I'm going to be driving the 49 with you, the spare car". He looked at me like I was off my rocker! I don't think he even knew that I raced. I was the Lotus importer, this guy picking him up at the airport. He must have thought I'd been drinking ... Anyway, I said "I've arranged it with Colin Chapman" and he said "Not bloody likely mate!" He was really quite upset. I thought that I'd better not say any more so I got on with the job of the press day. Anyway, off he went to Quebec and, of course, I went down there as well because I'd made this arrangement! So I arrived at the circuit and I am sure that he was not a happy guy. But he came around ...'

A decision had been made to try and improve reliability in the transmission area in the crucial final three races by going for a more robust gearbox, the Hewland DG300, on Hill's car. However, there was not enough time for the new boxes to be delivered and fitted to the car prior to its departure for Canada, so the gearbox followed on behind. It arrived on the Friday during first practice, and was fitted overnight in time for Saturday's final session. Although it was, in theory, a kit of parts that should just have been assembled and bolted on to the engine, this was not quite the case, as the records of Bob Dance (who, as Chief Mechanic, was in charge of the operation) show ...

What should have been a simple process turned into a saga of truly epic proportions, in typical Lotus style. 'The Old Man complained "Why have you been there all night, it's supposed to be bolt on?" I had made a list of why it didn't all go together. Some of the things had obviously never been fitted together. The lads back at the works just sent it all out, probably thinking "Get it out to them, they'll fix it". That sort of thing did happen, which was a bit of a nuisance for the chaps in the field.' There were sixteen items which Dance was not happy about and he wrote them down -

1. *Bottom suspension beam had never been fitted on the gearbox studs.*
2. *Top beam fouled the dampers.*
3. *Top tubular beam had the wrong hole centres.*
4. *Starter was solid when bolted up (wouldn't turn through some misalignment, or something).*
5. *Gearchange fouled an oil union.*
6. *Had to modify the exhaust bracket.*
7. *No unions with the clutch pipe.*
8. *Oil tank fouled the top union.*
9. *Oil tank pipe fouled the starter.*
10. *Oil tank pipe also fouled one of the brackets.*
11. *Oil tank fouled the slave cylinder.*
12. *Had to extend the support tubes.*
13. *Slave cylinder offtake was pointed into the tank.*
14. *Bottom beam fouled the starter by 1/8th of an inch.*
15. *Oil tank brackets 1 inch short.*
16. *Had to make new length oil pipes.*

'That was our pre-fitted kit! And we were working in a tent on a peeing wet, cold, muddy, night in the middle of nowhere with portable hand lamps ...'

The other notable point about the specification of the cars entered for Canada was that Gold Leaf Team Lotus were the only team to retain the revised exhaust system layout as used at Monza, which gave them extra power at the top end but reduced mid-range power, something which, on a twisty circuit like St. Jovite, appeared likely to count against them.

For Brack, this was the first time he'd even seen a Lotus 49, let alone sat in one and driven it. Consequently, he went through a rather steep learning curve in terms of how to drive a cutting-edge Formula 1 car: 'On the Friday practice day I recall that they let me out, saying "Bill, do 10 laps, take it easy". So I did and the session was over before I could get out again. Then I got another 10 laps in the next session. So I think I did 20 laps on the first day of qualifying. Of course, I wasn't even on the grid, time-wise. I'd never driven anything larger than a 1.5 litre and this car was just amazing. It was like a rocket to the moon, really something.'

Meanwhile, Hill was setting some rapid lap times, consistently up among the contenders for pole position, despite having to pit early on to replace a rear wheel which had cracked. However, practice was interrupted when Jacky Ickx had a major shunt at the fast downhill right-hander just after the pits. His throttle had jammed open and his car had shot across the trackside, vaulted a grass bank and landed near a lake. The result was not as serious as was initially feared, but it left him with a broken left leg. Thus, the closest of Hill's challengers for Championship honours had been eliminated from the fight with three rounds still to go. When practice was restarted, Hill went faster still and was only one tenth off the fastest time of the day, set by Chris Amon in his Ferrari. However, towards the end of the session, Hill's exuberance got the better of him and he spun off, damaging the nosecone and suspension slightly. Oliver was struggling to get up with the pace of the leading runners but was still joint eighth fastest along with the luckless Ickx.

When the cars reappeared on the Saturday for final practice, the patched-up nosecone had been fitted to Brack's car and the DG300 box was fitted to Hill's. After having experienced problems with his car's throttle sticking open in Friday's session, Siffert was making up for lost time and soon eclipsed the previous day's time of 1.34.7 set by Amon. His lap of 1.34.5 was equalled by Dan Gurney in a McLaren. After relatively few laps, during which he improved his lap time by 1.4 seconds, Oliver came into the pits with a crack in the rear subframe, effectively ending his practice session. Despite this, he still ended up ninth on the grid. Hill could not match his pace of the previous day, being more than a second slower, something which was eventually traced to his having Brack's rear wheels and tyres on instead of the ones intended for his car! The only consolation for the Lotus crew was that Hill's nearest rivals for the title, Hulme and Stewart, were both behind him on the grid.

It was left to Siffert to uphold Lotus honour; he improved his time by a further two-tenths for a spot on the outside of the front row, at the same time completely eclipsing the efforts of the works team with his private entry. The most likely explanation for this rich vein of speed was that Siffert had never been offered the revised DFV, with more power at the top end but reduced mid-range grunt, which the works cars were running. Quite simply, Siffert's engine was better suited to the characteristics of the circuit. Alongside Siffert on the front row were Amon's Ferrari and Rindt's Brabham. These two had both done the same time of 1.33.8, but Rindt secured the pole (plus the BOAC Trophy and a C$1000 prize) by virtue of having achieved the time first.

Brack was struggling to get anywhere near the pace, until he had a conversation with team-mate Oliver which helped him get nearer the limit: 'I thought I wasn't going to make the grid after that first day because I couldn't seem to get the car to go any faster, I was scaring the living daylights out of myself because it was so fast and handled so well! Anyway, I came in after the first qualifying session on the Saturday and said to Jackie "I can't get it through that corner, I'm going to have to change the gear ratios. You say that you go through that corner in third gear?" and he said "You don't want to change that ratio, just keep your foot in it Bill, the car will do it. You may not think so, but it will do it." And, sure enough, I could fly through the corners if I really forced myself to keep my foot on the accelerator. I went faster and faster and I made the grid.'

At the end of unofficial practice on the morning of the race, the familiar problem of driveshafts twisting was rearing its ugly head again on all three cars, with Hill's car's shafts the worst afflicted. A decision was made to swap them for the ones in best condition - from Brack's car - but things didn't look good for the race.

Warm and dry conditions greeted the drivers for the start of the lengthy (238.5 mile) race. As the flag fell, Amon jumped into the lead, hotly pursued by Siffert but after only a few laps, the engine in the Swiss driver's car had begun to smoke and it seemed unlikely he would go the full distance. Eventually, he began to drop back from Amon and, at the end of lap 29, he pulled into the pits to retire. From fifth on the grid, Hill had maintained his position until he passed Gurney on the 14th lap. Benefiting from Siffert's retirement he was up to third and, when Rindt came into the pits on lap 39, this elevated the Briton to second behind the seemingly invincible Amon.

Brack's race with his twisted driveshafts was a short but sweet experience: 'I was raring to go. The flag dropped and right away it was vibrating because I had Hill's shafts! But still I passed three or four cars. I lasted 18 laps before one of the shafts broke and that was the end of me.' In retrospect, he believes that he should have insisted on being able to drive the car before race weekend, as there was simply so much to come to terms with in such a short space of time. This was a piece of advice that he passed on to Gilles Villeneuve some nine years later, then an up-and-coming Formula Atlantic racer who had been offered the chance to drive for McLaren in the 1977 British Grand Prix: 'I said "Make sure you get some time in that car because it is a big change from Formula Atlantic to Formula 1" and I think that helped him.'

Oliver's steady progress through the field had brought him up to sixth place by lap thirty. However, Lady Luck was not smiling on him and, only three laps later, his race was run and he coasted to a halt with a broken crownwheel and pinion. It was left to Hill to uphold Lotus honour, but towards the middle of the 90 lap race, his pace fell off noticeably and he began to be caught by the two McLaren drivers, Hulme and McLaren. Surprisingly, he let them both past without any resistance and, eventually, on lap 57, after being passed by Rodriguez's BRM, he headed for the pits to try and find out what was wrong with his car.

Bob Sparshott remembers the pit stop clearly: 'Graham came in saying that the car felt like it was going through the ground one minute and flying up in the air the next. When we jacked up the back of the car you couldn't see a problem, because the engine was pressed up against the chassis, it went up normally. We were looking all around but we couldn't see it because the top engine mountings on the 49 were hidden. As we lowered it down, somebody said "'Christ, as the jack came out it went like that" [motions hingeing in the middle]. So we jacked it up again and lifted it in the middle and then we could see that the top engine mountings were broken. I was convinced we were going to push it away. The next minute, the Old Man was patting him on the head and waving him out, saying "You'll have to be careful, take it easy ..." and he was gone! And I thought this is it, the bloke's going to die! But he didn't, he hung on. The race was so strung out, it made it worthwhile, because we hung on to one or two points and I think we needed those points to win it [the Championship].'

Hill rejoined in sixth place circulating at a much reduced speed, but just fast enough to stay ahead of his nearest challenger, Vic Elford in the Cooper-BRM. By dint of the fact that there were a further two retirements, this translated into a very worthwhile haul of three Championship points for fourth place, some four laps behind the winner. Meanwhile, Jackie Stewart had pitted early in the race with a broken wishbone. After the Tyrrell mechanics changed it, he restarted and tore back through the field to take a well-earned sixth place, albeit some seven laps behind. Hill's other title rival, Denny Hulme, had benefited from the tremendous rate of attrition of other cars and the superb reliability of the McLaren to lead his team-mate home to a 1-2 victory. He had inherited the lead after the hapless Amon had been forced to retire when comfortably ahead with only 18 laps to go. His retirement was caused by final drive gears badly worn because he had to change gear without the clutch, which had failed early in the race. This was the second successive race in which Hulme had scored a maximum points haul, and this catapulted him to the top of the standings along with Hill. Suddenly, from almost nowhere in the early part of the season, the 1967 Champion had begun to put up a defence worthy of his title. The two leaders both had 33 points, while Stewart joined Ickx - now out of the title hunt due to his crash - on 27 points, but was still very much in with a shot at the title.

With two weeks between the Canadian race and the US Grand Prix at Watkins Glen, there was plenty of opportunity to put right the problems which occurred at

The looks on the faces of, from left, Bob Sparshott, Trevor Seaman (who was feeling ill anyway) and Chief Mechanic Bob Dance say it all. It is 10 o'clock at night, all the other teams' mechanics have long since departed but Chapman has just arrived to tell the lads that three engine changes are needed before morning! Maurice Phillippe, who stayed on to help out that night, looks on. (Bob Dance Collection)

It's dark, it's late and it's a scene of chaos ... it must be the Team Lotus garage! Midway through the engine change, Andretti's car has its engine in place, while the cars of Oliver (11) and Hill (10) await fresh units. (The Dave Friedman Photo Collection)

St. Jovite. The most pressing problems were the repairs that were required to Hill's chassis, R6. The car needed a lot of work since it had relied on the bottom engine mountings and the rear radius arms to hold the car together, causing quite a lot of damage to those mountings, as Bob Sparshott remembers: 'When we got to Watkins Glen - as well as all the normal hard work - we were faced with rebuilding Hill's car and finding out what had happened in the chassis. Because it buggered all the bottom engine mountings, they were all working on the chassis; completely buggered, they all had to be redone. A massive job! Plus all the other work, the routine changes and mods that were happening at Lotus all the time. And we had to run Mario there in the third car.'

Following his aborted debut at Monza, Andretti was finally to make his first Grand Prix start at the Glen. Chassis 49B/5, the one he had driven in practice in Italy, and in which Bill Brack had started the Canadian race, was readied for him. It seems that the mechanics were informed of the fact that they would again be running a third car rather late in the day. Furthermore, Bob Dance and his mechanics were not happy to be told of the need for three engine changes at 10pm when Chapman arrived in the tech-shed. Most of the other teams had finished their preparation, covered up their cars and left for their hotels. As Sparshott recounts: 'That was thrust upon us,

we didn't know that was going to happen in advance. A deal was done with Ford and, of course, he had to have a decent engine, because they demanded he had one. We were short of good engines and to get Mario a decent engine and still give Graham a decent one meant that all three engines had to come out of all three cars, with Jackie's the first car. We did a sort of "Round Robin" and put a fresh one in Graham's and put the best of the bunch - Oliver's engine, which was quite good - into Mario's. Oliver had to have the worst of the lot, which he didn't like, he was whingeing and moaning!'

The Lotus No. 2 was not a happy man at the prospect of having another race in what he now regarded as the spare car, chassis 49B/2, when he felt he should be driving the newer car which had been allocated to Andretti: 'I was served up a real load of old rubbish there, because Colin wanted Mario Andretti to drive my car. So they gave me what was left!' Although the passage of time may understandably have impinged on their respective memories, it is interesting to record Andretti's perception of the car he got for that meeting, which bore a remarkable similarity to Oliver's assessment! 'As a third car, I was getting all the crap parts, all the old parts. I didn't have a fresh engine or any of that stuff.' Which just goes to show that even in those days drivers always thought that their team-mate was getting preferential treatment ...

As a precaution to prevent a recurrence of the problems experienced by Hill at St. Jovite, the aluminium engine mounting plates were replaced on all three cars with steel ones. Hill's car retained the more durable DG300 gearbox he had used in Canada, while both Andretti and Oliver's cars used lighter FG400s. The 'Monza' exhausts were retained on the Hill and Oliver cars, while Andretti started the meeting with the older-style set-up giving less horsepower, although this would be changed overnight for the final session, with Andretti getting the second four-into-one system and Oliver the older layout, much to his disgust.

Lotus also switched to new type CV driveshafts for the US race, similar to those used at Monaco earlier in the year, in an effort to improve drivetrain reliability. This was all very well, but, according to Sparshott, it made for a lot of extra work: 'We were trying to perfect CV joint driveshafts, which was a new thing and, of course, being Lotus, they were never, ever, man enough for the job, they were always just about there. It was one man's job to maintain all of those joints and Bob [Dance] did it for the race on all three cars. That was a whole night's work. They had to be stripped out, measured, regreased and all the seals put back on. They were all glued on, it was a nightmare of a job. I'll never forget, dear old Maurice Phillippe [Lotus designer] helped us, because Trevor Seaman was ill, which left five people for three cars. Maurice said "I'll stay and give you a hand". He worked all through the night with us to make the six.'

'In the morning the drivers and the Old Man came back, all fresh and bubbling. We'd gone to have a bacon sandwich and we arrived back at the garage as they were all standing around looking at the cars. I'll never forget Jackie [Oliver]. He was looking down at his car, with his hands on his hips, to identify the colours on the springs, because one of the jobs that needed doing on his car was changing the front springs. It was a big job on a Lotus 49 with the inboard front springs. We had prioritised the work and the springs had been left, basically. Jackie didn't say anything to us, not "Good morning, I see you've been here all night, guys", nothing at all. He just walked over to the Old Man and said "Colin, they haven't changed my springs". Colin went to talk to Bob Dance, I suppose to say "Why haven't you done it?", took one look at Bob, who was totally wound up and got the vibes straight away. He just turned back to Oliver, put his arm round him and said "I want you to start practice on those, Jackie".'

The previous day in practice had been largely spoiled by rain in the early part of the session and the generally cold weather conditions. The Gold Leaf cars had not been particularly impressive, ending up fifth, eighth and fourteenth fastest in the hands of Hill, Andretti and Oliver respectively. On stripping the cars down that evening, it was discovered that Hill's clutch had a broken centre. Due to various problems, Oliver had not got much running. Consequently, for the final session on the Saturday, he was particularly keen to get out onto the track and improve his grid position. However, disaster struck after only a few laps, for his car speared off the track on the fast right-hander after the pits, striking the armco barrier twice before coming to rest in a rather sorry state - bad enough to eliminate him from the rest of the weekend's proceedings. Bob Sparshott takes up the story: 'Where we screwed up was that a truck came up in the middle of the night from the airport - as they did all through the night bringing parts for everybody - and a load of wheel halves arrived for us. They got shoved in the corner. Nobody said anything about these wheels, so we didn't know anything about them. The next morning, after all this work, Jackie wanted to get out quickly. I was still finishing my car off because I was down on manpower. Graham arrived and was not panicked at all and asked if he could help. I asked if he'd mind giving it a polish! You always had to polish the car before it went out, but I literally hadn't got time to do it. So Graham was polishing away and everybody else buggered-off out of the tech-shed to start practice and before we got out, they were back. Oliver had crashed the car, he'd stood on it and a wheel had broken on about the second lap.

'The first thing I saw was Chapman and Maurice Phillippe marching up to the tech-shed ranting and raving. Then the Old Man just dived into the wheels with Maurice and started frantically trying to find things to screw the wheels together. It was like a circus act! Then he went over to the Firestone tyre bay, which was also in the tech-shed. A bloke was fitting somebody else's tyres and Chapman just stopped him working and literally ripped the tyres and wheels off and threw them on the ground and put the Lotus wheels on and said "Fit those up, now!" Which he did. This guy was terrified because Lotus was a main runner and the Championship was very much at stake!

'Afterwards there was an inquest into why we hadn't put the new wheels on, which came to nought because we didn't know about them, or that they had suspected they were going to break.' Apparently, Chapman had realised after the events of Monza and

The end result of the tiring all-nighter was that the mechanics didn't fit the new wheels that had arrived that evening and one of the old wheels on Oliver's car failed, pitching him into the barriers. His car was too badly damaged to be repaired in time for the race, so he had to miss it, since the spare car was being used by Mario Andretti. (The Dave Friedman Photo Collection)

Challenge For The Championship

Canada, where Hill had experienced problems with breakages and cracks, that the wheels were not up to the job. He had ordered new ones and issued specific instructions that the Canadian wheels were not to be used. Somehow, there was a breakdown in communications, for these instructions had not been passed on to the mechanics. In Bob Dance's opinion, the whole episode was unnecessary, since he felt that the engine change on Mario and Jackie's car was of no great performance advantage. All it had done was to use up valuable time which could have been used building the wheels up, bolting the halves together …

Once the new wheels had been fitted to the remaining cars of Hill and Andretti, they rejoined the battle for grid positions. Everyone agreed that Mario looked comfortable and that his driving was very assured, but few expected him to seize pole position. However, after getting a tow in the slipstream of Surtees' Honda, he set a blistering lap of 1.4.20, 7/100ths quicker than Stewart's fastest time, set on Friday, and to the delight of the home crowd. It was an impressive performance, made all the more exceptional by the fact that, despite living and racing in the US, Andretti had never seen the circuit before, let alone driven or raced on it! He recalls: 'A lot of people said "Oh yeah, that's no problem he's on pole, it's his home course." I'd never seen it! I'd never even been to Watkins Glen, I had no idea which way the turns went so I was totally green there. The guy that really helped me that weekend was Maurice Phillippe. We set the car up the way I wanted and I put her on pole.'

Andretti's performance put a spring back into the step of Chapman and sparkle in his eyes. The American's brash confidence, bordering on arrogance, was impressive for one with so little experience of Formula One, as Bob Dance recalls: 'He told the Old Man: "You tell me when you want me to put it on pole". Those were his words. And he did it! The Old Man was rubbing his hands together and saying "It's just like having Jimmy all over again" …'

Soon after Andretti set his time, Hill came into the pits with a rear tyre off the rim, but was soon sent out with a new wheel. His team-mate's turn of speed seemed to spur the Briton on and he really wound things up to set a time of 1.4.28, just 1/100th of a second away from making the front row an all-Lotus affair for the second year running.

The other Lotus on the grid, Siffert's Walker/Durlacher car, had experienced engine troubles throughout practice. A down on power replacement for the unit used on Friday meant that he was unable to improve his times. As a result, he lined up on the sixth row of the two-by-two grid, 12th fastest.

Race day dawned overcast but dry and a huge crowd, estimated at 93,000, assembled to see if Andretti could carry his practice form over to the race. Despite not having any previous experience of race starts from a static grid in a Formula 1 car, he made a great getaway although, as Dave Sims recalls, it was not without cost. 'He burnt the clutch out on the line because he was still getting used to full tank standing starts.' As the field snaked up through the Esses, Andretti led from Stewart, who was closely followed by Hill. At the end of the back straight, the canny Scot caught his rival by surprise and

World of Wings: by this stage of the season virtually the whole grid had adopted some form of high wing. Here pole-man Andretti and his front-row companion Jackie Stewart wait for Tex Hopkins to leap into the air, drop the flag and set them on their way. (The Dave Friedman Photo Collection)

'It's like having Jimmy all over again' was Chapman's gleeful response when Andretti finally got to join the team for the US Grand Prix at Watkins Glen and promptly planted his car on pole. (The Dave Friedman Photo Collection)

- First Person -
Mario Andretti on the Lotus 49

It was awesome. To me the Formula 1 car was much easier to drive. It was more nimble and responsive, so it was a piece of cake! I had no problem adapting to it at all. The 49 was such a good car with the big wings on the uprights. It was forgiving and it could be driven with abandon, I mean you could just DRIVE hard. And that suited me pretty well. I felt so much at home because when we went testing I was immediately down to quick times.

The car was a winner, no question. Just one of those that any driver could get into and fall in love with - that simple. When a car is sincere, it is balanced, it feels good to everyone and you know that you can tweak it up here and there and make it exactly the way you want. But it's in the ballpark immediately, and that's the important thing – which is a statement you can make about that car.

Lotus 49

Andretti impressed on his race debut in the US Grand Prix, taking pole and running second before car troubles brought his race to a premature end. (The Dave Friedman Photo Collection)

out-braked the Lotus into the turn to take the lead. At the same time Hill, braking hard behind them, found his fingers trapped against the dash because his steering column had slid forward due to the clamp bolt not being tight enough. In extracting his fingers from this painful situation, he knocked the ignition switch off and the engine momentarily died. Although he quickly realised what had happened, this hesitation allowed Amon to slip by into third place.

After about five laps, the race order seemed fairly established, with Stewart not getting away from Andretti at all and a small gap to the chasing pack of Amon, Hill, Rindt and Gurney. However, one or two of the nose-clips on the same side of Andretti's car's nosecone flipped up and came undone and it allowed the underside of the nose on the right to droop down onto the track. In so doing, it was wearing away the foam wings so, after 14 laps, Mario was brought in to the pits for the offending piece of bodywork to be taped up. He roared back into the fray in 13th place and continued to circulate rapidly, overtaking Jack Brabham and moving up the order largely due to other people's retirements until he was 9th by lap 22. However, he could not find a way past the BRM of Rodriguez, which was particularly strong in terms of straight-line speed. By this stage he was also beginning to experience problems with his clutch slipping and this eventually got so bad that he limped into the pits to retire for good on the 32nd of the scheduled 108 laps.

Andretti's problems had promoted Hill up to third place and then, when Amon's radiator split causing him to spin, he found himself in second. However, he could do nothing to reduce the gap between himself and the flying Matra. The three Championship contenders were now in first, second and third places, for Hulme was close behind Hill. On lap 16, Hulme spun at the end of the straight, damaging a brake pipe. This necessitated a lengthy pit stop, which cost him eight laps and effectively put him out of the running for any points but he carried on, mindful of the high rate of attrition in previous races that had been most evident in Canada.

By half-distance, Hill was 40 seconds behind Stewart. However, with around 70 laps completed, the Matra began to smoke a little, so the Scot eased off slightly, allowing his rival to close the gap once more. A spate of retirements at around three-quarters distance saw Hulme move up to seventh place, with only his teammate McLaren standing between him and a point. However, before his fellow Kiwi had time to think about how (or whether) to let him reduce the seven laps deficit between them, Hulme spun wildly off the track at the last corner before the pits, deranging the right rear wheel and damaging the car beyond repair.

The biggest concern now, as Bob Dance remembers, was fuel consumption in what was an exceptionally prolonged race, even by 1960s standards. 'We fitted an auxiliary fuel tank behind the radiator, shaped to fit between the exit ducts'. Such additional capacity proved a prudent measure, for both Siffert and McLaren called at the pits for more fuel in the dying stages of the race. Despite this, Siffert went on to take a well deserved fifth place after a steady and uneventful run. Things were pretty tight with Hill's car, according to Dance. 'We finished with less than a gallon of fuel, when measured after the race with "The Guvnor" in close attendance!' But finishing, and finishing in the points, was what mattered. While Stewart took a comfortable victory - his third of the year - Hill finished in a safe second place to earn himself six valuable points to take his tally to 39. Stewart's nine points meant that he jumped up to second in the standings with 36, while Hulme, who had been joint Championship leader going into the US race, emerged empty-handed, leaving him back in third on 33 points. For the second successive year, the Championship was going to be decided at the final round in Mexico. Immediately after the race, Oliver's car was stripped and rebuilt as far as possible before the team left for the UK, while the cars were transported down to Mexico by road on an open transporter, a journey that took over a week.

After a four week gap, due to the fact that the Olympics were being held in Mexico on the date when the last Grand Prix of the season was traditionally held, the Grand Prix teams reconvened at the Magdalena Mixhuca circuit in Mexico City for the final and deciding round of the World Championship.

Once again, Lotus had contracted to run a third car, this time for the local driver, Moises Solana, who had acquitted himself so well at the wheel of the spare 49 in the US and Mexican races in 1967. Hill was entered in the car that had been his regular chassis since Monza, R6, while Solana was allocated chassis 49B/R5 and Oliver retained his usual car, 49B/R2.

The weekend saw Chapman the genius at his brilliant best, with his idea for a wing that feathered flat for the straights going from concept to execution overnight, as Bob Dance remembers: 'That was one of the Old Man's "We'll *just* make the wing pivot!" We were still working on it the next morning when he arrived at the circuit ...' That episode is also engraved in the memory of Bob Sparshott, Hill's mechanic: 'Chapman came up with the idea of making the rear wing pivot, because it was legal to do it. We did it, literally there, in the garage. We took the wings to pieces, moved the static pivot point to a dynamic pivot point - which was a fabrication job - and then reskinned the wing. We fitted a Bowden cable and made up a little pedal, because there was enough room in the cockpit to fit another thin pedal, and then there were springs which were bungee straps on the wing to counter-act the downforce. What Graham had to do was keep his foot on that pedal down the straights which feathered the wing flat and in the corner took his foot off and the bungee straps hopefully would pull the wing down.

'The Old Man reckoned that we'd pick up 200 or 300rpm. It was not a lot, but it was enough to give a little edge. But it took about 14 hours work by four people to do it. He came back in the morning, when we were riveting the wing skins up and started bossing everybody about and saying "How are you going to do this, how are you going to rivet that back on there?" So we said, "Well we're going to put a row of skin pins along there" and he was saying "No, no, you don't want to do it like that, that's no bloody good" ... and then he took one look at us all and shut up and disappeared!'

Sparshott tells a tale of co-operation from other teams that is hard to imagine in today's climate of bitter rivalry in Formula 1: 'On away trips in those days we only took a few things because we didn't have a lot of space. We didn't have enough rivets - we needed hundreds - so we went round all the teams and they lent us pop rivets and pliers to put them together! We had to borrow everything but nobody quibbled. You'd never do it now because it would all be worked out on a computer in advance. This was real seat-of-the-pants stuff. We needed a little edge and the Old Man came up with one.' Bob Dance also remembers another small touch which demonstrates Chapman's attention to detail. 'We had to cover up the rows of rivets with shiny black adhesive tape to improve the airflow over the airfoil surface.' This is why, in pictures of Hill's car in this event, there appear to be black stripes round the rear wing.

Everyone in the team was on edge, knowing that World Championship victory was within their grasp, as indeed it was for both Stewart and Hulme. The all-nighter to fix up the wing did nothing to ease the tension. For Friday practice, Hill was allocated 49B/R5 as well as his regular R6, with Solana and Oliver having to share driving duties in 49B/R2. For the majority of the session, he used 49B/R5, shorn of its wings and nose fins, a piece of gamesmanship designed to give other teams the impression that this was considered by Lotus to be the optimum set-up. It didn't take long for rival teams to appear with their cars in the same format, which they all discovered to be slower than when the cars had wings! When Hill did go out in R6, the wing was in the fixed position so as not to give the game away too soon about his 'secret weapon'.

Siffert, in the Walker/Durlacher car, had really got the bit between his teeth and was turning in some blistering laps, well inside the existing lap record. Eventually, he finished the session half a second clear of the next fastest runner, Hill, who had done his time in the winged car. This proved to be nearly three seconds a lap faster than he had managed in the wingless 49B/R5.

Mexico 1968: an extremely serious-looking Colin Chapman ponders how he is going to fit a fourth pedal alongside the clutch to operate the feathering rear wing. (Phipps Photographic)

Stewart was next quickest, while Solana caused a stir by turning in sixth fastest time, ahead of many more established drivers, although there was a question mark over whether, with all the switching between cars going on, this time had actually been done by Hill. The timekeepers seemed to be in a bit of a muddle, for they originally credited Oliver with a time almost three seconds slower than Solana, which did not reflect very well on the Lotus number two. However, this was revised to a time less than a second slower than the Mexican late on Saturday night, after final practice had taken place. The other Championship contender, Hulme, spent most of the first day of practice in the pits while the McLaren mechanics rebuilt his clutch, but still managed to set the ninth fastest time.

The weather for the final practice session was warm and dry and there was an air of expectancy in the pit lane as grid position would be crucial in dictating the extent of the battle between the three Championship rivals. After some initial exploratory laps in the wingless car, which saw him improve his lap time in that car by just under a second, Hill decided to concentrate all his attentions on the car with the feathering wing, and Solana was allowed to use 49B/R5, once the wings had been refitted, for the remainder of the session. This freed up Oliver to concentrate on honing 49B/R2 to his liking, but he had a disappointing session too, and was unable to better his time of the previous day before he spun at the Esses, tearing off the exhaust pipes against the earth bank.

Despite his best efforts, Hill could not improve on his Friday time either and would start from the second row, alongside rival Hulme, who had really got the McLaren going in the final session. It was all the more frustrating for Hill when Siffert, who already held pole position from his effort in the first session, calmly went out and knocked a further 0.3 seconds off his time to reinforce his place at the top of the time sheets. Meanwhile, Stewart, who had done his quickest time in the spare car (which he did not want to race), elected to take the start in his race car, which meant that he had to occupy the grid position of his Friday time, which put him a row further back than he otherwise would have been.

Solana was not at all happy with the performance of his car, having ended the session as the slowest man in the field. A fierce row ensued in the Lotus garage, the result of which was that it was agreed Solana could have the car he recorded his Friday time in, and Oliver would take over 49B/R5, a car which he had not driven at all at the meeting up to that point. By some judicious swapping of nose-cones (49B/R2 was using the old-style nosecone after the one with the top exit ducts had been destroyed in Oliver's US Grand Prix practice accident, while 49B/R5's nose had the top ducts), the team managed not to alert the organisers to the fact that the chassis had been swapped, and thus Oliver was able to take up his allocated position on the grid, instead of starting from the back as the rules required.

Just as it seemed that the problems in the Lotus pit had been solved, another row brewed up, over the eternally sore subject of fuel. Bob Dance explains: 'There was an argument between Bob Sparshott [Hill's mechanic] and the Old Man and myself about the fuel for the race day - the measurements didn't work out right. Graham, who was a pretty stable bloke, a good team man, said "Look this isn't the place or the time for arguing about it, let's just get on with it".' Bob Sparshott also remembers this vividly: 'We had a bit of a to-do in Mexico! Bear in mind we'd got a job list a million miles long, and one of the jobs on it was obviously draining fuel and doing all that stuff - if we could get the fuel out! The Old Man came in the garage and asked what the consumption was and Bob said "Well, I'm afraid it's not come out right." And he went bloody bananas! He started ranting and raving. I spoke out of turn and said "We can't do a check Mr Chapman. The only way we'd get a really good check is to stand the car up on end." He thought that was a real p**s-take! He just said "You b*****ds! Don't blame me if it breaks all the driveshafts on the line" and he stormed out of the garage'. The only pleasant thing Bob Dance remembers about this weekend is 'Ken Tyrrell, some of his mechanics and the McLaren mechanics, coming into our pit late on the Saturday night with a big cake and offering it round and wishing us luck for race day.'

At 2.15pm, the cars came out for a warming-up lap and lined up in front of an enthusiastic capacity crowd. Hill reported that a front wheel was out of balance, but there wasn't any time to rectify this and he had to start with it. When the flag dropped, Siffert squandered the advantage gained from his practice efforts by getting bogged down and the field streamed past him as he dipped his clutch to get things going. Hill burst through from the second row to take the lead ahead of Surtees in the Honda and a fast-starting Stewart, up from the fourth row. Siffert, smarting from his poor start, came through in seventh place but was rapidly making up his lost ground. On the second lap, Stewart passed Surtees, to set up a battle for the Championship with his main rival Hill, while Hulme was only two places further back and holding a watching brief. Solana's race took a decided turn for the worse when his wing collapsed on this lap. Although he drove several more laps with part of the wing hanging down at the side of the car, he clearly could not continue like that and, on lap 4, came in to the pits to have it removed.

There was potential for disaster on lap 3 when one of the wing-feathering bungees on Hill's rear wing came adrift, leaving just one to operate the system. This was immediately spotted by the eagle-eyed Chapman, who proceeded to give chief mechanic Dance a hard time about it for the rest of the race!

On lap five, Stewart passed Hill and Hulme slipped by Surtees, so the Championship contenders were running 1-2-3. However, it did not last, for on the next lap Siffert, continuing his meteoric rise up the order, dispossessed Hulme of third position and began to close on the two leaders, who had pulled out a bit of a gap. On lap nine, Hill repassed Stewart for the lead. A lap later Hulme's title bid came to a premature end against the barriers opposite the pits when a damper unit broke, sending him spinning off the track. Siffert was catching Stewart all the time and, on lap 17, he dived inside the blue Matra into the hairpin and set off after Hill. By this time, Solana had retired in a fit of pique, declaring that the car was not going well. Although the mechanics could

Challenge For The Championship

find nothing wrong with it, the fact that it had no rear wing but front nose-fins would undoubtedly have unsettled the car's handling. The reason for his retirement is recorded simply in Bob Dance's book as 'Driver Temperament'!

Before long Siffert had caught Hill and, on the 22nd of the scheduled 65 laps, he took the lead and began to pull steadily away. However, his glory was to be very short lived, for three laps later, he coasted into the pits with a throttle linkage that had popped out of its socket at the engine end. It took the mechanics an agonising two laps to locate the problem and rectify it and Siffert rejoined the race in last place, determined to try and salvage something from the weekend, even if it was only fastest lap. This he proceeded to do turning in a series of blistering laps, the quickest of which was almost a second faster than his pole position time!

Siffert's stop had promoted Oliver to seventh place from his initial starting position on the grid of fourteenth. When Dan Gurney dropped out of third place, he was in the points and running steadily if unspectacularly, just ahead of Mexican Pedro Rodriguez.

The tension in the pits was tremendous. With Hulme out, Hill and Stewart were effectively fighting for the Championship and they remained seconds apart, lap after lap. The mathematics of the situation were quite simple: Stewart had to win the race and, even if Hill finished second, the Scot would take the title by virtue of having scored more wins, while if Hill finished ahead of Stewart, in or out of the points, he was the Champion. On lap 38, the battle swung Hill's way for, instead of being close behind the Lotus, the Matra came past five seconds behind. The next lap the gap had stretched to 15 seconds and it was abundantly clear that Stewart had a problem. It transpired that the fuel pressure had dropped and this engine would not run above 6500 rpm, therefore severely limiting the car's pace. The helpless Scot fell back steadily down the order, his last hope of securing the Championship going when first McLaren and then Brabham passed him. They were followed by team-mate Servoz-Gavin and then the battling duo of Oliver and Rodriguez.

The retirement of Servoz-Gavin with a blown engine on lap 58 temporarily restored Stewart to the points, and Brabham's elimination through lack of oil two laps later raised hopes of a miracle. However, first Bonnier in the Honda and then the still charging Siffert passed him in the dying laps and finishing in the points became academic because Hill's Lotus was running like clockwork and showing every sign of taking victory. This it duly did, to the delight and relief of everybody in the team.

To complete a day of triumph for Gold Leaf Team Lotus, Oliver benefited from other people's retirements to move up to third in the closing laps and, although local hero Rodriguez managed to nip past him with only three laps to go, the Briton held his cool and re-took third a lap later to secure his best finish of the season. Unbeknown to the people back in the start/finish area, Rodriguez had almost unwittingly altered the outcome of the Championship in the dying laps. Having passed Oliver, he misjudged his braking into the hairpin and Hill, who was running just in front of them, had to swerve to avoid

Back from the brink: Hill single-handedly revitalised Team Lotus and went on to win the World Championship in partnership with Chapman, who saw him in a new light. (Ford Motor Company)

being taken out by the BRM. Fortunately, he had spotted the BRM in his mirrors arriving behind him in a cloud of tyre smoke! This would have been the ultimate irony, because it was at this spot in 1964 that Bandini ran into the back of Hill's BRM and damaged it beyond repair, thereby robbing the Briton of the Championship title.

Siffert's fine drive was rewarded with sixth place, the Swiss having driven faster than any other car on the circuit after his pit stop. He even unlapped himself once from Hill and, in a few more laps, would probably have caught and passed Bonnier's Honda for fifth. As well as the Driver's Championship, Lotus also won the Constructors' Championship handsomely, with a total of five wins - three for Hill and one each for Clark and Siffert.

Hill had a relatively uneventful race and despite the team's concerns before the race about fuel consumption, had no problems in this respect, no doubt due to more than the usual 'mechanic's gallon' having been put in, as Sparshott recalls: 'We finished that race with two or three gallons, I can tell you! We weren't about to bin the World Championship by running out of fuel.'

Despite finishing with only one of the bungees intact, the feathering wing survived the race, providing Hill with just the little edge (actual and psychological) that Chapman had intended. Had the second bungee broken, it might have spelt disaster for Hill, as the wing would have been permanently in the feathered (i.e. flat) position, with a resultant loss of downforce in the corners. However, in reality, the Lotus was so far ahead of its rivals this would have made little difference and, once Stewart dropped down the order, the finishing position of the Lotus became an irrelevance.

It was an emotional end to what had been a rollercoaster year, a year which had seen Team Lotus start off full of optimism, plumb the depths of despair with the death of Clark and Spence and then recover to win its third World Championship title. Much of the credit for the team's success can be attributed to Hill's ability to

The Championship-winning team: Graham Hill is in the car, which is a Formula 2 Lotus 48 as the Formula 1 cars weren't back from Mexico at this stage, with Chapman standing beside him. Front row from left: Trisha Strong (secretary); Andrew Ferguson (Competitions Manager); Jill Downing (secretary); Trevor Seaman (mechanic); Bob Sparshott (mechanic); Bob Dance (Chief Mechanic); Eddie Dennis (mechanic); Dale Porteous (mechanic). Middle row from left: Mike Watson (buyer); Mike Young (buyer); Martin Wade (designer); Maurice Phillippe (Chief Designer); Geoff Ferris (designer); Colin Knight (fabricator); Brian Leighton (cleaner); Dave Buller (fabricator); John Smith (fabricator); Roy Franks (fabricator); Mick Thorn (fabricator). Back row from left: Manning Buckle (accounts); Bob Black (accounts); Jeremy Pearson (Chief Accountant); Dick Scammell (Racing Manager); John Parr (inspector); Ted Fleet (fabricator); Nick Paravanni (fabricator); Ron Chappell (fabricator); Frank Cubitt (fabricator). This shot was taken in November 1968. (Express Newspapers/Bob Dance Collection)

create triumph out of adversity, as Bette Hill points out. 'He was determined that something that was as great as Lotus which had come from nothing, was not going to disintegrate because of their favourite driver dying. I know that in his heart of hearts, Colin knew that too. Colin was devastated of course, but I know that Colin at long last appreciated Graham for what he was.'

The second part of Walter Hayes' prophecy – that the Cosworth DFV might win a Championship – had now also been fulfilled, in only its second season of racing, and it had become the power unit that every team wanted to get its hands on for 1969.

As ever, there was little time to spare to dwell on this success, for in six weeks time the new season would kick off with the first round of the 1969 Tasman Cup and there were cars to be finished and shipped out to New Zealand before then ...

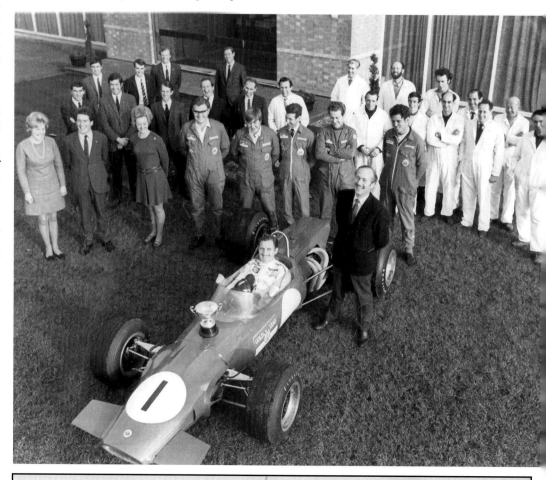

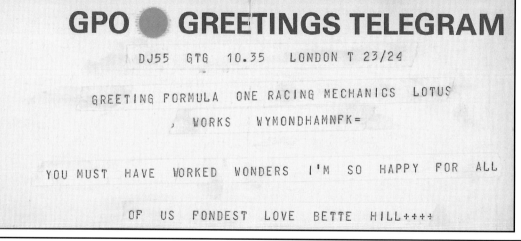

Bette Hill's telegram to the Lotus mechanics after Graham had won the World title in Mexico says it all - they did indeed work wonders, not just there but for the whole of 1968. (Bob Dance Collection)

Chapter 13
Winter Works

The winter of 1968/69 saw considerable upheaval at Team Lotus on the Formula 1 side, with changes made on the driving front and a shake-up of mechanics. Pretty much the only constants were the cars, which were updated to conform with 1969 regulations but otherwise unchanged. The team's sponsor, John Player was delighted by the coverage it received in the inaugural year of its relationship with the team.

The first change was on the driver front. Jackie Oliver had endured an eventful season in 1968: he crashed and wrote off his car in his debut race at Monaco; finished in the points in his second at Spa; crashed and wrote off another car in practice for his fourth at Rouen; and crashed in practice at both the Nürburgring in August and Watkins Glen in October. Although very few of these accidents were due to driver error – component breakages and failures being the primary cause – there was little doubt that he had been an expensive driver for Lotus to maintain when compared to World Champion Hill, who had rarely left the road. Therefore, and considering that he had won his seat in Gold Leaf Team Lotus largely because there were no suitable drivers with Grand Prix experience available, it came as little surprise when it was announced that he would be leaving Lotus and that Jochen Rindt would be joining as joint number one alongside Graham Hill. Oliver immediately signed on to partner John Surtees at BRM and would quickly find out the harsh reality of what Grand Prix racing was like with a team who were not exactly at the cutting edge of Formula 1 …

Rindt had also endured a frustrating season. Widely regarded as a driver who was 'on the up' during 1967, he was almost unbeatable in Formula 2 and had hoped that a move to the reigning World Champion team of Brabham for 1968 would put him in a position to challenge for race victories and possibly even the Championship. However, Repco's answer to the Cosworth DFV, the four-cam Type 860 motor, proved hopelessly unreliable, and the Austrian failed to finish in 10 of the 12 Championship rounds. When he did make it to the chequered flag, he was impressive, taking third places at Kyalami, in a 1967 car at the start of the season, and in the wet at the Nürburgring. His blinding speed and absolute commitment had not escaped the notice of Colin Chapman, with the result that he was signed up for 1969.

This would be the Lotus boss's first experience of working with a non-British full works driver, and the combination was to provide plenty of friction during the year. Born in Germany to a German father and Austrian mother, Rindt had been brought up in Graz, Austria, by his maternal grandparents, after his parents had been killed in an allied bombing raid when he was just one year old. Like Clark, he was never short of money, as there was a successful family business generating an income. This provided him with the wherewithal to buy the machinery which would enable his talent to blossom properly. It also resulted in a character who, to some, was extremely arrogant yet incredibly funny and self-confident, while to others, he just seemed bigheaded and plain rude.

The appointment of Rindt as Hill's team-mate marked the beginning of the deterioration in the Englishman's relationship with Colin Chapman. Although they appeared to have settled their differences over Hill's departure to BRM at the end of the 1959 season, the obvious favouritism the Lotus boss had displayed towards Jim Clark must still have rankled. Now here he was, after delivering a World Championship title in extremely difficult circumstances, having to suffer the ignominy of being given joint number one status alongside a driver who had yet to win a Grand Prix, let alone a World Championship.

A number of considerations influenced Chapman's decision to choose Rindt. Firstly, he yearned for a driver of natural raw speed rather than one like Hill, who had a very methodical approach and spent huge amounts of time tinkering with set-up before feeling confident enough

Lotus 49

The type 57/58 was built as a dual-purpose car for Formula 1&2. Built initially as a Formula 2 car with an FVA engine, its main feature was de Dion suspension front and rear. Clark was due to test it at Zandvoort the day after he was killed. The programme stalled but was later revived. (Dave Sims Collection)

to go for a time. Despite his sensational debut at Watkins Glen, Mario Andretti's lucrative USAC commitments precluded him from signing up to drive for the team on a full-time basis (although he would take in several races during the 1969 season) and Rindt seemed like one of the fastest young drivers in Formula 1 at the time, perhaps along with Jacky Ickx - who was also considered for the drive. Secondly, at three months short of his 40th birthday, Hill was no spring chicken, and it seemed plausible that the quality of his performances would go down in the coming years. Therefore, a younger, faster driver needed to be groomed to take over. Enter Rindt …

Not everybody seemed to have as much faith in Rindt's abilities as Colin Chapman. His most outspoken critic was the renowned journalist Denis Jenkinson, who was so convinced that the Austrian would never win a Grand Prix that he struck a bet with Rindt that he would shave off his trademark shaggy beard should this ever happen!

Despite having been Chief Mechanic in a World Championship-winning year, Bob Dance found himself on the receiving end of some typical Team Lotus man-management, being replaced by Leo Wybrott for 1969. Although Chapman had assured Wybrott that his predecessor had been informed of the decision, it transpired that the first Bob knew about it was when the official announcement was made … Dance explains: 'The Old Man obviously thought that he could find somebody better than me for Chief Mechanic the next year, so he decided that he would give me a choice of jobs. One was to go into another department of Lotus, or I could do the transmissions for the four wheel drive cars, so I decided I'd do that in 1969.'

Bob 'Bucko' Sparshott was another to leave the fold during the winter, to start his own company. Today, he continues to be involved in the sport, through his B.S. Racing Fabrications operation, in Luton. Having moved across to Team Lotus from Lotus Components in 1964 at the tender age of 20, he was still only 24 when he left.

Despite his relative youth, he felt that his time at Lotus had prepared him well, describing it as 'An unforgettable experience. Very challenging, but the net result was that it made you realise what you were capable of. It really drew stuff from right down inside you which, most times in life you don't have to do and you don't think you can do. It's amazing what Chapman used to drag out of us. But overall, it set us fair for a life after Lotus.'

The comings and goings of mechanics and drivers meant that there was a lot of shuffling around of people and new relationships had to be formed. Under Leo 'Squeak' Wybrott were Billy Cowe, Eddie 'Jake' Dennis, Dave 'The Beak' Sims, Derek 'Joe 90' Mower, Bob 'Lotti' Dance, Dougie 'Kildare' Bridge, Michael 'Herbie' Blash and, later in the season, Jim 'The Sponge' Murdoch. Cowe was returning to Formula 1 for the first time since 1966, and was to take over from Bob Sparshott as Graham Hill's number one mechanic, with Dave Sims working alongside him. Eddie Dennis was to be the lead mechanic on Rindt's car, working along with Bridge. Mower spent most of 1969 working on the Indy programme but did some races, working mostly on the four wheel drive team. Due to the similarity of the Type 63 Formula 1 and Type 64 Indy car designs, the mechanics on both tended to float between the two operations. Blash was the team's 'gopher' working on whatever was needed at the time, while Murdoch joined the team for the Dutch Grand Prix in Zandvoort. Additionally, Hywel 'Hughie' Absalom went to the first World Championship race in South Africa and the Race of Champions, but after that he went to work for McLaren.

The question of which cars to take to the 1969 Tasman series also had to be addressed in the latter half of 1968. The choice was between the known quantity of the Lotus 49, or a new project which had evolved in secret and had yet to race. During 1968 Lotus developed the Type 57/58 series, which was basically a dual-purpose Formula 1/Formula 2 chassis designed to take both the DFV (57) and FVA (58). The design was akin to a slimmed-down version of the Type 56 Indy turbine cars which had come so close to scoring the first win at the Brickyard for a turbine-engined car. It followed the wedge-shape principle which was a favourite theme of Chapman and Phillippe at the time, and which culminated in the design of the all-conquering Lotus 72 for the 1970 season. However, the major difference with this car was that it was designed with front and rear de Dion rear axles, something of a favourite with Chapman, but a design which had never before been made to work successfully on a Lotus of any kind.

Originally, the Type 57/58 had been tested briefly at Hethel by Graham Hill in Formula 2 format, with an FVA engine. It had been completed the day that Jim Clark died, and his death and its affect on Chapman really stymied the design from birth because the programme lacked direction. In the end, it was revived with the return of Leo Wybrott, who had been charged with preparing the cars for the Tasman series, with the intention that, if the car looked suitably quick in testing, it would cut its competitive teeth in the New Zealand and Australian races.

Like his boss, Kiwi Wybrott had been deeply af-

Winter Works

Wedge profile pre-dated Lotus 72 by two years... (Dave Sims Collection)

fected by the death of Clark. 'I came back after the [1968] Tasman series and I'd only been back a short time when Jimmy was killed. At that stage I'd spent two and a half years of summers with him and then the Tasman series, nearly three months, together. We were really close-knit, there was just me, Jimmy, Roger and Dale, so we'd go everywhere together. We had a very close relationship. Certainly a lot of my enthusiasm for the sport went when Jimmy was killed.' After the accident, he left the team, ostensibly to get married but also to start a new life with his wife in the south of France. When this didn't work out as planned, they returned to the UK and he rejoined Team Lotus in August of 1968.

A 2.5 litre DFW engine was installed in the car and Wybrott did the initial testing at Hethel prior to Graham Hill coming up to try it out. The car didn't seem too bad to him but, as he recalls, the team's number one driver was distinctly underwhelmed by its performance: 'From the limited knowledge I had, it seemed to go quite quick. You have to assume that, as we were young and wild, it probably was going quite quick but not from the viewpoint of a racing driver. Graham drove it for quite a time. It wasn't just one or two laps to say that it was a bucket of rubbish, he did persevere a bit with it on the day.' After such a disappointing evaluation, and with the complex Type 63 four wheel drive car on the horizon for the 1969 season, it was decided to abandon further attempts to develop this design. Although a national newspaper managed to get a 'spy' shot of the car testing at Hethel, this was the closest anyone would ever get to seeing it: it never ran in public.

Instead, the decision was taken to rely on the trusty old 49s once more, since they were by this stage well-developed and, in Grand Prix, still quite capable of trouncing the opposition on a regular basis. Because the Tasman series ended only two weeks before the start of the first round of the World Championship in South Africa (which had been moved from its traditional New Year's Day date), it was not practical to take any cars to the Tasman series that would be needed for the race at Kyalami. As a result, two cars were built up specifically for the Tasman series, these being 49T/R8 and 49T/R9. Chassis R8 was built new and was actually the ninth Lotus 49 constructed, following on from the seventh (Rob Walker's car, R7) and the eighth, which was built to replace the one written off by Oliver at Rouen. The eighth chassis subsequently assumed the identity of R6, in order to avoid having to apply for a new customs carnet.

The second Tasman chassis, R9, was effectively the prototype Lotus 49, 49/R1, rebuilt after Oliver's Monaco crash and renumbered. The reason it is possible to say this with certainty is that this car was the only other tub apart from 49/R2 to have a second skin sewn on to it at the front of the tub when the new internal bulkheads were added during 1967, and this feature is clearly visible on a number of pictures of R9. All cars built from 49/R3 onwards incorporated this bulkhead from new, and were built in 16swg aluminium sheet, so a second skin was not required.

The cars that emerged were hybrid 49/49Bs with special deep cut-outs in the tubs for mounting the lower rear radius arms, due to the fact that they were using the Cosworth-Ford DFWs but still with the old ZF box and fir-tree rear suspension mountings. This was instead of the standard 49B set up of rear suspension subframe and Hewland gearbox that had been adopted early in 1968. Because they were not using the rear oil cooler set-up, the

The radiator was located at the rear of the car, allowing a chisel nose and pure wedge shape. This is the FVA-engined version with a Hewland FT200 gearbox. (Dave Sims Collection)

Lotus 49

Type 57/58 had no nose fins - all the downforce was created by the shape of the body. (Dave Sims Collection)

The project was revived as a possible Tasman car when Leo Wybrott returned. This is the car running at Hethel, with Wybrott at the wheel and a 2.5-litre Tasman DFW engine installed. (Express Newspapers/Leo Wybrott)

oil tank was back in its original place in the nose behind a combined oil/water radiator and original spec rocker arms were used, at 90 degrees to the tub rather than swept forward as on the 49Bs. It seems that reasons of economy were the cause of this apparent retrograde step. As Jim Endruweit recalls: 'Tasman cars were usually built from left over bits. You scratched around and found what you could and cobbled something together.' This is confirmed by Dale Porteous, one of the mechanics who went to the 1969 Tasman series, who remembers them as being: 'Both built out of old bits. They didn't want to take the 49Bs out there.' Having successfully pioneered the concept of feathering wings at the Mexican Grand Prix, Lotus fitted both these Tasman cars with the same system, activated by a fourth pedal in the cockpit.

A third tub was built to this same Tasman specification but was destined to be a show car for Ford and never turned a wheel in anger. The basis for this car is believed to have been the tub which Oliver wrecked at Rouen in July 1968, the original 49B/R6. It was rebuilt at the same time as the Tasman cars were being prepared and was reskinned, the tub being to the 1969 Tasman design with the deep cut outs for the radius rods because this was how the jigs were set up at the time. As with the Tasman cars, it featured the old-style fir-tree suspension and did not have the B-spec cross beam. In addition, it featured a ZF gearbox rather than the B-spec Hewland, although a mock oil tank and rear cooler were slung over the gearbox 'a la 49B' and a high rear wing was fitted. This car was delivered to Ford late in 1968, bearing the chassis number 49/12 and made its first public appearance at the Racing Car Show at Olympia early in January 1969.

Chapter 14
Tasman Troubles

The first race in the 1969 Tasman series was at the Pukekohe circuit in New Zealand on January 4th. The cars had been flown out early so that they had arrived by Christmas, the mechanics following a few days before the first race. With new Chief Mechanic Wybrott staying at home to supervise preparations for the forthcoming World Championship campaign, the team which flew out to New Zealand was led by Dougie Bridge, with Dale Porteous and Trevor Seaman. Once again, to bolster numbers, Dale's brother Roger joined up with them when they arrived.

As in 1968, the two Lotus cars would start as favourites for the Tasman title. However, Chris Amon, who ran Clark so close for the title the previous year, was back again in a Ferrari. He was once more at the wheel of a V6-engined Dino Formula 2 car, but this year was accompanied by a team-mate in the form of Briton Derek Bell. Other strong contenders were expected to be Piers Courage in an ex-works Brabham BT24 with a Cosworth DFW engine and twin wings entered by Frank Williams and the unique Alfa Romeo V8-engined Mildren, driven by the gritty Australian, Frank Gardner. This semi-monocoque car had been designed by Len Bailey and was built in the UK at Alan Mann's Byfleet workshops then shipped out to Australia.

In practice, Rindt laid down a pointer for the rest of the season to his World Champion team-mate, when he out-qualified him to line up on the front row alongside pole man and local hero, Chris Amon. This was despite it being the first time he had sat in a Lotus 49! Clearly, he was going to be quick when he'd got used to the car, and it was small wonder that Hill was grim-faced at the end of the final session. To add insult to injury, Hill had suffered a number of minor niggling problems which had restricted his practice time and so it was with some relief that he finally managed to pop in a quick lap to end up third on the grid, alongside Courage.

In the race, Amon made a copybook start, but was passed on the second lap by the flying Rindt. After running fourth behind Courage, Hill – already suffering from a faulty rev counter, a similar malady had already forced Rindt to have his changed on the grid before the race – pulled into the pits before quarter-distance with broken front suspension. Right from the start he had already had to contend with another problem, as Porteous relates: 'The fuel tanks were perished. Graham sat in a plastic bag, in a pool of fuel and I stood with my feet under the car, draining the petrol up on the start line, so no-one saw. And he started the race like that!' The reason for the problems with the rubber bag tanks was that they had been shipped dry and the long journey by sea out to

A holidaying Denny Hulme joins Hill in looking concerned about the problems which were keeping him in the pits while new team-mate Rindt was putting his car on pole. (Chevron Publishing Group)

Lotus 49

Spot the Lotus: the Gold Leaf Team Lotus wings dwarf those of the rest of the field as they head down towards the first corner at Pukekohe. (Euan Sarginson)

The two cars sent to the Tasman series were hybrid 49/49Bs, with the cut-outs in the tub for the rear radius arms but the old-style front rocker arms, rear uprights, tubular sub-frame for the suspension and ZF boxes. However, they both had the feathering rear wings as used in Mexico, with bungees on the leading edge to pull them down into "maximum attack" position. This shot is taken in the tents which served as makeshift garages at Levin. (Euan Sarginson)

New Zealand had adversely affected them.

On the 18th of 58 laps, Rindt spun on some oil at the hairpin, letting Amon through into a lead he would hold to the chequered flag. Due to various problems with his car, Rindt was unable to fight back. For three-quarters of the race, he had been battling with no clutch and then the feathering system for the wing failed, leaving it permanently in a high-downforce setting. He eventually finished some 20 seconds behind winner Amon, but had the consolation of setting a fastest lap and a new outright lap record.

A week later, the series reconvened at Levin for the second of the four rounds in New Zealand. Amon was looking to do well on this, his local track only a few miles from his home in Bulls. However, it was Rindt who set the early pace in practice and he ended up fastest, a second inside Jim Clark's 1968 lap record. Lotus were still experiencing problems with the fuel tanks. After asking for some fresh bags to be air-freighted to Wellington, a 60 mile journey to pick them up the night before the race proved fruitless as they had not arrived, so the mechanics had to patch up the old ones for another race as best they could.

In the preliminary race, the starter caught many of the drivers off guard. The result was that Rindt found himself down in fourth behind Hill, and his team-mate was not very accommodating when it came to letting him past. Out of the hairpin on one lap, Hill's right rear wheel clipped the left front nose fin of Rindt's car, an incident

Tasman Troubles

The hotshoe versus the wily old Champion. During the preliminary heat at Levin, Rindt found himself behind Hill after confusion at the start. Hill was uncompromising in defending his position, much to the chagrin of his team-mate... (Euan Sarginson)

which led to terse words between the two afterwards.

Bell won the heat, ahead of team-mate Amon and the two Lotus cars. However, the result of the race did not materially affect grid positions for the main race, which were decided by a curious combination of practice times and fastest laps recorded in the preliminaries. The only person to benefit from this was Hill, who, as a result, moved up from sixth to fourth on the grid.

Amon led pole man Rindt away again at the start but, before the first lap had been completed, the Austrian had snatched the lead. He held this position until spinning on only the 3rd of the 63 laps, which relegated him to sixth. A few laps later, as he chased furiously to catch up again, something caused the car to leave the track.

Synchronised starting: Rindt and Amon smoke off the line neck-and-neck at Levin. This shot clearly shows the deep cut-outs which had to be built into the 1969 Tasman cars because they were using the old-style rear suspension and the lower radius rod was consequently higher up than on a standard 49B. (Euan Sarginson)

Lotus 49

In a corner of the tent at Levin, the reigning World Champion ponders the prospect of having to race alongside Jochen Rindt for the rest of the year! (Euan Sarginson)

These two shots, taken at Wigram, show the differences between the 49B sent out for Rindt (R10) and Hill's purpose-built Tasman chassis (R8). Rindt's car has the nose ducting, swept forward front rocker arms, fabricated cross member holding rear suspension and rear oil tank/cooler arrangement while Hill's is using the original 49 rear suspension and his car still has the oil tank in the front. Note Rindt's 'stretched' rear wing struts - due to using a wing made for the narrower rear suspension and lack of cable running up the right strut to control feathering wing (R5 did not have a fourth pedal fitted). (Euan Sarginson)

Mechanic Trevor Seaman remembers that 'they suggested when it got back to Hethel that the master cylinder caps weren't screwed on properly because it was very easy to cross-thread those. But we'd also had problems ever since the start with the airflow through the nose, that would actually suck out the brake fluid.'

Rindt went off at Castrol bend, slid up the earth safety embankment, and the car flipped over, trapping the driver inside. For a few seconds nobody did anything and the petrol-soaked driver lay helpless, strapped in his stricken car, the risk of fire only too obvious to most onlookers. However, a spectating saloon car driver by the name of Dick Sellens took the proverbial bull by the horns and tried to right the car. With assistance from others, he managed to do so and free the shaken but unharmed Rindt. While the driver emerged unscathed from this incident, the same could not be said about the car, which was badly damaged. Five laps later, the other Lotus was out with a broken driveshaft, to complete a sorry afternoon for the team. Once again, the only – albeit small – consolation was that Rindt had again proved the quickest man on the course by setting fastest lap.

Although the damage to 49T/R9 was repairable under ordinary circumstances, thousands of miles away from base in New Zealand, this was not a practical proposition, according to Porteous: 'It wasn't too bad, it was just too bad to be rebuilt out there because we didn't have enough parts.' After frantic discussions with the works back at Hethel, a decision was taken to air freight a fresh car out to Christchurch, while the damaged tub was sent back to the UK.

The car sent out to replace Rindt's damaged tub was chassis 49B/R5, the original 49B, which was in exactly the same specification as when it had been driven by Moises Solana in the Mexican Grand Prix some three months earlier. It had yet to be updated to 1969 specification, so still had the old-style roll hoop. However, it was quite different to the one it replaced, since it had the swept forward front rocker arms, the oil tank and cooler mounted at the rear, a Hewland box and the steel subframe holding the rear suspension, while the cut-out for the lower rear-radius arms was, in true 49B style, much shallower and the mounting point lower down. In addition, it had a larger rear wing, which was wider and deeper than the standard Tasman variant.

Having two cars of completely different specification created an absolute nightmare for the mechanics. Since the newer and, in all probability faster, machine had been sent out for Rindt, it also put Hill's nose out of joint, for he was left with his regular machine, which one wag from a rival team described as 'a museum piece' …

The next race on the schedule was the Lady Wigram Trophy at the Wigram airfield circuit. Rindt quickly got to grips with both the circuit and the new car, setting the fastest time in practice over a second inside the existing lap record. This gave him pole position for the preliminary heat. However, he was again left standing at the start, and had to fight his way up from fourth, passing Courage and Amon, until he was on the tailpipes of his team-mate. A bitter battle once again ensued. Although the Austrian slipped past once, Hill was giving no quarter and slammed back past him. As with the previous race at Levin, words were exchanged between the two after the finish.

Although he won his heat, Hill was the unfortunate victim of one of his own suggestions regarding setting the grid for the feature race, since he had proposed that lap times recorded in the heats should determine positions. Therefore, although the Englishman had won his heat, Rindt had set the fastest lap time, and would start from pole, with Hill and Courage for company on the front row. When the flag dropped, Rindt made a great getaway and was never headed. Courage got the jump on Hill, who took a while to regain this place, by which time his team-mate was a long way ahead. Lotus scored their first one-two of the series, with Rindt again taking fastest lap, although this time jointly with Amon, who closed right up on Hill in the final few laps. While Amon maintained his lead in the series, with 22 points, Rindt was now not far behind, on 15, while Hill had actually managed to finish a race and posted his first points of the series to put him fourth.

The final round of the New Zealand leg of the Tasman series was held at Teretonga Park in Invercargill, the world's southernmost racing circuit. By this stage, Hill was getting pretty fed up with having lengthy gearbox changes every time he wanted to change ratios, for the ZF box was the original fixed ratio proposition. Consequently, a Hewland box was flown out from England, complete with driveshafts and other ancillaries, and this arrived in time for the mechanics to finish fitting it in time for the first official practice session. Due to the longer length of the bellhousing on the Hewland car, this conversion resulted in the driveshafts ending up being angled back slightly, but this did not seem to be a problem and the car ran in this even more 'hybrid' form for the rest of the series.

The leading cars were quickly under the 61 second lap record of Jim Clark set in 1968, the high wings really coming into their own. In fact, they went so fast that they went off the lap speed table in the race programme! At the end of practice, Piers Courage was given the fastest time, of 57.2, which was a whopping 6.2mph faster than the record average, in the space of just a year. Rindt, Hill and Amon were all within 0.2 seconds of this, promising some close racing. In the preliminary heat, Rindt won easily from Courage and Amon. Hill should have taken fourth but ran low on fuel towards the end and ended up a disgruntled sixth behind Bell and Gardner. The finishing positions from the heat dictated the starting grid for the feature race, so he was going to have some catching up to do.

All Rindt's work in gaining that vital pole position came to nought when he broke a driveshaft as the flag dropped and was rammed by Bell's Ferrari. This at least eliminated two of the cars ahead of Hill, who came round fourth at the end of the first lap behind Amon, Courage and Gardner. After Courage got past Amon on the third lap the race was rather processional, although Hill managed to catch and pass first Gardner and then Amon to take second behind a delighted Courage at the chequered flag. With Rindt not scoring, this enabled Amon to maintain his lead in the series, moving up to 26 points,

By the fourth race of the Tasman series at Invercargill, Hill had gained a Hewland gearbox in place of the ZF originally fitted to his car. The shorter length of the box and bellhousing meant that the driveshafts were angled backward to locate on the old-style hubs but this did not seem to compromise performance! This shot very clearly shows the control cable for the wing feathering system running up the wing strut. The bungees pulled the wing back down into maximum downforce setting when the driver took his foot off the cockpit pedal. (Euan Sarginson)

Lotus 49

One problem with the feathering wing system was that it placed variable loadings on the wing struts and, quite often, they were unable to cope with the increased strain and simply bent back. This is Graham Hill, in practice at Lakeside, with the bemused brothers Porteous (Roger - left and Dale - right) looking on. (Nigel Snowdon)

Natural-born racer: the uncompromising Rindt made an immediate impact as soon as he sat in a Lotus 49, outqualifying Hill in his first race and always seemed to have the upper hand thereafter. Here he is pictured on the grid at Lakeside. (Chevron Publishing Group)

while Courage's win promoted him to second in the standings with 22 points. Hill remained fourth but had closed up on his team-mate.

The circus moved to Australia for the next round of the series, at the Lakeside circuit, which was to be the 34th Australian Grand Prix. The Gold Leaf Team Lotus entries were held up by customs, which set them at a disadvantage to the Ferraris, which had an extra day of practice at the circuit. Then the motor in Hill's car blew up when it was started up in the garage the night before the meeting was due to begin. This meant the last spare had to be installed, leaving them with no other engine to fall back on in the event of problems.

When they appeared, the Lotuses had missed the first official practice session and only just arrived in time for the second one. Hill's car was fitted with a larger 49B-style rear wing like Rindt's for the first time but this did not last long, for it collapsed at the end of the session. Without any spares, this was replaced with the original wing that he'd used in the New Zealand rounds. Amon emerged quickest ahead of Teretonga victor Courage, with Hill third, ahead of the other Ferrari of Bell. Rindt languished down in fifth as a result of a persistent misfire which could not be cured. More significantly, it was not possible to change the engine as the team had no spare, so he would have to start the race with it.

As they lined up on the grid, an oil leak was found on Hill's car, which resulted in some hasty tightening up of oil lines as the one minute board was shown. At the start, Amon streaked into the lead, ahead of Hill and Courage. On only the fourth out of 67 laps, Courage and Hill clashed as the former tried to pass the latter, the Brabham spinning out and breaking a wishbone. Hill continued, although he had lost a place to Bell and his wing appeared to have been damaged because it was waving around much more than it should. Meanwhile, Rindt was having a miserable race with a down on power engine, which eventually started to make expensive-sounding noises, forcing the Austrian into the pits to retire. A few laps later, Hill's wing finally gave way and fell sideways across the back of the car. He had already been troubled by a broken rev-counter (again!) and fluctuating oil pressure, so this really was the final straw. After a few laps driving with the wing like this, he dived into the pits to have it hacksawed off, dropping him down to fourth. He could do nothing about catching the third placed man, Leo Geoghegan (in an ex-Clark Lotus 39 with a Repco V8) in the remaining laps, since the handling was now atrocious and he was losing oil as well.

Amon coasted to a win which, after five of the seven rounds, virtually guaranteed him the Tasman title. Only Courage could exceed his points total in the remaining races. Hill's fourth place lifted him to joint third in the series with his team-mate but this was hardly representative of the quality of the performances of the two drivers, for Rindt had been by far the faster of the two. With only one serviceable motor in the whole team – the one in Hill's car - it was clear that something had to be done if a two-car entry was to be fielded for the next round of the series in a week's time. With this in mind, Cosworth's Mike Costin was despatched to Australia on the Sunday night of the Lakeside race with a brief to salvage and repair as much as possible: 'I didn't get there until the Tuesday, practice was on Friday. By the first session, I'd only just finished. It was a case of stripping down four engines and seeing which was the least worst and chopping and changing and rebuilding an engine.

'The real problem was that the camshafts had snapped in such a stupid position you wouldn't believe it. So mainly it was a case of fitting new valves - because some were bent - and rebuilding the engine. We'd got plenty of people who could rebuild engines but within our [*i.e.* Cosworth's] defined system, not when you haven't even got an engine stand! You couldn't turn it around and look at it and so on. So it had to be somebody who could make the judgement as to what was going to be done. It wouldn't have been any good Keith [Duckworth] going, because he'd have taken too long!'

The next race in the series, round six, was at the Warwick Farm circuit. Rindt was unable to go out on the Friday, when there was unofficial practice laid on, because his car was still being put together with its 'bitsa' engine. However, during the two three-quarter hour official practice sessions on the Saturday, he again demonstrated his incredible ability to learn a circuit very rapidly, getting well under the circuit record within only ten laps. At the end of the first session, he was joint fastest with Amon (who knew the circuit very well), some 1.7 seconds inside the record. By the end of the second session he had lopped an incredible 5.2 seconds off the record and Amon was left 0.7 seconds adrift.

Race day dawned bleak, with rain pouring down in torrents. At the start, Rindt got bogged down and it was Amon who led the field into the first turn. However, by the end of the first lap, Rindt was in the lead, the length of the start/finish straight ahead of his closest pursuer, his team-mate Hill. Amon and Courage had tangled as they fought for second place and both were eliminated by suspension damage. This settled the Tasman series in the New Zealander's favour, for Courage was the only person who could have beaten him to the title and could not now score sufficient points in the one round remaining after Warwick Farm to overhaul him.

By the end of lap two, Rindt was 11.5 seconds ahead and continued to pull away from his pursuers at the rate of two or three seconds a lap for the first 15 laps, or so. Early on, Hill had begun to slow due to getting water in the electrics of his engine and pitted at the end of the 11th of the race's 45 laps. The problems took him ten laps to rectify, so that when he rejoined he stood little hope of scoring any points. Rindt continued on his triumphant way, his engine not missing a beat, and he crossed the line some 45 seconds clear of second man home Bell. Hill, who finished eleventh, had the consolation of setting fastest lap as the track dried a bit towards the end of the race, but his race had certainly not gone to plan, as Costin recalls: 'Graham thought "Bloody engine, never been on a test bed ... old Mike's just had to stuff it together in the garage, without any equipment to do it". Anyway he went out and Rindt beat him! Old Graham was boot-faced! He didn't like Rindt beating him anyway, but especially when he'd thought he was going to be alright because no way could that engine that Mike had stuffed together be any good!' Rindt's win took him back up to second in the Championship, leapfrogging Courage, whose miserable form on the Australian leg of the series continued.

Between Warwick Farm and the final round of the series at Sandown Park, near Melbourne, Costin continued his work on the DFWs, replacing the inlet camshaft on Hill's engine as a precautionary measure. A further engine was flown out from the UK so that the team actually had a spare unit. Sandown Park was one of the faster Tasman circuits and so the Lotuses were well to the fore. In a last-gasp effort, Rindt pipped Amon to pole position, with Hill back in third, once again out-qualified by his new team-mate. Right at the end of practice, tell-tale oil-smoke from the exhausts of the World Champion's car signified the need for yet another engine change ...

At the start, it was Rindt's turn to jump into the lead, but the determined Amon had snatched the lead

Rain-master: in a brilliant display of wet-weather driving, Rindt just drove away from the field at Warwick Farm to record a well-deserved victory. Note that, by this stage, his car had a wing feathering system again. (Paul Cross)

This close-up action shot captures the reigning World Champion in full flight in practice at Warwick Farm. (Peter Mohacsi)

Lotus 49

Deja-vu: twelve months on and a Lotus 49 is taking the chequered flag at Warwick Farm, the driver giving the thumbs-up signal as he crosses the line. This time it is Jochen Rindt, not Jim Clark, at the wheel. (Paul Cross)

back before the field crossed the line to complete the first lap. Even at this early stage Hill was heading for the pits, with the same problem as Siffert had experienced in the previous year's Mexican Grand Prix. The throttle cable end had come out of its socket, and he lost nearly three laps getting this rectified. This was turning out to be another troubled meeting for the World Champion, since at the end of the warm-up lap he'd had to ask the mechanics to disconnect his rev limiter, as the car would not rev above 9000rpm.

Fairly early on, it became clear that it was to be a two-horse race, for Courage had broken a halfshaft on the opening tour, and Brabham, Gardner and Bell were losing ground rapidly on the leading pair. For the next thirty laps or so, barely a second separated the New Zealander and the Austrian. Although Rindt managed to get alongside the Ferrari several times, he never regained the lead and eventually fell back slightly to finish seven seconds behind Amon, whose victory made up for losing the race by less than a car's length to Jim Clark the previous year.

Amon's win put him fully fourteen points clear in the Championship, ahead of second man Rindt, while Hill finished fifth on 16 points. As the Champion lifted the Tasman Cup, it was time to reflect on the relative performance of the top five drivers, and particularly the disparity in performance between the Team Lotus pair: Amon had driven superbly, taking four wins (two each in New Zealand and Australia); Courage had gone well in the New Zealand rounds but only managed eight racing laps in Australia; Bell had played a strong supporting role in the second Ferrari; and Rindt had blown off the reigning World Champion comprehensively, despite never having driven a Lotus 49 prior to arriving for the first round of the series. Taking this into account and the fact that they had 'exchanged words' about each other's driving after several of the Tasman races, the scene appeared set for an explosive World Championship season in the Lotus camp. Such a situation did not bode well for team morale in the coming months ...

Chapter 15
Wing Worries

In the wake of the rash of fatalities during the 1968 season, new FIA regulations introduced for the 1969 season required the addition of a full fire extinguisher system and a much more substantial rollover hoop to all Formula 1 cars seeking to compete in World Championship races. As a consequence, all three existing Team Lotus 49B chassis were re-fettled and updated during the winter of 1968 to conform to these new regulations.

Two of these cars were given new chassis numbers to reflect the fact that, when they reappeared at the first Grand Prix of the new season, they would be substantially different to how they had been when they were first built. These were 49B/R2 and 49B/R5. Chassis R2 was the tub in which Clark had won first time out at Zandvoort in 1967 and which Oliver had campaigned during the latter part of the 1968 season, the car having been converted to B-spec in August of that year. The other car, the prototype 49B, chassis R5, had been raced by Hill for the bulk of the 1968 season before becoming a number three/spare car. Chassis R6, which had only (in its second reincarnation, at least) been introduced at the Italian GP in September 1968, was updated to 1969 spec, but retained its existing chassis number.

At the end of the season Chapman had agreed to sell one of these cars to the American driver Pete Lovely. The two remaining chassis were to be retained for use by Gold Leaf Team Lotus in 1969 while they were waiting for the new Type 63 to be completed. Lovely had asked for an ex-Graham Hill chassis and originally understood he was buying R5. 'I did the deal with Colin at Mexico City in 1968. I went down as a spectator to the race and I asked him after the race if he'd be interested to sell one of the Lotus 49s. He said that he would. He asked me which car I would like. Graham Hill was an old buddy of mine - I drove for a very short time in Formula 1 with Team Lotus in 1959 with the Lotus 16s and knew Graham from those days. In 1958 we'd gone to Le Mans together when he drove the 2-litre car with Cliff Allison and I drove the 1500cc with Jay Chamberlain - so I said "I'd really like to have Graham's car" which was chassis number 5. Colin said "Yeah, that'd be fine".

'I say "Graham's car" because of the fact that Graham actually drove chassis number 5 and 6 to win the World Championship. The one that was offered was number 5 and I said, "OK, I'll go for that one". We shook hands on price [US$36,000 - approximately £15,000, or around £140,000 at the values of the late 90s] and the whole thing in Mexico and then he said, "By the way, we've got the car entered in South Africa for Mario Andretti to drive and we'd like to do that" which was in early March. They said that they would bring the car up to the 1969 specifications which was the higher roll-over structure and onboard fire extinguisher system. So I said that was fine with me.'

Although Lovely had agreed to buy chassis R5, what he ended up taking delivery of was the second chassis, with the unique and distinctive filler cap and access hatch arrangement that had been carried over from the days of Jimmy Clark, and it was this car which Andretti drove in the season-opening South African Grand Prix. One possible explanation for this is that the car he originally expected to take delivery of had been pressed into service following Rindt's Levin shunt (perhaps because it had been the only ready-to-race car in the workshops at the time) and could not now be returned from Australia in time to be readied for the South African Grand Prix. Interestingly, although at this stage it had been given a new chassis plate stamped R10, none of the modifications for the forthcoming Grand Prix season had yet been carried out, and the car appeared in New Zealand in exactly the same specification as it had raced at Mexico, save for the substitution of a DFW engine in place of the DFV.

Apart from the addition of the new roll-over bar and fire extinguisher system, the only other work done to

Lovely's car before the 1969 season was the fitting of a fourth pedal/wing-feathering arrangement as first used on R6 in Mexico 1968. When it was sold to him, his chassis was renumbered 49B/R11, due mainly to the fact that eleven is a pretty significant number in the life of Pete and his wife, Nevele – which is eleven spelt backwards! 'When we sat down in Colin's office in December, he said "We could renumber it because there will be these modifications made". Now I was born on the 11th of April 1926. My wife was born on November 11th at the eleventh minute of the eleventh hour of the eleventh day of the eleventh month, and so elevens are pretty big in our life! So I said "Well, how's about number 11?". And he said "Well, yeah, eleven is open because we've got chassis numbered up to 10. We could do it if you'd like. We'll just take the chassis plate off and stamp a new number". So that's the way it went.'

Aside from changes to meet the new regulations for 1969, the other main specification change for the Team Lotus 49Bs was the addition of a high wing mounted on the front rocker arms, in addition to the usual nose fins and suspension-mounted rear wing. Like the rear wing, the front wing could be feathered along the straights through use of the fourth pedal in the cockpit. This modification was applied to the cars of Rindt, who was to drive 49B/R9, repaired after its Tasman roll and Hill, who would start the season in his Championship-winning car from Mexico, 49B/R6. In addition, it featured on the third works car, for Andretti, and the privately-entered Rob Walker car of Jo Siffert. This latter car differed from the works machines in that slim three gallon fuel tanks had been added alongside the tub of the car in an effort to increase its range and keep its centre of gravity low. Finally, the works cars and the Walker car had now all switched to the more robust Hewland DG300 gearbox, which had been fitted to Hill's car for the last few races of 1968 in the interests of reliability.

Kyalami in South Africa was the venue for the first round of the World Championship, on Saturday March 1st – only 13 days after the end of the Tasman series – having moved from its traditional New Year's Day date. This race boasted the largest ever number of Lotus 49s to feature in a single Grand Prix, with five entered. These included the three Gold Leaf Team Lotus entries, plus the Rob Walker car for Siffert and John Love's 49/R3. The latter was an original specification 49 (*i.e.* it had not been converted to B spec) but was by this stage fitted with a rather ugly mandatory roll hoop, a Hewland FG400 gearbox and front and rear adjustable wings. The wings were controlled by an ingenious system relying on compressed air, as Love recalls: 'I made a hydraulic-type system that ran off the gearbox at the back so that, when I went down the straight, it took the downforce off the wings and just left them in the neutral position. I had a cockpit lever which was a hydraulic-type thing and I just moved it with my left foot. As I went in on the brakes at the end of the straight, I depressed the pedal and the things went down to where I'd actually set them for the corners. So I was trying to go quicker down the straights and also have the downforce on the corners.

'It was our own design – myself and a chap by the name of George Pfaff who was a brilliant machinist and Gordon Jones, my mechanic. We machined all the cylinders ourselves. We made them out of a propeller from a Vickers Viscount aeroplane which had crashed! We knew the airport guys and they gave us a bent prop. It was made out of a special dural, so it was very light and very strong. It was quite an innovation. Colin Chapman looked at it and shook his bloody head. He said "Christ, what are you guys trying to do?!" I said "Well, we're copying you and trying to go just a little bit better!"'

Practice took place over three days – Wednesday, Thursday and Friday – with a three hour session each day from 2.30pm to 5.30pm, giving a total of nine hours of qualifying! As a result, teams were in no hurry to set their times on the first day, preferring instead to experiment with various set up changes. Team Lotus spent most of the day trying out different wing combinations, while Siffert didn't get out at all as the front wing was being fitted to his car.

On the second day, most drivers improved. However, Hill's chances were spoiled when he broke the fourth pedal inside the cockpit which controlled his pivoting wings. As a result, his wing became fixed for the rest of the session, which ultimately turned out to be the fastest out of the three. However, this problem probably prevented an incident similar to the ones experienced by his team-mates. Within minutes of each other, both Rindt and Andretti came into the pits with rear wings which had collapsed. On the long fast corners that are a feature of the Kyalami circuit, the wings had been leaning over and, in some situations, rubbing against the inner wall of the rear tyres. This combination of the rubbing on the tyres, the constant flexing and the increased pressure applied to the supporting struts when the wing was put into its maximum downforce position on entering a corner eventually caused the struts to collapse.

Fortunately, neither driver crashed when their wings collapsed (although Andretti spun at high speed), but wing failure was an ominous portent of things to come later in the season. To stop the wings flexing too much, the height of the struts was reduced, down to the level of the front wing and the team experienced no further problems in this area for the rest of the weekend. However, a shortage of suitable struts meant that Andretti ran with just a rear wing for the rest of the meeting, while struts which had been lent to Rob Walker's team were repossessed. As a result, Siffert only had a rear wing for the rest of the weekend. However, this was the least of his problems as the car would not run cleanly as a result of fuel feed troubles and he did not manage to record a single representative lap.

When the session finished, crafty 'Black Jack' Brabham had posted the fastest time in his newly DFV-powered Brabham, but Rindt was only two-tenths further back. Andretti was a very creditable sixth fastest, with the other Gold Leaf Team Lotus car right behind in seventh. Love was not quite able to match the pace of local rival Basil van Rooyen. This was to be the session which determined final grid positions, for the vast majority of the practice session the next day was ruined by rain, which began to fall after only 30 minutes. Fortunately, by this time Siffert had already managed to get out and set a decent time, which was joint third quickest on the day, but only enough to put him 12th on the grid. Towards the end of the session, the track began to dry again, but not quickly enough to offer the prospect of improving on the previous day's times. However, Rindt did manage to break a camshaft in his engine, which curtailed his day's work.

The morning's untimed warm-up was an eventful one for Lotus drivers. Rindt's engine had been changed overnight but he couldn't get it to run cleanly and it eventually stopped out on the circuit. Meanwhile, Love was having electrical trouble and his black box transistor unit burned out, forcing a new one to be fitted. However, not having been able to get to the bottom of why this was happening, his team were not confident of getting very far in the race.

On the warming-up lap, Rindt's engine was still playing up, which did not bode well for the race. Stewart jumped into an early lead when the flag dropped and was closely followed by Brabham, Rindt and Hill. Love made a tremendous start from the fourth row and came by at the end of the first lap in eighth place ahead of Siffert, who had also made a lightning getaway, and Andretti who, in contrast, had not had a good start at all. After Brabham had to stop to have a collapsed wing removed, the race settled into a pattern, with Hill and Rindt pursuing Stewart and the Siffert/Andretti train moving rapidly up the order. Rindt's engine was misbehaving and he gradually began to lose places, while Andretti passed Siffert to move into third and closed the eight second gap to Hill in only ten laps.

At one stage, it looked like the young American would waltz right past the reigning World Champion and start to put pressure on Stewart, but on his 32nd out of the race's 80 laps, he coasted to a halt, unable to find a gear. Rindt, by now down in seventh place, retired just after half distance, the mechanical fuel pump having failed and the electrical pump – only designed to provide pressure for starting the engine - was unable to deliver sufficient quantities of fuel. So Stewart ran out the winner, with Hill making a solid start to the defence of his Championship with a steady if unspectacular second. The private Rob Walker entry of Siffert finished in fourth, having been passed by Hulme's McLaren as the Swiss's brakes worsened.

Less than two weeks later, the European season started with the Race of Champions at Brands Hatch. This didn't give much time to get the cars back from South Africa and prepare them. As a result of the rush, the Team Lotus mechanics simply painted the car which was to be sold to Lovely the same colour red as the top half of the Gold Leaf Team Lotus livery, as the American recalls with a laugh. 'It was the cheapest way for them to get rid of the Gold Leaf Team Lotus colours that were on it before handing the car over to me!' This caused much head-scratching among spectators at the meeting because it looked like a Ferrari (and, indeed, one had been entered for Amon but did not arrive) but they did not recognise the driver …

Lovely was to take delivery of the car at the circuit on the Friday morning of the meeting. As he recounts, liaison between customer and supplier was minimal to say the least: 'It was sitting alongside the transporter when we got there with its funny red paint job and the wheel wrench and four spare wheels and tyres and nothing else! And the crew, all the Team Lotus mechanics, said "We're off to the pits, see you out there" because we were in the old paddock up on the hillside there and you had to go through the tunnel to get to the racetrack. Anyway, away they went. And, gee, I hadn't the foggiest notion how to even start the engine! So I kinda looked at the switches. I turned on the electric fuel pump and that sounded like something to do and turned on the ignition and cranked it for a while but it wouldn't start.

'It was a cold day and we had a rental car, which was an Austin 1100 four-door sedan, so we borrowed a rope and my wife drove the Austin around the paddock with me tied on the back, trying to get it started. And it would go "splutter bang, splutter bang". And finally Ken Tyrrell flagged us down and said "Did you put it on full rich?" And I said "No, how do you do that?". He called Allan McCall over and Allan reached in there with a couple of screwdrivers - I didn't even know what he was doing as I was still sitting in the car – and he put it on full rich. By golly, the next trip around the paddock behind the 1100 it started and ran! I thought oh, that's good. And then it ran on full rich for the next three races because I didn't know how to change it - nobody ever showed me!'

The works Gold Leaf cars had turned up minus their front high wings, reverting to the high rear wing and front nose fins set-up which had proved so successful in 1968. However, Siffert in the Walker was still persisting with the dual-wing set-up, although in the dips of Brands Hatch, this sometimes caused visibility problems. A novel Indy-style qualifying system was used to determine the grid, although fog played havoc with this and the eventual grid was a hybrid combining the times of those who actually completed Indy-style runs and times from the free practice for those who didn't! Hill was on really good form and returned a time on the Friday which

With the Tasman cars still on their way back home, Rindt started the 1969 European season at the wheel of R9, the chassis he had rolled at Levin. It was sent back to the UK after the accident and rebuilt as a full 49B. The only clues to its identity were the presence of the high 'Tasman-style' rear cut-outs for the lower rear radius rod mountings and the overlapping skin on the front of the tub where the tub had been rebuilt with baffles following its Silverstone '67 shunt. Here the Austrian waits to go out of the pit lane at Brands Hatch. (Ford Motor Company)

Lotus 49

A few minutes later, Rindt appeared out of the mist and gloom, having suffered another wing failure to add to the one he'd had in South Africa. In this case, rather than falling backwards, the left strut buckled under cornering load and collapsed. (Ford Motor Company)

Pete Lovely drove a steady race to sixth place in his first outing in a 49B in the Race of Champions at Brands Hatch. (Ford Motor Company)

would be good enough to give him pole position for the race. On the other hand, Rindt ended his session needing an engine change as the bearings had gone. Then, in Saturday's free practice, his high rear wing collapsed, as it had done in South Africa. Due to a lack of spare parts, there was little else that could be done other than to shorten the struts again, while some supporting stays were jury-rigged on Hill's as yet undamaged car, attached to the roll-over bar.

While Rindt lined up on the second row as fourth fastest qualifier, Siffert had got his car working well round the twists and turns of Brands Hatch and had managed to get on the front row, alongside Hill and Stewart's new, bulbous Matra. Having got his car running, Lovely drove sensibly on his introduction to 3-litre Cosworth DFV power and managed to do a respectable time which put him ahead of Ickx and Rodriguez on the grid.

The race itself was a relatively dull procession, with Stewart disappearing into the distance and lapping all but the second, third and fourth-placed finishers. Hill and Rindt ran in close formation. The Austrian was unable to find a way past his team-mate, despite being obviously the quicker of the two, and Hill was unwilling to give him any room to find a way past! Just as the Lotus pit hung out a sign to the World Champion suggesting he let the Austrian past, Rindt got badly baulked by Jackie Oliver's BRM, losing a lot of time. However, the speed differential between the two Lotus drivers was amply illustrated by Rindt storming back up onto Hill's exhaust pipes and, in the process, smashing the lap record to set a time almost one and a half seconds faster than the Englishman's pole time. However, further confrontation was avoided when Rindt's oil pressure began to fall and, fearing a repeat of his practice engine malady, he pulled into the pits to retire. Hill continued to finish second, while Siffert had an uncharacteristically quiet race to finish fourth. By dint of the fact that he kept running and out of trouble, Lovely finished his first ever race in a 49 in sixth place, albeit four laps down on the winner.

Two weeks later, a slightly larger field assembled for the BRDC's *Daily Express* International Trophy at Silverstone. Following their wing collapses at both Kyalami and Brands Hatch, Team Lotus had finally resorted to bracing their wing struts using flexible rods attached to little 'ears' welded on to the roll-over bar. Aside from this modification, which appeared on both cars, the two Gold Leaf 49Bs had the same specification as the previous two races. Siffert was at the wheel of his regular Walker car (still fitted with the dual high wing arrangement) while Pete Lovely was taking in this race prior to flying his car back to the United States.

Team Lotus had elected not to take part in the first day's practice on the Friday. This turned out to be a rather daft decision, when Saturday's proceedings were ruined by rain. As a result, both Rindt and Hill were forced to qualify in the wet, and ended up eighth and tenth on the grid. The only consolation for Rindt was that he was the fastest in Saturday's session, and managed to claim two £100 prizes for being quickest in two of the 30 minute periods that the session had been divided into. His former team-mate Jack Brabham claimed the other two and joked that 'Jochen and I have earned more in the last two hours than we did in the whole of last year's Grand Prix season!' a reference to the troubled year they had endured with the four-cam Repco engine in 1968. Although the track had dried towards the end of the session, it was still very slippery and so it was amazing to find that Rindt's time was 1.2 seconds under the old lap record! However, it was some three seconds slower than the fastest time on Friday, hence his lowly grid position. Siffert qualified sixth, having set his best time in the dry session, while Lovely ended up 12th after a poorly-timed glance at his gauges during Friday morning's practice resulted in an 'off' at Becketts which bent his car's suspension and prevented him running again until Saturday.

The wet weather returned with a vengeance on Sunday. Just as the cars came out for their warm-up laps, there was a downpour which left the track awash with water. The works Lotuses, which hadn't yet come out onto the track, had some last-minute suspension tweaks aimed at improving their wet-weather driveability, as it seemed fairly certain that the conditions would prevail for most of the rest of the day. In Hill's case, he had the front anti-roll bar disconnected, a mistake which would cause him to be well off the pace in the race. Pole-man Stewart elected to start in his proven 1968 car rather than the new Matra MS80, which had proved a handful in the wet.

This meant he had to start from the back of the grid, which at least offered spectators the prospect of a charge through the field to liven things up a little.

As the flag fell, Jack Brabham jumped into the lead, followed by team-mate Ickx and the Frank Williams-run BT26 of Piers Courage. At the end of the first lap, Stewart had already made up four places, while Rindt and Hill had each gained two positions. On the second lap Lovely, who had no rain tyres and therefore had been forced to start on dry-weather tyres, understeered off into the bank at Becketts and virtually repeated the accident he'd had in practice, damaging the suspension but doing no harm to himself. Rindt's engine was starting to suffer from the rain and began to splutter and he dropped rapidly out of contention. Stewart meanwhile, was fairly scything his way through the field and was up to sixth before even five laps had elapsed, although he was still losing ground to leader Brabham.

Suddenly, Rindt's misfire cleared up and he really began to race. From having been 11th at the end of lap 5, he gained a position every lap, so that by lap 10 he was in sixth place and, more importantly, gaining on leader

At the Race of Champions, Hill's car was reinforced with support braces running from the roll-over hoop to the wing struts for the race itself to try and prevent rearward or sideways collapses. Due to a shortage of struts, Rindt's wing was shortened but not strengthened with braces. In addition, he had to run with his wing in a fixed position rather than using the feathering system which Hill had at his disposal. He still managed to give Hill a hard time during the race... (Ford Motor Company)

Jo Siffert was one of the few drivers to persist with the double high wing arrangement at the Race of Champions. Other drivers found the undulating nature of the circuit meant that the front wing was often in their line of sight and was distracting. Note the polished cam covers and chromed exhausts on this car, which was always immaculately prepared. (Ford Motor Company)

Lotus 49

Rindt was spectacular as ever at Silverstone, entertaining the crowds with his sideways style as he drove through the field to second place. (Ford Motor Company)

Jo 'Seppi' Siffert was tremendously popular among his fellow competitors and fans alike. Here he is caught in a contemplative mood during practice for the International Trophy. (Ford Motor Company)

Brabham at the unbelievable rate of two seconds per lap! The Austrian was simply awesome, hands sawing away at the wheel fighting to control wild oversteer. His meteoric progress contrasted starkly with that of his team-mate Hill, who once again was being overshadowed by the 'new boy'. His car was handling atrociously in the wet conditions (not helped by his pre-race adjustment) and on lap 17, he suffered the ignominy of being lapped by Brabham.

All eyes were now on Rindt rather than Stewart, and he passed the Scot on lap 20 out of 52 to take fourth. Five laps later, he came upon a gaggle of cars, consisting of Rodriguez and Hill, already lapped, fighting for seventh, plus second and third placed men, Ickx and Courage, trying to find a way past them. In one fell swoop, he passed all four cars going into Stowe, and came round the next time in an amazing second place. He was still 27 seconds behind Brabham, with the same amount of laps remaining so, in theory, given his rate of progress to this point, the task of catching the Australian was quite possible.

However, Brabham had been driving with something in hand and, try as he might, Rindt could not close the gap fast enough. The final drama was provided in the dying laps when Brabham's engine began spluttering because he was running out of fuel. From 12 seconds with two laps to go, the gap fell right down to 2.2 seconds by the time the chequered flag came out. Another lap and the Austrian would have scored a remarkable victory but it was not to be... Hill and Siffert, both afflicted by dreadful handling (although it didn't seem to have affected Rindt) came in two and five laps down in seventh and 11th places respectively – hardly their most impressive showings!

There was a five week gap between the International Trophy and the start of the first European round of the World Championship, the Spanish Grand Prix at the Montjuich Park circuit in Barcelona. This enabled Gold Leaf Team Lotus to concentrate on the build programme for the replacement for the 49Bs, the four wheel drive Type 63. This was essentially a scaled down model of the Type 64 Indy car, and the two shared many components and design features. The Indy car, due to the pressing nature of the May date for the 500, was finished early in April, while the intended debut for the Formula 1 car was the Monaco Grand Prix in May. As a result of the team focusing on the new cars, the Gold Leaf Team Lotus cars emerged from the transporter ostensibly in unchanged specification, save for the addition of larger surface rear wings to give greater downforce on the twisty turns of Montjuich Park.

This was the first time that the Grand Prix had been held at Montjuich Park, the previous year's version having been hosted by the tight and twisty Jarama track, which had not been particularly popular among the drivers. Set in the heart of Barcelona, Montjuich had been a venue for several important races before the war and had hosted a number of minor races since the Royal Automovil Club de Cataluña had reopened the circuit in 1966. These included Formula 2 and Formula 3 events, but it had been upgraded substantially for the Grand Prix. The most obvious improvement had been the addition of double-height armco barrier round much of the circuit, a revision which was to prove to be of immense value as the weekend's events unfolded.

Chapman was still experimenting with the size and width of the high rear wings on his cars. Given the number of wing collapses so far, this was something which Rindt was becoming increasingly uneasy about. However, the racing driver in him kept him behind the wheel as they were clearly performance-enhancing. Even at the circuit, Chapman was supervising further work on extending the wings.

Dave Sims, who was working on Rindt's car that weekend, remembers the occasion well. 'Chapman said "I want to extend the width of the rear wing and its actual shape, take it back, to get a bigger surface area." So we got packing cases made of polystyrene, and cut them to the shape of the wing. If I remember rightly, it was another six inches he told us to put on each side, and then he extended the whole thing back too. And then we had some thin aluminium and put it round the polystyrene shape of the wing to make it look stronger, but it was very,

Wing Worries

very thin. Astonishingly, it stood up. Stewart come up and said "This is ridiculous, you can't do that". And the Old Man said "Yes we can, there's nothing in the rules to say we can't." There were no rules then on wings. Then someone from Ferrari come up and said "Jesus." The Old Man said "Take no notice, come on, let's get it done". So we did it on both cars. Of course the thing was horrendously quicker, especially through Barcelona. The extra downforce helped with the traction and everything. But, it was a severe load to put on the struts.'

Practice was held on the afternoons of Thursday, Friday and Saturday. Rindt's session was ruined after only three laps on Thursday when he hit a stray dog and damaged his suspension. Ferrari driver Amon, still with a chassis-mounted rather than suspension-mounted rear wing, was the driver to beat, setting fastest time, while Hill was quickly on the pace, ending up second fastest. Siffert's Walker car was the only other 49B in the field, but was off the pace because it was still suffering from a pressurised oil system which had caused the team's brand new 9-series DFV to blow while testing at Thruxton a few weeks previously.

On the second day of practice, Hill and Amon set joint fastest times, with a ratio change for the Englishman having got his car just right. Rindt was seven-tenths off the pace and complaining that his car was handling very unpredictably. Eventually, the mechanics found a problem with one of the front shock absorbers, which was a consequence of the Austrian's encounter with the stray dog the previous day. For the final day's practice the rear wings of both the works cars had been extended at each end. A trailing lip (like a modern-day Gurney-flap) had been added and the front nose fins tweaked to provide additional downforce to balance out the greater loads on the rear of the car. It was all rather 'seat-of-the-pants' stuff, the trailing lip being held on by adhesive tape and adjustments to the front fins finished off with a pair of pliers! All the team knew was that, as Dick Scammell says 'every time we extended the wings the cars went quicker. And in the end we ended up with monstrous pieces taped on the end. I can remember Colin saying "Cor, it's even better! Go and see if you can make a bigger one."'

The improvements to the performance of the car were, in Rindt's case, significant enough to take pole with a stunning lap half a second clear of his nearest rival, Amon. Meanwhile, Hill was able to go out and knock a further second off his Friday time, to put him alongside his team-mate on the front row. The Walker team eventually got their engine to work properly, although the car's handling was still not to Siffert's liking, and he managed to put it on the third row.

Overnight, further revisions were made to the wing extensions, which had drooped somewhat. What emerged was undoubtedly the largest wing ever seen on a Formula 1 car! The all-night session was not without incident, one in particular which serves to illustrate the mercurial and temperamental nature of Colin Chapman, as mechanic Herbie Blash recalls. 'We were extending the front fins and rear wings. In fact, I think the rear wings were overhanging the outer rear tyre walls! You couldn't move two cars in the pits without there being a gap of a foot or something. Anyway, I got this big sheet of aluminium –

The start of a dramatic weekend: Leo Wybrott, assisted by Eddie Dennis, rolls Rindt's car off the transporter at Barcelona. On the Sunday evening the two cars would be thrown in on top of one another... (Ford Motor Company)

I was working on the front nose fins – and cut it up. Colin Chapman arrived at about five o'clock in the morning and looked for "his" sheet of aluminium. He'd been thinking about it overnight and he wanted to modify the rear wing once again. And I'd gone and cut up "his" - and our - one and only sheet!

'He went totally mental! He made everybody load the cars up and said "That's it, we are not racing." He shouted at me and said "You come with me". So we walked outside and there was a little all night café there and he kind of pushed me in through the door and I thought "Oh my God, he's going to kill me!" I'll never

A penny for your thoughts: 'Chunky' Chapman looking pensive in the paddock at Barcelona, framed by the wing which would cause so much trouble that weekend. (Tim Considine)

Lotus 49

It's Saturday morning and Rindt rather apprehensively studies the previous night's modifications, which have seen a lip added to the trailing edge and the wing extended. The white tips are where the wing ended, the bare aluminium being the additional sections added overnight... (Tim Considine)

Don't lean on it too hard, Jochen! The Austrian continues to gaze in disbelief at Chapman's latest mods while he chats to the Lotus boss and photographer David Phipps. (Tim Considine)

forget it. He got me a Coca-Cola, sat me down and told me this story of how a carpenter had come along and cut too much off a door and how you must think before you do things and that was it. Then it was back to the garage and he was saying "Hey, what are you doing loading up those cars? Come on, get those cars off ..."'

At flag-fall, Rindt jumped into an immediate lead, followed by Amon, a fast-starting Siffert and Hill. Even on full tanks, Rindt was returning some impressive lap times and was pulling away at the rate of around a second a lap. Then on the ninth lap, disaster struck as Hill, who had managed to get past Siffert, lost all control of his car

Taking a leaf out of McLaren's book, the Rob Walker team fitted endplates to the front and rear wings of their car for Barcelona, long before they appeared on the works cars. Siffert is caught here during practice, exiting one of the hairpins on the descent from Montjuich Park into Barcelona itself. (Ford Motor Company)

Wing Worries

Taken in practice, this shot shows Hill swooping down the descent at Barcelona with the new extensions on his wing already sagging badly. Note the effect of the roll of the car on driveshaft angles and the braces supporting the wing struts. (Ford Motor Company)

just after the hump following the pits straight. He was a passenger as first it hit the inside guard rail on the left and then cannoned off across the track to make contact with the armco on the other side. Eventually the car slid to a halt in a rather crumpled three-wheeled mess, its wing folded in a V-shape and hanging down behind the car. It transpired that the wing had folded upwards in the middle as the rear of the car went light over the hump. According to Dick Scammell 'the load on the tips was so great that it upset the balance of the wing and lifted the middle up.' This in turn caused the wing to tip back, creating massive lift which pulled the rear wheels right off the track and made the car to go out of control.

His mechanic, Billy Cowe, had seen the beginning of the accident from the pits and feared the worst. 'We could just see him lose the car as he crested the hill beyond the pits. I, and another mechanic, rushed off with a hacksaw thinking he might be trapped in the car because in those days you couldn't get the driver out because of the steering column and things like that. But lo and behold when I got there, Graham had got out and he was looking over the wreckage, to see if something had broken or a bolt had come out. I said no, it was the wing that had collapsed, watch the next time Jochen comes round and I'll show you, you can see his one doing the same. And his car did exactly the same thing that lap. Graham said "God, we've got to stop him". So he sent the other mechanic [Dave Sims] back to the pits to tell Colin

They're off: the start of a fateful race for Rindt, Hill, Lotus and high wings in general. (Tim Considine)

147

Lotus 49

This familiar sequence shows how Hill's Barcelona accident unfolded. In this shot, his wing has just collapsed over the brow of the hill after the pits, pitching him left into the barriers, rear wheels spinning furiously and only feet from the crowd. (Ford Motor Company)

Having rebounded past the photographer across to the other side of the track and hit the barriers, Hill's car has just come to a halt. He is shielding himself from a bouncing wheel which has detached itself from the wreck. (Ford Motor Company)

Miraculously unscathed, a shaken Hill steps out of his car and gets ready to vault over the barriers to safety. (Ford Motor Company)

to bring him in, and he was going to try and get Jochen's attention to try and stop him.'

However, it was too late: the next lap, in almost a carbon copy of Hill's accident, the Austrian's wing collapsed and he too lost all control of his car, hit the inside armco, and rebounded across the track. Where Hill had simply slid down the armco until he came to a rest, Rindt slammed into Hill's stricken wreck, which had only been pushed in slightly towards the barrier. As well as further severely damaging his car, this launched him into the air and onto the top of the armco barrier, as Cowe who witnessed the whole thing, recounts. 'His car did exactly the same thing in exactly the same spot. Graham's car was at the side of the armco and Jochen's car came crashing into that. It went across the top of the armco, and I was underneath it at this time, because there wasn't much protection other than the armco and a bit of chicken wire. Jochen's car teetered the right way up on top of the armco and it was a 50/50 chance whether it went back onto the circuit, or into the crowd. Fortunately it went on to the circuit and only Jochen was involved. If it had tipped into the crowd it would have been a massive shunt.'

As it came back onto the circuit, Rindt's car turned over and he slid down the track upside down until he came to a rest, with petrol pouring out over him and onto the tarmac. Added to this, fuel from Hill's stricken car was making matters worse, according to Cowe. 'The fuel from Graham's car was rushing down the gutter and I was terrified. That was why I had been standing by Graham's car, to stop people smoking near it.' It was a powder keg and one spark would have sent the unfortunate Austrian and his car up in flames. Fortunately, common sense among the crowd and marshals prevailed and, under the supervision of Hill, Rindt was extracted from his car, mercifully without serious injury. Hill was aided by marshals and the Lotus mechanics, more of whom had come down to the scene having seen the beginnings of the accident from the pits. Not only was it amazing that he had not been seriously hurt but it was also incredible that it had been possible to get him out without any cutting gear, since the car was bent into a banana shape, as Sims recalls. 'How Rindt got out of that, I'll never know ... We tried to get in it when we got it back to the factory, but we couldn't, it was impossible. Not even a little kid could have got into it.'

Miraculously, Rindt escaped with a broken nose and cheekbone, facial cuts and a hairline fracture to his skull. His helmet and the roll-over bar had both been worn away where he had slid along upside down, so he was very lucky not to have suffered more serious injuries. To cap a poor race weekend for Lotus, Siffert went out with a blown engine while lying a strong second, while Hill's Championship rival Jackie Stewart took his tally to two wins out of two starts to amass a maximum possible 18 points in the title chase.

Team Lotus had two seriously damaged cars – one in need of an extensive rebuild and the other a complete write-off. They had less than a fortnight until the start of the next meeting – the Monaco Grand Prix – to sort the situation out, this being on top of the demands already being placed on the Formula 1 section by the build programme for the Type 63 four wheel drive car and on

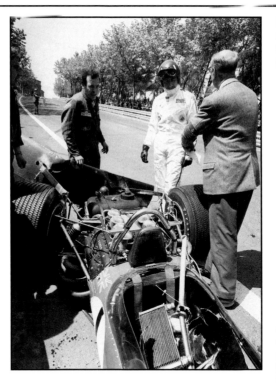

Cowe and Hill inspect the damage to R6, trying to find the cause of the accident. (Ford Motor Company)

As in previous accidents involving 49s, the tub stood up remarkably well to the impact, with the damage being mainly confined to the corners and the radiator/nose area. Here mechanic Billy Cowe - who rushed to the scene with a hacksaw in anticipation of having to cut the driver out - and Hill are looking worried, possibly having seen the beginnings of a crease in Rindt's wing. (Ford Motor Company)

Team Lotus resources generally by the Indy programme running hopelessly behind schedule.

As they packed up the transporter at Barcelona, most of the Lotus mechanics assumed that they would only be able to field a singleton entry in Monte Carlo, if that. However, Chapman had other ideas, as Billy Cowe recalls. 'I flew back that night with Graham in his plane and we were all a bit shaken up. A couple of guys stayed behind and took the truck back. The purpose of us flying back was to get cars ready for Monte Carlo.' Fortunately,

as beyond repair. Given that it was descended from the prototype 49, chassis 49/R1, it had not had a bad run for its money …

Herbie Blash remembers the trip back from Spain in the transporter well, particularly the surprise reception they got on their arrival at Hethel. 'We'd decided that we were sure to miss Monte Carlo. There was just myself and Eddie Dennis to bring the truck back from Barcelona, which was a hell of a long way. And it was just really one of those cruising trips where you stop off here, sleep there, not the typical Formula 1 journey where you are sleeping in the transporter and driving day and night. When we arrived at the gates, there they were waiting for us, in an absolute panic! I think we arrived back on the Friday or something and they had to leave for Monte Carlo on the Monday. And we literally never got home. We stepped off the transporter, worked day and night, day and night, got back on the transporter and went straight off to Monte Carlo!'

It became apparent that Rindt would probably not be fit to race at Monaco, so Competitions Manager Andrew Ferguson was instructed to arrange a suitable alternative. Richard Attwood had attracted plenty of attention with his drive to second place in the 1968 Monaco Grand Prix at the wheel of an unwieldy and uncompetitive BRM, and he was also the circuit's outright lap record holder, so he was quickly lined up. Attwood recalls: 'Chapman obviously recognised that I was a bit of a Monaco specialist, and I think he thought that I would go well in the car.' Such one-off outings at short notice were starting to become something of a habit for Attwood. 'I hadn't driven since before Monza [1968] in a Grand Prix car. It was very much a last minute thing. There was no time to test or anything, I just pitched up and drove. I came straight in from the airport and got in the car. I just had the official qualifying and then it was into the race. But it was worse the year before, where I hadn't driven a car for two years or so and had never driven a 3-litre Grand Prix car!'

The crease developing in Rindt's wing can be clearly seen, just to the right of the strut on the left hand side of this shot. It was caused by the additional pressure on the outer tips of the wing through them being extended. (Ford Motor Company)

On a wing and a prayer: Only a matter of feet away from spectators, Rindt is a helpless passenger as his car flies, all four wheels off the ground, towards the barriers. (Antonio Leal)

the task was made a bit easier by the fact that one serviceable car was immediately available, that being the one sent out to the Tasman series after Rindt's Levin accident. This car was, rather appropriately, allocated to Hill since this was descended from chassis 49B/R5, which was the very first B-spec 49 and the chassis Hill had driven to victory in Monte Carlo 12 months ago and then used for most of the rest of the 1968 season.

Incidentally, the car Rindt had crashed at Barcelona was the same one as he had rolled at Levin four months earlier. He had contrived to have two accidents in the same car, neither of which were his fault and both of which resulted in the car turning over and him being trapped upside down with petrol pouring over him. Clearly, on these occasions at least, luck had been very firmly on his side. However, after Barcelona, there was no hope of the wreck being rebuilt and this was the only Lotus 49 ever built that would be consigned to the skip

As pressing as sorting out a driver was the problem of arranging a second car. The one crashed by Hill at Barcelona was repairable but not in the available time. Instead, the mechanics had to go down to a freighter at Southampton and burrow down through tonnes of rotting fruit in order to retrieve the car driven by Hill in the Tasman series. While the chassis driven by Rindt (a pukka 49B, R10) had come back by air so as to be available as a spare car for the Grand Prix team, the Hill car had come back from Australia by boat, since its speedy return was not deemed necessary due to its outdated, hybrid specification. However, it was now the only serviceable 49 at Team Lotus's disposal, hence the unpleasant task which needed to be undertaken to retrieve it.

There was no time to bring the Tasman chassis up to full B-spec, so high wings were added at the front and rear, together with the bracing rods from the wing struts to the new mandatory roll-over bar which had to be added, while a fire extinguisher system was also installed. Although a Hewland box had been fitted midway through the Tasman campaign, the car still featured the original Mark 1 Type 49 rear suspension with the fir-trees, plus it

Wing Worries

With fuel pouring out onto the track, mechanics Dave Sims and Billy Cowe help right Rindt's car as officials and Hill (obscured) attempt to free the dazed Austrian from the wreckage. (Ford Motor Company)

still had the combined oil/water radiator set-up, with the oil tank in the nose behind the radiator, and the original front rocker arms at 90 degrees to the tub.

Both cars utilised 1968-specification DFV engines, since the two 9-series units in the cars crashed at Barcelona were too badly damaged to be repaired in the time available.

After his Barcelona shunt, Rindt, who had been quite critical of Lotus's safety record in the past, wrote an open letter about the dangers of wings to the FIA and copied it to journalists. Several magazines refused to embroil themselves in affair by publishing it but Autosport did so, in full -

Wings – Jochen Rindt's view
"This is an open letter to all people who are interested in Formula 1 racing. I want to demonstrate a few points about the aerofoils which at the moment are used on most of the F1 cars, in order to convince the so-called experts that they should be banned.

Basically I have two reasons why I am against them:
1. Wings have nothing to do with a motor car. They are completely out of place and will never be used on a road-going production car. Please note, I mean wings and not spoilers, which are incorporated into the bodywork. You can say they bring colour to racing, and I cannot argue against that; but after all F1 racing is meant to be a serious business and not a hot rod show.
2. Wings are dangerous, first to the driver, secondly to the spectators. When wings were first introduced to F1 racing at Spa last year they were tiny spoilers at the front and back of the Ferraris and Brabhams. They had very little effect except at high speed, when they were working as a sort of stabiliser. This was a very good effect and nobody thought any more about it until Lotus arrived for the French GP at Rouen a month later with the first proper wing. Suddenly everybody got the message about what could be done with the help of the air; but unfortunately nobody directly concerned gave much thought to what could happen if the wings went wrong, and what effect they would have on racing.

First of all, it is very difficult to design a wing which is going to stand up to all of the stresses, because who knows how big the forces are. If you make the wing stronger, it is going to be heavier and therefore produce bigger forces on the construction; you make it lighter and it all goes the opposite way. This is not my wisdom, it all comes from one of the most successful racing car designers. Nevertheless I am sure that after some time – and a few more accidents because of wing failure – this problem could be solved.

Now some personal experience gained by racing with the wing: The wing obviously works via the airflow over it, and this situation changes rapidly if you happen to follow another competitor; he has the full use of the wing and you yourself have to put up with the turbulence created by his car. This could mean that the man in front is actually going slower than you, but you cannot pass him because, after getting near to him, your wings stop working and you cannot go so quickly. This fact spoils racing to quite a large extent. On the other hand the turbulence can be so great that your car starts behaving very strangely and completely unpredictably.

This, I think, explains Oliver's accident at Rouen last

year, and I personally have been in similar trouble very often, but luckily I have always managed so far. You will understand that these two facts stop close racing, which is one of the most exciting things to watch. Therefore it is in the interest of the spectators and the drivers to ban wings.

Let us have a look at the wing if something goes wrong with it. And they do go wrong quite often, but so far nobody has been severely hurt. My accident in the Spanish GP has been the biggest one so far and, through a lot of luck and the safety precautions taken by the Spanish organisers, nothing serious happened. Naturally I will always be grateful to the Automobile Club of Barcelona for lining the circuit with double guardrails and for providing such efficient marshals.

To explain the reason for my accident, I was happily driving round the fastest bend on the track when my wing broke and changed its downthrust into reverse. The back end of my car started flying, and I nearly flew over the double guardrail on the left side of the track. Fortunately I was flying about 10 inches too low and got bounced back into the road. I have got a picture to prove it. Can you imagine what would have happened if the car had flown into the crowd? By next year we will probably have wings big enough to do so, and all the owners of the circuits will have to think about new crowd protection. You can also get lift instead of down-pressure if you spin the car at high enough speed and start going backwards.

Altogether I have come to the conclusion that wings are very dangerous, and should therefore be banned.

Jochen Rindt
Begnins, Switzerland

Rindt's Barcelona accident had brought relations between him and Chapman to an all-time low, as Jabby Crombac, a good friend of Rindt's (he had been, along with Hill, a best man at the Austrian's wedding) recounts: 'Things sort of peaked after Barcelona because, quite rightly the car should not have had these extensions on the wings and so Jochen was really furious with Colin. During the period he was in hospital in Spain, he circulated the letter against wings, which the CSI used. And of course, this letter incensed Colin, because Colin knew that when it came to a technical development like that, he had the edge over the others, so he knew that banning the wings was banning a way for him to be better than the others. He really wanted to keep the wings and for his driver to campaign against the wings made him absolutely furious. So they went into a sort of four-sided discussion from then on. Colin would not speak to Jochen, so whenever he wanted to tell Jochen something, he would call me and say "Please, call Jochen and tell him this". And at the same time, Jochen would not speak to Colin. His manager was Bernie [Ecclestone], so whenever he wanted to say something to Colin, he would call Bernie and say "Would you please tell that to Colin". It was a square relationship for a while!'

Added to the fact that they were at loggerheads over the issue of wings, Rindt's command of the English language was such that he found it difficult to communicate articulately and effectively with Chapman, and what he said often came out in a blunt or rude way, even when it was not intended as such. Crombac describes it thus: 'It was very difficult. Jochen was a very close friend of mine, so I knew him very well and I had absolutely no problem with him. But I could see him behaving in a rude way very often. He was very aggressive. Also, he didn't have a full command of the English language. So he was apt to say things which normally you would not dare say openly in the face of someone. But he didn't realise it was so offensive.'

For all the friction between him and Colin Chapman, Rindt was tremendously popular among the mechanics. Dave Sims recalls him with great affection and admiration. 'He was a classic guy, one of the most underrated drivers ever. He was quick, astonishingly quick. He would just say "I can drive it, I don't care what you do to it. I'll drive it".' Sims also remembers that Rindt used to look after his mechanics, particularly showing concern about the amount of hours they worked. 'He was always saying to Chapman, "Your mechanics, you're going to kill them, it's ridiculous." He really used to stand up for us.'

Herbie Blash remembers Rindt as someone who was easy to work with, a stark contrast to his team-mate. 'Basically Jochen was so simple to work for. He was so laid back it was untrue. Whatever car you gave him, he'd be able to get the best out of it and he'd drive accordingly. Whereas Graham, to be honest, was a pain in the butt! He wanted every single little thing. He'd sit in the car and one day he'd want the steering wheel a quarter of an inch closer to him and the next day he'd want it a quarter of inch away from him. We used to have increments of an eighth of an inch on the roll bars but he'd want to go a sixteenth. And the problem was that, he'd do all these things and he wouldn't actually go any quicker. Graham would get out of his car and he'd sit down and he'd go from one to twenty … and then he would fill in his job list! Jochen would get out and say "Leave it, that's fine".'

Another point which Blash believes didn't help the relationship between driver and team boss was the importance Rindt attached to the amount of money he was being paid. 'Jochen became a bit of a financial man as well, and I think his demands didn't really go down too well with Colin. But of course, that was when Formula 1 was just changing and you'd have to say that Colin brought about one of the major changes by bringing in a commercial sponsor.'

Even though all the teams turned up at Monaco with high wings on their cars, there was a feeling among them that it was only a matter of time before they were banned, as Formula 1 was not the only international racing series to be affected by a spate of such collapses – it was just the one with the highest profile. However, they were not expecting a ban to come quite as quickly as it did.

Prior to practice, officials of the organisers, the Automobile Club de Monaco, had been canvassing teams to find out if they had any objection to the wings being removed from all cars in the interests of safety but one team, Ken Tyrrell's Equipe Matra International, objected. Consequently, the ACM contacted members of the FIA's Commission Sportive Internationale (CSI) who were present in Monaco to see whether they could organise a meeting to sit in judgement on the issue that evening. In the meantime, Stewart had emerged fastest from the day's practice session, ahead of Hill, who was familiarising himself with his old mount once more. The Englishman was also making his first appearance wearing the latest style of full-face helmet - painted in his traditional London Rowing Club colours - a consequence of his rather lurid accident at Barcelona, where he had been lucky not to have been injured by flying debris as his car disintegrated.

The main Lotus incident of the day concerned Siffert, whose 49B had been weaving wildly under braking for the Gasometer hairpin and eventually the inevitable contact was made with the barrier on the outside of the braking area. The resulting accident ripped the nose section off the car, causing it to disintegrate, and slightly damaged the radiator. Since the team had no spare nosecones and Lotus did not have any either, the mechanics were forced to patch up the damaged one with layers of fibreglass. Consequently, it looked decidedly non-standard.

The CSI members duly met, with five representatives in attendance (the British delegate was one of those not present) and decided that, forthwith, and to include the Dutch Grand Prix, all wings that were not part of the bodywork of the car were to be banned. They were able to do this by invoking the 'safety clause' which allows them to alter the rules immediately and without notice if they feel there is a major safety issue which needs to be addressed, in much the same way as the FIA was to react some 25 years later to the death of Ayrton Senna. Crucially, the influential, articulate and persuasive Chapman was not present when these decisions were taken by the CSI, for he was attending Indy 500 qualifying and would not be in Monte Carlo until the Sunday of the race. Nose fins were still permissible, but this effectively forced all teams to remove their high wings overnight and run 'naked'. In addition, all times set in the Thursday practice session were annulled and would not count towards final grid positions.

Stewart remained fastest throughout the remaining practice sessions despite the dramatic changes in the handling of all the cars following the removal of their wings. Amon continued his rich vein of good form to take the other front row slot, followed by Stewart's team-mate Beltoise and Hill. Siffert was a strong fifth quickest having reverted to an 8-series unit after problems with the latest spec DFV, while Attwood took a little while to accustom himself to the 49. He ended up 10th on the grid, despite being more than one and a half seconds inside his existing lap record, showing the pace of change in Formula 1 racing in just 12 months!

Various teams were looking at how they might incorporate some kind of wing-like device into their bodywork in order to claw back some of the performance advantage they had lost with the banning of the high wings. Consequently, on race day, a number of cars appeared with differing appendages doubling as engine cowls, gearbox covers and air deflectors. These included Amon's V12 Ferrari, Bruce McLaren's M7C and Hill's 49B. The cowl on Hill's car had been fashioned from some sheet metal commandeered from the inside of the Gold Leaf Team Lotus transporter, as Leo Wybrott recounts: 'We unriveted the panels that covered in under the benches of the transporter and they were part of the wing tray that we made up because, after Barcelona, we didn't have enough aluminium.'

The weather for the race was warm and dry. A typical Monaco race of attrition followed, with Stewart surging into a seemingly unassailable lead until being sidelined by a driveshaft failure. Hill assumed the lead, having passed Beltoise on the third lap and then inherited second place when Amon's Ferrari went out, and held it to the end of what was rather a dull race. Jackie Oliver, who had made his Grand Prix debut 12 months earlier with Lotus, achieved the dubious distinction of involving himself in an accident which caused him to retire on the first lap for the second successive Monaco GP. Fortunately, this year he was in a BRM, so the Lotus Team could simply watch and nod knowingly …

Jo Siffert, despite a sick-sounding engine, managed to keep running at a good pace throughout the race and ended up in third place, completing a good day for the privateer entrants, since Piers Courage finished second in Frank Williams' Brabham. Richard Attwood drove a sensible race and benefited from the high rate of attrition as well as several well-judged passing manoeuvres to move up through the field and eventually took a well deserved fourth place. He probably would have finished higher up had it not been for a problem in the early stages of the race which lost him a lot of time: 'The problem I had was that, very early on, the gearlever knob fell off. I spent several laps trying to retrieve it because I worried that it would get under the brake pedal, or something, and then I wouldn't have any brakes. Also, what was left on the gear lever was so rough that I wouldn't have been able to finish the race with it like that anyway.

'Of course, the only time I could try and grab or retrieve it was under acceleration and then of course I was braking again and it went forward. I don't know how many laps it took but it took quite a while and I lost quite a lot of time from that. I don't know if it would have made any difference to the outcome of the race but, obviously, I would have been a lot nearer to Siffert. It's a bit silly really, isn't it – a detail that shouldn't happen. It just unscrewed itself I think. It didn't break, otherwise I wouldn't have got it back on again. It was a very unnerving feeling to think that any second it could jam somewhere and I'd have no brakes.' Attwood's performance completed a good day for Lotus with three cars in the first four. Not bad for a marque that supposedly built fragile cars and considering the heavy workload before setting off for Monaco!

Hill's victory in Monaco had also revived his Championship hopes and, with it, team morale. With Stewart not scoring he was only three points behind the Scot, in second place in the standings. Rindt, on the other hand, had still to score any points, despite being quite clearly the faster of the two works Lotus drivers to date.

An indication of the incredible pace at which Colin Chapman led his life, and the perfectionist standards by which he judged others, is given by Dick Scammell's account of the hours following the Monaco Grand Prix. 'It was rather a hectic weekend for Chapman because he arrived race morning and helped us and then, after the

Lotus 49

Hill's Tasman car, R8, was pulled off a freighter and rapidly fettled up for Monaco, where Richard Attwood deputised for the injured Rindt. He drove a solid race, notwithstanding having to grope about in the cockpit for the gear knob early in the race, and eventually finished fourth. (Ford Motor Company)

Wily winner: Hill's fifth victory at Monaco (and his second successive one, having won in 1968 too) owed much to the unreliability of others but also to his tremendous experience of getting cars to the finish at Monaco. Here he is pictured climbing up the hill from St. Devote towards Massenet. (Ford Motor Company)

race, he said "Right we're all going back now, we'll be in bed by midnight." And we proceeded to drive back to Nice, to the airport, where his aeroplane was waiting, with its pilot. We got stopped by the police because we'd gone down the wrong side of the road! He told me to get out and deal with them. My wife was in the car and when I took more than about 10 seconds to sort it out he was sitting there saying "What's that bloody fool doing?" I think we got stopped three times and got away with it.

'When we arrived at the airport, we climbed aboard the aeroplane and he decided he was going to have a sleep in the back. We hadn't eaten all day and we managed to get three and half sandwiches and a bit of fruitcake between us! He said "You sit up front with the pilot and help him". Then he went to sleep. We got as far as Paris and there was ice and everything. In the end he woke up and he came up front. One of the things he couldn't stand when people flew his aeroplanes was that they had to synchronize the engines absolutely spot on so that you didn't get this humming noise. The poor old pilot was fighting his way across the Paris zone and Colin's in there twiddling all the knobs!

'We flew on and he called up Hethel and said "We'll be coming in, get the customs man over from Great Yarmouth." Strangely enough, the customs man would not turn out from Yarmouth at that time on a Sunday night! So then we went to Stansted and they [the customs] decided to have all the luggage out. Finally, we got back to Hethel and Colin said "I'll see you in the morning at eight o'clock". And by this time it was about half past two or three o'clock in the morning. I got in at about two minutes past eight and he was in his office. He said "Where do you think you've been? Just don't ever be late

Wing Worries

Master of Monaco: Hill rounds Station Hairpin on the way to what was to be his only Grand Prix win of the 1969 season. After Barcelona he switched to a 'full-face' crash helmet. The 'tea tray' wing over the engine was fashioned the night before the race from aluminium sheet commandeered from the inside of the Gold Leaf Team Lotus transporter. (Aldo Zana)

Siffert swings around Station Hairpin in Rob Walker's R7. The team always went its own way in terms of specification and development compared to the works team. In this shot, the car is still using an old-style nose rather than a top-ducted one, has its extinguisher mounted over the engine instead of behind the oil tank and has an auxiliary pannier fuel tank on the side of the tub for long range fuel capacity. (Nigel Snowdon)

again! Sit down, we'll go through the job list." So he wrote the job list out in his little book and he got back in his own aeroplane, flew down to London Airport and got on the aeroplane back to Indy! I was just amazed by it. That was the thing about him. He could do things you couldn't even think of doing in the time.'

Due to the fact that the Belgian Grand Prix at Spa was cancelled, there was a five week gap until the next round of the World Championship. This might have represented a welcome respite for Team Lotus had they not had Hill's Barcelona car which needed rebuilding and the Type 63 cars still to be completed. To further complicate matters, six days before the Dutch race, the team was committed to appearing at a Speed Day at Thruxton with the aim of setting a new outright lap record at the circuit. Adding insult to injury, Hill failed to do this by some significant margin. In the wingless Monaco-winning chassis, 49B/R.10, he recorded a time of 1.16.8, compared to Jochen Rindt's best Formula 2 time (with wings) of 1.14.0.

Chapter 16
Trials & Tribulations

Early in the week preceding the Dutch Grand Prix, Leo Wybrott finally took the Lotus 63 out for its first proper run. As he explains, it turned out to be rather an inauspicious start to its life. 'We'd worked some serious hours to finish it. The bodywork wasn't completely finished, but I went out to test it anyway. Colin was out there, Dick Scammell and all of the people. I'd done a series of laps to see it all went the right way. We used to have a little area that we called the pit area that they took the van out to, and all the bits and pieces, and I had been in there and they checked it all over and it all seemed OK. It was quite late in the afternoon. After another series of laps, a reasonable amount to get the thing up and warm and working, I came in again and it was all checked over and I know they checked the wheel nuts, looked around it, all the sort of stuff you did with a new car.

It is June 18th 1969 and John Miles poses with Chapman at the press launch of the new Type 63 four-wheel drive car which the team mechanics had worked so hard to complete. (Ford Motor Company)

'Then I went off again and just went quicker and quicker. It was getting quite late in the evening. Down the long straight my goggles misted up due to the damp in the atmosphere. I was going fairly quickly, certainly well up near 9000rpm and I thought "Well, that's it" because it was dark enough that the change of air temperature was causing the goggles to fog up. I braked at the end of the straight. That was the time, we worked it out, when the wheel had become loose enough that the nut came off. There are two sharp right handers at the end of the straight, then you accelerate up and there is a left kink, which is almost in front of the main office building, and there was a moat all the way along the front of it.

'It was the right-hand rear wheel and, being under power all the time, it would have stayed on. But when I lifted for the kink, the thing just went out of control and I had no idea what was wrong with it. There was a funny little builders hut on the left-hand side of the circuit, where they had obviously not finished what they were doing, and a big pile of shingle and stones. It went into the shingle and stones, launched itself and went between the circuit and the moat and knocked down some little trees. It could have easily gone into the moat but it didn't. Then it went across the circuit into the infield and into the corn. I got out and was looking around it and then everyone turned up. Colin was very concerned about the car and not at all concerned about me, because I was standing there!

'They found the wheel well into the field, and then they discovered that the other one was almost off as well! What had happened was that the wheels had been designed with no support between the taper cone where the wheel nut goes up against and the mounting plates – they were hollow inside. Another of Chapman's ingenious designs. Once they'd got hot and started to really work hard, they had flexed and allowed the wheel nut to come loose. The car wasn't all that seriously damaged but it did have a huge hole in the floor of the tub and a foot

Trials & Tribulations

The 63 was supposed to be the answer to every Formula 1 driver's prayers. In fact, it was an unmitigated disaster. (Ford Motor Company)

On his return to Grand Prix racing at Zandvoort, Rindt sported a brand-new Bell helmet, the effect of which was rather spoiled by his name being scrawled in pen across one side and 'This $pace to let' on the other. Mechanic Eddie Dennis helps Rindt adjust his mirrors - amazing how a Dutch florin comes in handy sometimes! (Ford Motor Company)

of gravel in the foot well and Graham was coming to test-drive it the next day!

'So we took it back to the workshop, got it up on the stands and looked at what had to be done. I think we scraped out all the shingle and stuff, there was a huge pile of grass and shingle and rubbish underneath it. Then the fabricators must have had to repair the floor. We decided we'd have plenty of time in the morning to clean up and finish it off and we went home. As it turned out, Graham turned up really early in the morning and he came in and saw the car, all of the shingle and everything. Meanwhile, overnight, Colin had flown up to Magnesium Electron who made the wheels, got the people up and they worked all night and sleeved the wheels so that there was a solid fixing between the inner and the outer flange, in readiness for Graham to test it the next day. Graham did actually test-drive it the next day and it broke a front driveshaft at the end of the straight! That happening and what happened to me was one of the final factors in Rindt's decision not to drive it. I don't think Graham was very enamoured about it either.'

Just three days before the Dutch race, the 63 was revealed to an expectant press (and one blissfully ignorant about the week's goings on!) at Hethel. Gold Leaf Team Lotus Formula 3 and GT sports car driver John Miles was given the task of demonstrating the car, although it was not intended at the time for him to race it at Zandvoort. The 63 was a complete contrast to the 49. Where the chassis of the 49 had been a simple bracket to take the DFV, the chassis of the 63 was a highly complex piece of engineering. Where the 49 had been two wheel drive, this car utilised four wheel drive, which necessitated the DFV engine being reversed, the gearbox being located behind the driver's seat back and the propshafts running fore and aft down the left-hand side of the chassis to the differentials and drive axles.

A second car was finished in time for the Dutch race, with the result that the team left for Zandvoort with

Lotus 49

Hill tried the 63 briefly at Zandvoort but was not impressed. Rindt refused to even sit in it... (Ford Motor Company)

four cars in total – two 63s and two 49s. Rindt was making his return to the team following his Barcelona accident, and an abortive trip to the USA to try and qualify the Type 64 car for the Indy 500. He sported a brand new white Bell full-face helmet, the effect of which was somewhat spoiled by the name 'Rindt' having been scrawled on one side in black felt tip pen and 'This $pace to let' on the other side! Mario Andretti – fresh from victory in the Indy 500 - had also been entered in a third car. However, he was scratched from the start list following a dispute over starting money (which was resolved) and then a problem due to a USAC race being postponed, which prevented him attending. The objective had been for Andretti to drive the 63, leaving Hill and Rindt to concentrate on the 49s.

This was the first race where constructors' interpretations of the new CSI regulations on wings were seen. Some manufacturers stuck to the letter and the spirit of them, whereas others stuck to the letter with appendages that were perhaps outside the spirit. The main problem for the scrutineers was that the wording of the new regulation was ambiguous. For the first practice session on Thursday afternoon, both the Lotuses appeared with upswept engine covers similar to the one fashioned for Hill overnight before the Monaco race. However, after seeing the wings of rival teams, Chapman flew back to Hethel and brought back some fabricators, who produced some wing sections which could be mounted over the oil tank and gearbox.

After setting a competitive time during the first of the three scheduled practice sessions, Hill put in a few laps in the prototype 63, chassis 63/1 but the springs were all wrong and the nose was bottoming under braking. After several all-nighters to try and finish it, the second Type 63 chassis arrived separately to the other three cars and was finished in the paddock. However, Rindt refused to countenance the possibility of even sitting in the car, let alone driving it - much to the dismay of the mechanics who had worked night and day to complete it. Herbie Blash remembers the occasion as 'the only time I've seen Jochen really nervous. Chapman was trying to talk him into driving the four wheel drive. And there was no way he was going to.' Bob Dance remembers overhearing a conversation between Jackie Stewart and Rindt back at the garage in Zandvoort, where the Austrian said: 'What does Colin think I am? I am not a monkey! I am *not* driving it.' Clearly, feelings were running high at this stage ...

The strain of running so many cars was almost too much for the mechanics, according to Blash. 'That was a very hard time for the team because it was very small, and to have two such totally different cars was pushing us to the limit. The four wheel drive was one hell of a lot of work, it was so advanced for its time.'

The two four wheel drive cars were subtly different from one another, with the exhausts on 63/1 going up and over the rear suspension and there being a flat tail on the car, whereas 63/2 had the exhausts going under the suspension and an upswept engine cowl. For the second practice session on Friday morning, Hill elected to drive 63/1 but with the upswept cowl and revised exhaust system. It would have been much simpler to swap cars, but Rindt had already raided the four wheel drive car allocated to him for its engine, the unit in his 49 having not been running properly in the first session. As a result, 63/2 was left lying in pieces in the garage barely a day after it had been completed!

Rindt's dogmatic attitude towards the 63, while understandable since it was nowhere near race-ready, further incensed Chapman, as Jabby Crombac recalls: 'Colin had spent a fortune designing and building this four wheel drive car and he thought that it should be given a try. One of the reasons Jochen had joined Lotus was he had been promised a four wheel drive. And suddenly he changed his mind about it.'

Hill persevered with the 63 for most of the Friday morning session. However, this proved to be the fastest practice session of the three and even then he only managed to get down to a time that would have put him towards the back of the grid. Consequently, the decision was made to put the 63 to one side and concentrate on honing the 49Bs for the race. Rindt had really got his car going well, and set a blistering time, aided by the new rear wing which didn't really seem to be a part of the bodywork but was still passed by the scrutineers, much to the disgust of rival teams! His time of 1.20.85 remained unbeaten in the afternoon session on the Friday, due to a change of wind direction. In this session, Hill was unable to improve and, in fact, only went a few tenths quicker than he had done in the morning with the four wheel drive car. However, his Thursday time was still good enough to secure a place on the front row, the two Lotuses sandwiching Stewart's Matra.

Siffert, in the Walker/Durlacher car, suffered initially from too much downforce from his upswept engine cowl, which was pushing the car down on the bump stops and making for some very exciting motoring. After some adjustments to the angle of attack of the nose fins, he found the handling was improved but could still do no better than 10th on the grid. Meanwhile, both Matra and Ferrari lodged protests about the wings being run by their rivals, with most of their anger directed at the Lotus arrangements, although Brabham also came in for criticism. However, the protests were rejected and Lotus would once again be starting a race with superior cars as a result of Chapman's ability to interpret regulations creatively, while still remaining within their letter.

At the start, Hill made the kind of getaway that most racing drivers only dream of, hurtling down towards Tarzan corner fast enough to cut right across Rindt's bows to claim the lead. Hill's move prompted nothing short of internecine warfare between the two Lotus drivers, with Rindt, who once again had clearly been the faster of the two in practice, desperate to get past so that he could demonstrate his superiority over both his team-mate and his rivals. The two cars made contact two or three times on that first lap, the old hand and the young pretender driving as if their life depended on crossing the finish line first after one lap, not 90. At the end of that lap, Hill came by with a small lead and Stewart tried to get round Rindt on the outside going into Tarzan. The Austrian was having none of that, elbowing the blue Matra aside and setting off after Hill. On the third lap, he finally managed to barge his way past and quickly pulled out a lead over his pursuers. Two laps later, Stewart was up into second place and the race order appeared to be settling down.

With a comfortable lead, Rindt again seemed to be on the way to his first Grand Prix win but it was not to be. On lap 16, a driveshaft universal joint let go as he passed the pits – some flying debris from which injured McLaren team manager Teddy Mayer – and he pulled off at Tarzan, handing Stewart the lead. It seemed plausible to assume that the wheel-banging with Hill which had

ensued in the opening laps had not done either car any good and probably went a long way to contributing towards the failure which put Rindt out of the race. Certainly, mechanic Dave Sims believes this was the most likely explanation. 'The driveshaft would go if they banged wheels. The constant velocity joint we used was very, very small and if you tapped the wheels when it was under power, that would just put all the loads in. That's what we said happened and I'm sure it did, they did have a coming together. Jochen didn't give a stuff anyway. He didn't have much respect for the "old hands".'

Eleven laps later, his team-mate came rushing into the pits complaining of poor handling and suspecting a problem with the steering or suspension. However, nothing could be found, so he rejoined in eighth place, effectively out of contention, eventually finishing a lowly

The garage used by Lotus at Zandvoort was in the town, some way from the circuit. As a consequence, the cars were either towed or driven in on the roads. On the first day of practice, before 'proper' rear wings were put on the cars, Eddie Dennis is about to depart with an Aston Martin as a tow-car! (Autocar/Richard Spelberg Collection)

Lotus 49

Into the first corner at Zandvoort and Hill stakes his claim to the lead rather forcefully over team-mate Rindt. Before the lap was out they would be banging wheels... (Ford Motor Company)

seventh, two laps down on victor Stewart. Siffert was the only remaining Lotus driver in with a chance of a good finish, and he was driving like a man possessed. Having disposed of Hill to take third place on lap 14, he then inherited second when Rindt dropped out, a position he held until the chequered flag. This was his best result since the British Grand Prix the previous year and, coming hot on the heels of his third at Monaco, really lifted the Walker/Durlacher team.

Stewart's form and the Matra's reliability was starting to look ominous for Team Lotus in terms of the prospects of retaining their World Championship crown. Stewart had scored three wins from three finishes, and would have certainly won at Monaco but for his broken driveshaft. By stark contrast, the Lotuses were looking less robust and the rivalry between the two drivers was sapping team morale as well as diverting their attention from the real job in hand, beating everyone else in the field rather than just each other. After the Dutch race, Denis Jenkinson suggested that the team should be renamed Gold Leaf Lotus, dropping the word 'Team' from the name, such was the obvious division between the two factions.

Although there was a natural antipathy between Hill and Rindt, this was further fuelled by the perception on Hill's part that Chapman clearly favoured the Austrian, as Billy Cowe points out. 'I wouldn't say they got on tremendously well. It was Colin Chapman's fault, to a degree, because he thought that Jochen was the man to back, as it were. Not to put too finer point on it, it was very difficult in those days for Team Lotus to prepare two equal cars from a development point of view and Colin would have preferred Jochen to have any new goodies.'

It was after this race that the New Zealand mechanic Jim Murdoch, who had just joined the team, earned his nickname 'The Sponge' for his ability to drink literally anything, as Leo Wybrott recounts. 'We had our normal bash after the race down in the town at whatever the place was and Jim drank the contents of a flower vase after throwing the flowers out! I'll never forget the reek of the water!'

The French Grand Prix took place just over two weeks later at the La Charade road circuit near Clermont-Ferrand. At just over eight kilometres and more than three minutes to the lap, it was a challenging venue, rather like a mini-Nürburgring. The last time the French round of the World Championship had been held there was in 1965, so it was clear that the existing lap record, held by Jim Clark, would easily be beaten.

For this race, it was decided to recruit a third driver who could take the wheel of the Type 63 four wheel drive car. It was felt that the car required a different driving style and that someone new to Formula 1 would find it easier to develop than either Hill or Rindt. Also, the fact that the car was still a clearly far from competitive proposition probably contributed to the decision, since neither Hill – still lying second in the Championship standings – or Rindt wanted to dilute their chances of a race win by driving it. As a result, John Miles was drafted in to make his Grand Prix debut, Mario Andretti's USAC commitments once again precluding him from taking the wheel.

The Gold Leaf 49Bs were essentially unchanged from Zandvoort, save for a revised rear wing. In first practice, this caused great instability on both cars, so it was quickly replaced with the type of wing run at the Dutch track. Even so, both cars were strangely off the pace. Rindt was suffering from sickness and double-vision, which was partly a throwback to his Barcelona accident, but also just generally due to the undulations of the twisty and hilly circuit. After the first session, he ditched his Bell helmet in favour of an open-face model borrowed from his friend, Piers Courage, which seemed to be a slight improvement. Hill just couldn't get his car handling to his liking, experiencing similar problems to those in the race at Zandvoort, but unable to effect any

Siffert drove probably his best race since Mexico 68 to finish 2nd behind Jackie Stewart in the Dutch Grand Prix. (Nigel Snowdon)

changes which significantly improved the situation. In the final session, Rindt found some more time and ended up third on the grid, but Hill was languishing down in eighth, on the fourth row, only just ahead of Siffert's private Walker entry. Miles, making his debut on a difficult circuit, took things steadily and qualified fourteenth in the four wheel drive car.

Neither driver's confidence was helped by an incident on the morning of the race, when the steering columns on the works cars were found to be cracked. Leo Wybrott takes up the story: 'They were Triumph Herald type sliding length steering racks. It had one tube inside the other and there was a clamp arrangement that clamped the two tubes together once you'd decided what length you needed. It was the outer tube that had cracked quite a considerable way round on both cars. We welded them up and I had strict instructions from Colin not to mention it to the drivers. Anyway, somehow it got out – one driver found out from his mechanics and told the other.'

In the race itself, Rindt was running fourth when the sickness and double-vision he had been suffering all weekend got too much and he pulled into the pits to retire. Hill ran in eighth place for many laps until a prolonged pit stop for Hulme elevated the Englishman up to seventh. However, he then suffered the ignominy of being caught and passed by Vic Elford in a privately-run McLaren M7A. Elford was a very respectable and accomplished driver at that time, but the Antiques Automobiles McLaren was running an old 8-series DFV and this was only 'Quick Vic's' second outing in the car. To rub salt into Hill's wounds, he was then lapped by his Championship rival, Stewart, who was on his way to another dominant victory. It was only the retirements of Amon and Rindt that promoted Hill into sixth place for a solitary point. Stewart's win had taken his tally to 36 points. Against this, Hill's total of 16 didn't look too good and it was beginning to appear as if the title was slipping away from him, even at this early stage in the season.

After the lacklustre performance of the cars and drivers in France, it seemed that things could not get much worse. However, team morale and fortunes reached an absolute nadir at the British Grand Prix at Silverstone. While Rindt and Hill were determined not to drive the 63 until it was shown to be competitive, Chapman was equally determined that, having built them the much-demanded four wheel drive car, they should actually make some sort of effort to drive it. To force the issue, Chapman claimed he had reached a provisional agreement with John Love to sell him Hill's regular 49B, chassis R10. Love doesn't recall discussions ever reaching this stage, so it was probably a bit of Chapman gamesmanship! Either way, the net result was that R10 was left at the works in Hethel ... In addition, the 49 used by Richard Attwood at Monaco had been brought up to full 49B spec – although still with the high Tasman-style rear cut-outs for the lower radius arm mounting points – and sold to the Swedish driver, Jo Bonnier.

Chapter 17
Boiling Point

A laid-back looking John Miles discusses the handling of his Lotus 63 with mechanic Dave 'Beaky' Sims at Silverstone. (Dave Sims Collection)

When the Gold Leaf Team Lotus transporter rolled into Silverstone on the Thursday afternoon, it contained not two but one 49B - Rindt's regular chassis R6 - and two 63s … one for Hill and the other for John Miles. Bob Dance summed up this inauspicious start to the weekend thus: 'Another typical Lotus action. We were late arriving again because we'd done an all-nighter and we had to transport the cars to the circuit on the day of practice in a great rush!' The only noticeable modifications carried out to the 49B for this race were the introduction of a small scoop on the top of the nose which guided air down into the driver's footwell via a ventilation hole which had been cut in the front bulkhead - in an effort to improve cockpit cooling in that area - and the addition of endplates on the rear wing, which was to cause problems later on… The nosetop scoop was only to be found on Rindt's car for this race, since it was the only 'official' works 49B entered.

Hill, who had used his free morning while waiting for his team to arrive to engage in some lappery at the wheel of a Brabham BT26, was unimpressed. Although he did some laps in the 63, it was still nowhere near

sorted. His Austrian team-mate, as Dave Sims recalls, was even less taken with the idea of driving the 63. 'Jochen just said "No way". He wouldn't even get in it! He led the charge against it and Graham followed suit.' As a result, the Lotus boss was forced to back down and undertook to try and retrieve the 49B he had just sold to Bonnier. For this event at least, it was too late to prepare R10 and get it over to Silverstone in time.

Meanwhile, Rindt got on with the job of putting in some fast lappery at the wheel of his car, ending up third fastest behind Stewart and Denny Hulme in the McLaren. Just in case the mechanics were not already stretched to breaking point, Miles suffered a recurrence of the fuel pump problems he had suffered in France with his 63, which necessitated the removal of his car's engine.

On the Friday morning, Bonnier's van turned up with his car and a deal was put to him whereby the works team would borrow his car back for the meeting, and they would put him in Hill's 63. Ever the gentleman, Bonnier agreed. His 49B differed from Hill's regular mount in that, because of the high cut-outs, the capacity of the seatback fuel tank was reduced, requiring an auxiliary tank to be added in the nose, where the oil tank on the Mark 1 49s used to go. Hill had only done a couple of laps in his 'new' car before the front wheel bearings failed due to incorrect assembly, damaging the hubs. Chapman made a quick dash back to Hethel in his plane to collect some replacements and these were fitted in time for the afternoon session.

The problems of running four cars were really taking their toll on the overworked Team Lotus mechanics and it is easy to understand how little mistakes could have occurred. As an example, Rindt didn't get any laps under his belt in the morning's untimed session, due to what Leo Wybrott jokingly now refers to as 'The Golden Rivet Affair'! 'When the chassis were made solid aluminium rivets were used and they were dollied [crushed] in place with a rivet gun and a dolly inside. Most of the nice dome-heads were on the outside so if they weren't properly riveted they fell out. But there were some that, when you built the monocoque, you did from the inside out, so that the fuel bag had a smooth surface inside, the inner skin.

'It was one of those that hadn't been dollied and had just worked loose and dropped out, eventually chafing a hole in the rubber fuel bag. We taped over all the rivets in the tank bays but, there were so many fuel leaks, that nine times out of ten the tape was a soggy mess anyway. Obviously this caused a big deal of hassle. Somehow, someone handed it to Colin, who was sitting in the front of the transporter with Dick Scammell. He called me over to the side door of the truck and started to lecture me and rant and rave about this rivet. You can imagine the state that we were in, having worked all night to get there; it was completely spontaneous, but I just looked at it and threw it over my shoulder into what was the shingle and gravel behind the pits at Silverstone. He went absolutely apes**t! I think it was Scammell who must have calmed him down. I didn't do it intentionally to be aggravating, but that was just the way it was!'

By the afternoon, repairs had been effected and the Austrian recorded the second fastest time behind Stewart.

As it transpired, this was good enough to give him pole position, as Stewart crashed the car he had set his quickest time in, damaging it too badly to be repaired for the race, and had to commandeer Beltoise's MS80 and qualify in that. This he duly did, but his times in that car were only good enough for second position on the grid.

In contrast, Hill's bad weekend was showing no signs of improvement. He missed the first hour of the final session with a broken water pipe, then ran out of fuel. Consequently, it was no surprise to find him once again languishing down the grid, on the fifth row between Jackie Oliver's BRM and Vic Elford's McLaren. Siffert's Walker Lotus was never on the pace and he qualified mid-grid, albeit three places ahead of Hill.

The race turned into a two-man battle between the two drivers who were clearly head and shoulders above

Rindt, in an open-face helmet borrowed from his friend Piers Courage, discusses car set-up with Chapman and his mechanic, Eddie Dennis. (Ford Motor Company)

- First Person -
Dick Scammell on working with Colin Chapman

He just ignored the fact that not everyone could keep up with him. He used to come into me and say "Guess who had a good night's sleep last night?" I knew the answer: "You". "Yeah, I got three hours!" But I'm glad to have known him. You couldn't help but be inspired by him. He almost had a power over people, and he used it. He could sum you up at a flash too. I used to get myself to the state where I thought "that's it, I'm finished, absolutely finished" and I used to storm over to his office to tell him where to stick it. I'd knock on the door, he'd open it and I'd walk in and he'd say "Hey Dick, I've been meaning to have a word with you. Don't sit there at the table, come over here and sit down in the easy chair. I've really been intending to have a chat to you about what a good job you're doing!" What could I do? You'd have cut off your right arm for him, he was one of those sorts of people. He was totally intolerant, but I suppose that's what these people who achieve so much are like. They can't be worrying about people around them, they just get on with their own life.

Start of an epic race: Rindt and Stewart tear away, neck and neck, towards Copse corner. (Nigel Snowdon)

the rest of the field in terms of out-and-out speed – Rindt and Stewart. The Austrian held the lead for the first six laps, then Stewart came past in front, led for another nine laps and then it was Rindt ahead again. The crowd were kept entranced by this battle, the two protagonists racing nose to tail the whole time. However, trouble was brewing for Rindt, for the left-side endplate on his rear wing had come adrift and began rubbing on his left-rear tyre. Not wishing to risk a high-speed blow-out, he reluctantly headed for the pits, having been passed by Stewart on the same lap. His stop took 34 seconds. Bob Dance recalls it well: 'Dick [Scammell] got a rocket from the Old Man for not being able to tear a rear-wing endplate off with his bare hands! "Tear it off!" he was shouting and Dick actually got really stroppy and swore back at him.' Such had been the speed differential between him and Stewart and the rest of the pack, that Rindt was able to comfortably rejoin in second place, leaving Scammell to nurse a badly cut hand and Chapman glaring at him for not 'just' tearing the endplate off!

Even that wasn't the end of Rindt's sorry tale, for with just seven laps remaining, he came into the pits for more fuel. The sheer pace of the battle with Stewart had taken its toll on the fuel consumption of the 49B, according to his mechanic, Eddie Dennis. 'He didn't have the [auxiliary] tank. And because Rindt drove an absolute balls-out race with Stewart, he ran out [much as Clark had done in 1967 at Monza]. To put fuel in those

The two cars were rarely any further apart than this during their struggle. Here Stewart is well crossed-up in his efforts to keep up with Rindt. (Ford Motor Company)

Boiling Point

cars wasn't quick because the funnel went on with a very fine thread. I remember putting the funnel on, and Leo Wybrott waiting to pour the five gallons of fuel over the top and the panic of getting the fuel in. But that could have been avoided if we'd fitted the extra tank.'

A lap later and in came Hill, who had worked his way up to fifth despite awful handling caused by a nose fin having been knocked askew early in the race and a down on power engine. His problem was also a lack of fuel, and the stop pushed him down behind Courage and Elford and just out of the points in seventh. A lap after Hill's stop, the third 49B in the field, that of Siffert, also had to stop, so clearly fuel was very marginal for the Lotuses. This was a point that hadn't escaped the spectating public, according to Bob Dance: 'I remember in the evening when we were clearing up, I was putting some fuel from a can into the tank of the 10/12cwt Ford van and a spectator said "You should have put that in the racing car instead of in the van, and then you would have won!" I thought "thanks very much!" But he was right, all the same ...'

The frustration of the mechanics with Colin Chapman's preoccupation with not putting too much fuel in his cars eventually manifested itself in that oft-quoted term, the 'mechanic's gallon', as Billy Cowe describes. 'Lotus cars running out of fuel wouldn't be an indication that they were low on capacity because we used to lose more points running out of fuel than we ever got winning races! It wasn't that the capacity wasn't there, it was just that Chapman didn't like putting in too much fuel. At one time, the mechanics always used to put an extra gallon in because we knew we were always tight on fuel consumption. Then Chapman found out about this and so he used to issue the fuel figures with a gallon off. And it got to an almighty showdown one day when we actually put in how much was on the ticket and it obviously ran out way before the end of the race – maybe that was Silverstone!' Bob Dance recalls that Chapman's argument about putting in too much fuel was perfectly logical. 'It was all to do with the requirement to keep his cars down to weight. He used to say "A gallon of fuel weighs 7lbs. We spend hundreds of pounds on weight saving and you mechanics go and stick in an extra gallon or two of fuel!" So you could see his point.'

Although Chapman was notorious for being obsessed about the amount of fuel put in his cars, it was not always him occasionally being a bit 'borderline' with his calculations which caused his cars to run short, as Eddie Dennis points out. 'In those days, [fuel] bags never fitted the tank tremendously well. You always had to be careful after you drained tanks. Although the bags were supposed to be clipped up at the sides, they could fall, and fold themselves and that was why, possibly after the likes of Silverstone, you pumped out everything and then you measured everything in, you didn't ever assume you had x amount of fuel in. So if someone filled it up and said "Right it's full, it's got 38 gallons in", you didn't take that as being correct, you pumped out and actually physically put in the fuel you required for the race, so that you knew that what you had put in was correct. It happened on one or two occasions - people assumed they had 10 gallons left in the tank when in fact they might have only had five. Bags used to pull themselves away from the tank bay walls when a vacuum was created as the fuel was sucked out.

'There were also problems in the early days when we had the foam inside the fuel tanks as a safety device. It held up fuel and it didn't allow the car to drain all the

While Rindt fought for the lead, Hill grappled with horrendous oversteer caused by a nose fin which had been knocked into a high downforce position. Bonnier's car, which Hill drove, was R8, the same one that he had raced in the Tasman series and Attwood had driven in Monaco. Since then it had been updated to 49B spec but still retained the distinctive Tasman-style cut-outs for the lower rear radius arms. (Ford Motor Company)

Jo Bonnier made his first appearance in R8 at the German Grand Prix. It was an inauspicious debut, with the Swede slower in practice than many of the Formula 2 cars and his race ending with a leaking petrol tank. (Ford Motor Company)

fuel into the seat tank, where the collector pot was. It didn't absorb the fuel, just retained it. Any pick-up system on racing cars relied on acceleration or braking to surge the fuel backwards or forwards. And, if it couldn't surge [go back] in a rush, you didn't get it into the seat tank and you'd soon run out of fuel because the pot only held half a gallon or so. And that often was a lot of the problem with some cars – they had the fuel but they didn't pick it up.' Later that year, as Dennis explains, the fuel pick-up system on the 49s was revised and, as a consequence, significantly improved. 'We had a better system on the later cars, whereby the collector pot was under the driver's seat – instead of in the seat tank – and it picked up on acceleration and deceleration very well and drained exceptionally well too. We had some proper little flat valves which wouldn't let anything back. Plus, the fact that you'd taken that catch tank and everything out of the seat tank, gave you another half gallon or three-quarters of a gallon in the system. We had that system on the 49s in late 1969 and 1970. That was definitely a better system.'

It seemed now that, barring a major disaster, Stewart would be crowned the 1969 World Champion. He had scored 45 points so far or, to put it simply, maximum points in five out of six rounds. Hill's performances in the last three races had been disappointing and compared very poorly with his team-mate's tigerish drives. Rindt's 'drive it as it comes' approach to car set-up also contrasted starkly with the constant fiddling the Englishman was renowned for. Despite this, Hill was tremendously popular among the mechanics, as Bob Dance points out. 'Away from his race car Graham was a really good bloke: he and his wife always looked after the mechanics very well and he was the life and soul of the party.'

After the British race, Bonnier's 49B was refitted with an 8-series DFV and painted red with a white stripe down the front. His first Grand Prix with the car was to be the German round, held at the fearsome Nürburgring. Despite being entered by Ecurie Bonnier, when it appeared it had Scuderia Bongrip (Bongrip was Bonnier's own company) on the side, but was looked after by the Parnell team! Bonnier was able to do this as team manager Tim Parnell and his mechanics had time on their hands because their BRM was back at the works and had not appeared since Monaco.

The works cars featured revised front rocker arms which, instead of being slightly angled upwards, were now straight, thus increasing suspension travel, particularly important at the Nürburgring with its dips and jumps, which really gave the car and driver a pounding. The 'sale' to John Love having been called off, Hill was reunited with his old faithful double-Monaco GP-winning car, R10, while Rindt was in his by now regular car, R6. Both cars featured revised fuel pick-up systems in an effort to help them use all their fuel (on inspection after the British Grand Prix they had both been found to have five gallons left in their tanks). A third car, 63/2, had been entered for Mario Andretti, the American having been brought in to show the regular drivers just what they were missing by not driving the four wheel drive car!

Boiling Point

The Nürburgring circuit always offered spectacular viewing for spectators and equally dramatic racing for the drivers. Here Hill becomes airborne in R10 during practice. (Ford Motor Company)

In practice, Rindt was the best of the 49Bs, although he was unhappy with the unpredictable nature of the handling of his car. Belgian Ickx put his Brabham on pole, his pace reflecting his love and knowledge of the tortuous circuit in the Eifel Mountains. Another fan of the 'Ring, Jo Siffert, secured fourth on the grid, while Hill was again struggling and down in ninth, albeit having beaten the magical eight minute barrier. Andretti had been afflicted by a rash of camshaft failures in Cosworth DFVs, which restricted his running time and ended up 12th on the grid. Even so, he set some quite impressive times for a first-time visitor to the Nürburgring, and was optimistic about his chances in the race.

Due to problems fitting the engine, Bonnier's car didn't arrive until Friday afternoon and he didn't manage to get a flying lap in until the Saturday session. Even then, he found the cockpit too cramped for a man of his size, to the extent that he had to change gear by reaching across with his left hand, as he could not reach the lever with his right hand! Consequently, it was no surprise to find that he was the slowest of all the Formula 1 runners.

The only Lotus 'incident' during practice occurred when a ball joint joining the bottom front wishbone to the upright on Siffert's car broke, fortunately without causing an accident. Having already decided to change a down-on-power engine, the Walker/Durlacher mechanics were therefore in for a busy night, as they now had a suspension to rebuild as well.

Rindt's weekend was not going well. First a front suspension pin was found to be loose and then his engine wouldn't run above 8500rpm. However, a quick change of the spark box was effected and he was wheeled onto the grid in the nick of time. Similarly, Jo Bonnier found that his car had a leaking rubber fuel bag, but was unable to repair it in the time available and so faced the prospect of sitting in a puddle of fuel for the race.

At the start, Stewart jumped into the lead, followed by Siffert and Rindt, while Ickx completely fluffed his getaway and ended up down in seventh place. He had soon passed everybody but the Matra and this he eventually succeeded in doing at the start of lap seven. From this point on he never looked in trouble, particularly as Stewart was having gear selection troubles. Rindt's engine maladies returned, it again refusing to rev above 8500 and sounding very rough. Eventually, the problem got so bad that he came into the pits. The mechanics couldn't do anything to solve the problem, so after doing another lap, he retired. Hill was another experiencing gear-selection woes, the car refusing to stay in gear unless held and then later he was unable to select fourth gear, while Bonnier had already given up his unequal struggle after only four laps, having got fed up with sitting in a pool of petrol. Even worse, Andretti had lost control of the four-wheel drive car as it bottomed out in one of the circuit's many dips, badly damaging it, but doing no harm to the driver. However, debris from the accident caused Vic Elford to crash heavily and the injuries he sustained effectively finished 'Quick Vic's' Formula 1 career.

Siffert was running a strong third when he had a massive shunt approaching the corner leading up to the Karussel, from which he was lucky to emerge unscathed. A front rocker arm had snapped in two - perhaps a legacy of strain caused by his practice incident - causing the car to plunge off the track into the surrounding trees and bushes, fortunately not hitting anything substantial.

As a result of the high rate of attrition, Hill was able to nurse his stricken car into fourth place, the last of the Formula 1 cars still running, to secure a further three points for himself. However, Stewart's second place cemented his hold on the Championship, giving him a 29-point lead over closest challenger, Ickx. Hill's run of poor form had seen him slip to fourth in the standings, behind 'Mr Consistency' Bruce McLaren, who was making a habit of finishing in the points and this time had finished a well-deserved third.

Chapter 18
Team Spirit Returns

With a five week gap until the next Championship round, Lotus was presented with a real opportunity to re-group, get some serious testing in with the 63, and return to the fray in Italy significantly more competitive than they had been in recent races. Two weeks after the German race was the non-Championship Gold Cup race at Oulton Park and Chapman scored a remarkable victory by finally persuading Jochen Rindt not only to sit in a 63, but also to race in one! It seemed that, having had a 'clear the air'

After putting his 49 on the front row for the Gold Cup at Oulton Park, Jo Bonnier had a big shunt caused by a front suspension failure. It badly damaged the car and shook him up to the point where he returned the car to the works and never drove it again. (Ted Walker/Ferret Fotographics)

168

Team Spirit Returns

Chapman finally persuaded Rindt to drive and even race the 63 at the Gold Cup meeting at Oulton Park, where he finished a steady if unspectacular second. (Ford Motor Company)

chat, the Austrian and the Lotus boss were enjoying considerably improved relations. Having qualified fourth, he eventually finished a steady rather than spectacular second to give the four wheel drive car its best ever result.

Jo Bonnier should have started from the outside of the front row, having really got his newly-acquired 49B, chassis R8, going well. However, late in practice, a bottom front wishbone came away from the upright on the approach to the fast Island Bend, and the car slammed off the track and into the bank. Bonnier was knocked unconscious in the impact and his crash helmet was torn from his head. The Swede was taken to hospital and was unconscious for several hours but recovered well enough to appear at Oulton the next day as a spectator. The car was quite badly damaged, with the worst of the impact having been taken on the left side of the tub.

Wheel-to-wheel combat: the start of the Italian Grand Prix and the wingless cars of Rindt and Stewart lead the field away. Denny Hulme is trying to thread his way through the gap between Stewart and Piers Courage while Jo Siffert is making a run down the outside. Hill is mid-pack while Miles is bringing up the rear in his misfiring 63. (Aldo Zana)

Ultimately, this was to be Bonnier's second and last appearance in the car, for it returned to Lotus and stayed there. Although this was ostensibly for repairs, Bonnier had clearly been very shaken by such a dramatic component failure and it seems he had no appetite for driving the car again. Two weeks after the crash, it was reported in *Autosport* - clearly as a result of a conversation with the driver - that: 'It looks as if he won't be driving the Lotus 49B again this season, and the car may revert to Lotus for the time being.'

The four wheel drive car was still not even near to being on the pace, so the team stayed on at Oulton Park to carry out further testing, as Bob Dance remembers. 'We raced on the Saturday and stayed and tested on the Monday. What we found was that, the more torque we transferred to the rear wheels, the better the car handled. Originally, we started off with the torque split as 60% to the rear wheels and 40% to the fronts but it ended up at 80:20.

'It was at that point that we decided that we might as well revert to two-wheel drive with all the weight and friction penalties associated with the four wheel drive car! At one stage the Old Man was lying on the ground on the inside of the first corner [Old Hall] to get a good look at it, to see what it was doing in the corners.'

Dick Scammell also remembers Rindt being very derogatory about the 63 after driving it. 'We were all going to sit down and decide what we were going to do with it. We were all sat there waiting for him and he came in said "I'm off for a cup a tea, because whatever you do to that car you can't make it any worse!" And he just left! He used to say, "Once you've settled on entering the corner, you've got to have it right because there is nothing you can do with the car. If you have got it wrong, you know a long time before you go off the road that you are going to go off because you can't actually do anything with it!" A horrendous car!'

Following Siffert's Nürburgring accident and the problem found on Rindt's car on race morning, the two regular works 49Bs turned up at Monza with new pins fitted to the bottom of the front uprights and they also retained the flat rocker arms which had been introduced to give more suspension travel. New, smaller, rear wings had also been fabricated especially for this race to give a low downforce set-up. As well as Hill and Rindt, a third car, the original 63, was entered for Miles to drive again.

In the first of two three hour practice sessions on Friday, Rindt set the quickest time, with Hill back down in seventh, one place behind Siffert, while Miles was plagued by a chronic misfire in the 63. Rindt once again demonstrated his blinding speed by taking pole on Saturday, while Hill suffered a broken camshaft early on in the session. With no spare car and insufficient time to change the engine, he was forced to sit out the rest of practice and would, once again, start from way down the grid. The spare engine was put into his car for the race, while another engine was borrowed from the Walker team in an effort to solve the misfire problem with the four wheel drive car, which persisted throughout Saturday.

On race day, both works Lotus cars appeared devoid of spoilers and wings in a bid to maximise straight line speed. As they appeared to be particularly good round the corners, it was felt to be a worthwhile trade off to lose some of this advantage in return for more top-end velocity. The race was an absolute classic, in front of packed grandstands and a crowd of 100,000 racing-mad Italians. The early laps saw the lead change between Stewart, Rindt, Courage and Hulme, with Siffert, McLaren and Beltoise keeping a watching brief. Hill, despite losing a right hand tail-pipe from his car's exhaust system, was making up lost ground and eventually joined the leaders. With ten laps to go, the leading group was down to five, with the two Lotuses of Rindt and Hill ranged against the two Matras of Stewart and Beltoise and a singleton McLaren driven by the team's owner and founder.

With only four laps to go, disaster struck for Hill when his car's left hand driveshaft broke. He had been driving very well and looked a strong candidate for victory, which made it all the more disappointing for the Englishman given his dire performances in the previous four races. The other four drivers were left to fight it out, Stewart timing his run to perfection to lead out of the final corner and just out-drag Rindt, team-mate Beltoise and McLaren. The official margin of victory was eight-hundredths of a second and only 19 hundredths of a second covered these first four cars! It was the closest finish ever seen in a Grand Prix and, more importantly for the Scot, it guaranteed that he had won the 1969 World Championship in the best possible way – from the front, as indeed he had done all season, with the exception of Germany, where he had suffered a rare defeat at the hands of *Ringmeister* Ickx. Rindt had come agonisingly close to his first Grand Prix victory and Jenks' beard must have been positively quivering with anticipation as the cars shot towards the finish line! It was not to be this time, but his time was surely to come soon.

Of the other two Lotus cars in the field, Siffert was slowed by a sick engine after having been up with the leaders in the early stages, and eventually the engine suffered a rather expensive looking piston failure. Miles completed a dreadful weekend by only managing four laps in the four wheel drive car after yet another DFV camshaft failure – something which Cosworth was putting down to a bad batch of shafts.

With the European season ended and the Championship title decided, the Grand Prix circus took off for the final three rounds in the North American continent in a relaxed frame of mind. Gold Leaf Team Lotus set off for these races with the 49Bs in unchanged specification, save for the re-introduction of nose fins and the conventional sized rear wings, plus a four wheel drive car, the one Andretti had crashed in Germany, which had been repaired.

The Canadian Grand Prix, held at the Mosport Park circuit, some 50 miles north of Toronto, was the first of these North American races. As well as the two Gold Leaf Team Lotus 49Bs and the Walker/Durlacher example, a fourth one joined the field for this race, and for the US and Mexican rounds. This was R11, the 49B which had been driven by Andretti in the South African Grand Prix in March and then delivered to the American Volkswagen dealer, Pete Lovely. The car had been re-sprayed since it last appeared in an international Formula

1 race and featured a low wing, but was otherwise little changed.

In practice, Rindt was again well to the fore, ending up on the outside of the front row, third fastest behind Ickx's Brabham and the Matra of Beltoise. Hill was higher up than he had been for some time, on the third row and seventh fastest, one place ahead of Siffert. The Swiss had a heavy crash after catching his foot under the brake pedal, the impact ripping off both wheels and suspensions on the left side of the car and putting several dents in the tub. However, mechanics Tony Cleverley and Stan Collier were able to get it rebuilt in time for 'Seppi' to take the wheel in the race. Lovely was the slowest of the 49Bs back on the seventh row, clearly finding the Grand Prix pace somewhat overwhelming, while Miles put in a competitive performance to slot the 63 onto the inside of the fifth row, putting many a two wheel drive car to shame. Even Rindt deigned to have a couple of tries with the four wheel drive, either suggesting that he felt useful progress might have been made with the car, or that he wanted to satisfy himself that it still wasn't up to a competitive level yet. Either way, he managed to have an almighty spin in the car and was not as quick as team-mate Miles, although of course the Englishman had considerably more experience of driving the car.

After making a lightning start, Rindt was able to enter the first corner in the lead, a position which he maintained until part-way round the sixth lap, when Stewart passed him. Three laps later, Ickx had slipped by too, but the Austrian was promoted up to second again when the young Belgian and the new World Champion had a coming together, which left the Scot in the ditch. At two-thirds distance, Jack Brabham dislodged Rindt for second and went on to score Brabham's first one-two since 1967, with Rindt finishing third, the only car on the same lap as the two BT26s. Hill's wretched luck continued with yet another camshaft failure. Things could only get better in the remaining races, couldn't they? As in Italy, he was the last retirement, although amazingly this was on lap 43 out of 90 and no-one else dropped out for the rest of the race! Siffert was another retiree, going out two laps before Hill with driveshaft failure, while Lovely was the only other Lotus finisher apart from Rindt, coming home nine laps behind the winner in seventh place but some three laps behind the sixth place finisher.

A fortnight after the Mosport Park race, the teams assembled just across the border in the northern part of New York State at the Watkins Glen circuit, for the 11th Grand Prix of the United States of America. While this was no longer a significant race in Championship terms, it was a race that everybody wanted to win because the size of the race purse was so much more than for any other race. The overall prize fund was US$200,000 (about £84,000 at 1969 exchange rates but £784,000/US$1,254,000 at late 90s values) with the winner taking a massive US$50,000 (around £21,000 in 1969, or £196,000/US$314,000 at late 90s values)!

Clearly if Rindt was going to break his duck, as he had looked like doing on so many occasions during the season, this was the time to do it. At a little over a minute to the lap, the Glen was rather short compared to other

Grand Prix circuits, but still challenging nonetheless, with the fast swoop of the Esses leading on to the front straight, returning via the cambered Loop corner, leading onto the appropriately named Chute which propelled the cars onto the Back Straight, through a challenging right sweep called Fast Bend, and round a left curve which led to a 90 degree right turn and back onto the start/finish line. On past form, it was a circuit which suited the Lotus 49s well, with Clark and Hill having occupied the front row in 1967 and Andretti and Hill having qualified first and third in 1968.

As the cars had come direct from Canada without returning to the factory, they were in unchanged specification. However, there was a alteration to the driving strength, with Mario Andretti returning to the scene of his impressive Grand Prix debut in 1968, this time at the wheel of the four wheel drive 63 in which he still appeared to retain a lot of faith. As it turned out, this meeting would do much to erode his and Chapman's belief in the concept, and probably sounded the death knell for four wheel drive in Formula 1 racing.

Practice was divided into two four-hour sessions, one each on the Friday and Saturday. Friday's practice was wet but it offered the first good opportunity of the year for the four wheel drive cars to demonstrate their roadholding superiority. However, as Andretti recalls: 'The car was diabolical in the wet! I was licking my chops, thinking "Man, you know this is good, finally they are all going to envy me" and they were all saying "Oh jeez, I want to switch" [from a two wheel drive to a four wheel drive car]. And I said "Not today!". So, I went out there and when I came back in I said "Yup, I want to switch!".' The four wheel drive 63 was around two to three seconds a lap slower than the two wheel drive 49Bs – in other words, pretty much the same margin as in the dry.

The Saturday session was dry and, after some problems with a misfiring engine and overheating tyres, Rindt really got down to some serious lappery, taking

Despite his poor start, Hill made it across to the leading bunch and both he and Rindt took turns in challenging Stewart who, as this picture shows, was trying very hard to stave them off. (Ford Motor Company)

As he had done many times before in the 1969 season, Rindt led the Canadian Grand Prix early on but fell back before the end. He is shown fending off Jack Brabham, who at this time was discussing with the Austrian the possibility of him returning to the Brabham team for the 1969 season. (Nigel Snowdon)

pole position – and the US$1000 that went with it – for the fifth time that year, albeit only three hundredths of a second ahead of his nearest rival, Denny Hulme in the McLaren. Champion Stewart was third a further twelve hundredths back, while Hill was higher up the grid than he'd been for some time, alongside the Scot in fourth and one place ahead of the mercurial Jo Siffert in his Walker 49B. A frustrated Andretti languished back on the seventh row, only just over a second quicker than Pete Lovely in his privately owned and run 49B. An overnight engine change for Rindt's car was all that was required in the Lotus section of the Glen's Tech Shed, so the team were feeling quietly confident about their prospects come race day.

At the start, Rindt jumped into the lead, hotly pursued by Stewart in a re-run of their Silverstone battle. The Matra driver enjoyed a brief period of nine laps in the lead, after which the Austrian was never headed, reeling off the remaining laps quite comfortably, aided by the fact that Stewart went out on lap 36 out of 108 with engine problems. The same could not be said about his team-mate who, having started third, lost ground steadily almost from the beginning and by lap 12 was languishing in seventh position. In *Life at the Limit*, he explained that he was using some experimental Firestone tyres and, because they had been concerned about the risk of chunking on the inside of the shoulder due to overheating, had asked the team to take some camber off the rear wheels. This compromised the car's roadholding – hence his poor race pace.

Regardless of the reasons for his being off the pace, by the 31st tour he had been lapped, serving only to emphasise the superiority Rindt had enjoyed all year over the Englishman. However, much worse was to come. Hill spun on oil which had been dropped on the far side of the circuit and stalled the engine. As the car was undamaged (and the cause of the spin seemed innocuous), Hill push-started the car on a downhill gradient and jumped in but of course, was unable to buckle up his belts. As he push-started the car he noticed that the tyres were bald and showed signs of chunking, which went some way to explain the ease with which he had spun when he had seen the oil and was taking care. A couple of laps later, the car was beginning to handle very oddly, so he gesticulated to his mechanics to have new wheels and tyres ready for him next time round.

Unfortunately, he never made it back to the pits for, going down the Front Straight, the right-rear tyre deflated, causing him to spin out of control and hit an earth bank, which pitched the car into a series of rolls. As he was not strapped in, the hapless Hill was thrown out of his car, the act of exiting the cockpit with such force breaking his right knee, dislocating his left and tearing a lot of ligaments. The incident cast a shadow over Rindt's inaugural victory, which had been so long coming and was well-deserved, especially as he had outdriven Stewart fairly and squarely when they had been racing head-to-head. It also meant that, after several narrow escapes earlier in the season, Denis Jenkinson would finally have to eat his words and shave his beard off!

More importantly to the mechanics, they got a 10 per cent share of the prize money won by the team, which helped to make the whole North American trip a very lucrative one for them. In all, the six mechanics – Wybrott, Dennis, Cowe, Blash, Murdoch and Sims - shared around US$6000 (£2510 then, but the equivalent of around £23,500/US$37,600 at late 90s values!) as Billy Cowe recalls: 'We did three races in all, the Canadian, American and Mexican GPs and that took about six weeks. We were on expenses, which we probably didn't save a lot of! We also got the Watkins Glen prize money paid out there. I was a young lad then and I went shopping in New York and bought some trendy clothes and when I came home I still had plenty of money in my pocket!'

Although Hill's car was not written off in his Watkins Glen accident, it was too badly damaged to be repaired without going back to the factory so, while the rest of the team headed for Mexico and the final round of the Championship, R10 was sent back to the works. There had been some talk of putting local hero Pedro Rodriguez in a Gold Leaf car for his home race but it was never clear what car he would drive if this happened. What was obvious following Hill's accident was that there would be no 49B for the Mexican to drive. Up until a week or so before the race he was apparently still in the frame for a drive in the 63 in place of Miles, but in the end these plans did not come to fruition.

The Magdalena Mixhuca circuit was a curious mixture. It started with a long, flat-out straight, at the end of which was a sweeping bend leading onto another shorter straight and then a quick left-right, leading down to a hairpin. It then made its way back to the start/finish area via a series of right-left-right curves which led on to a long 180 degree bend, slightly banked, which slingshot the cars once more down onto the main straight. To make things more complicated, the circuit was also at high altitude, which caused additional problems for nearly all the teams in terms of overheating and fuel mixture. As a result, for this race (and for the other high-altitude race at Kyalami) a different cam was used in the fuel metering unit which altered the characteristics of the mixture.

With Hill in hospital and no spare car, Rindt was the sole works 49B runner in the field, but he was joined in the team once more by John Miles in the four wheel drive car. There were two other 49Bs in the field, Siffert at the wheel of his regular Walker/Durlacher entry and the American Pete Lovely in his privately owned version.

This race was all about tyres and, if you weren't with the right manufacturer, you quite simply weren't going to be in the running. The tyres to have on this rather abrasive track were Goodyears, with the Firestones at a clear disadvantage throughout the weekend.

In the Friday practice session Rindt was still tired, having arrived just before the start. Consequently, he was over three seconds off the pace of Jack Brabham, who recorded a scintillating lap 1.3 seconds inside Jo Siffert's 1968 lap record. This proved to be the crucial session, with the first five on the grid for Sunday's race setting their times on Friday. This worked against Rindt, who improved his time by more than two seconds on the Saturday but this was only good enough to elevate him to the outside of the third row in sixth place. Thus, for the first time that year, Siffert's private entry had qualified faster than the quickest Gold Leaf Team Lotus works car. Miles was the best of the four wheel drive cars entered but back on the sixth row, while Lovely, who had missed Friday's practice, lined up in the penultimate spot on the grid, ahead of the young Canadian driver, George Eaton, in the third BRM.

Continuing the resurgence in that team's fortunes, the front row of the grid was an all-Brabham affair but, more importantly, it was also an all-Goodyear one. At the start, Stewart made a tremendous getaway, bursting through from the second row to take the lead, a position he held for the first five laps. Meanwhile, Denny Hulme was recovering from a poor start which left him in fifth place at the end of the first lap. After disposing of Rindt on the second lap, he spent four laps behind the wily Jack Brabham before finding a way past, took Stewart on the next lap and then, after three laps of running behind Ickx, nipped past the Belgian and was never headed again. After his early lead, Stewart fell rapidly down the order and ended up battling for fourth place with Rindt for the crown of top Firestone runner. This ended on lap 21, when the left front suspension on the Austrian's car started to collapse and he pulled into the pits to retire.

After the highs of Watkins Glen, and the strong performances Rindt had put in during the 1969 season, it was a slightly disappointing way to end the year. However, he had finally managed to break his 'duck' after having shown all the hallmarks of a potential Grand Prix winner for several years and the prospects for 1970 looked very good. Without doubt, 1969 was the year in which Jochen Rindt matured.

At last! Rindt finally took a long-overdue and well-deserved first Grand Prix victory at Watkins Glen. (The Dave Friedman Photo Collection)

Chapter 19
Trading Places

Despite offers from the likes of Brabham and March to join forces with them for 1970, and even the possibility of starting his own team with Robin Herd and Bernie Ecclestone, Rindt decided to stay on with Chapman and Lotus. However, as Jack Brabham points out, this was after he had agreed to return to his former team. 'We actually signed Jochen up for the 1970 season and when we got to Watkins Glen, Colin offered him a lot more money to run for him and Jochen came to me and explained the situation. So I said "OK, I'll drive another year and you go and do your thing with Lotus" which he did. Unfortunate really, because had he driven for us he would probably have won the Championship another two or three years. It's just fate, isn't it? We would have enjoyed having Jochen drive for us and we had the car that was going to be competitive

End of the road: the US Grand Prix was symbolic in many ways for both Hill and Chapman. Hill's accident effectively finished him as a top-line Formula 1 driver, while the race also marked the last time Chapman would field the Englishman in a works Lotus. In the shadow of Clark in '67 and Rindt in '69, '68 was the only year when Hill felt he was allowed to blossom and he pulled off a World Championship title. (The Dave Friedman Photo Collection)

but, unfortunately, the money swayed him.'

There were other reasons, aside from money, behind Rindt's decision to stay at Gold Leaf Team Lotus. Firstly, he had become a winner and, having got to know the Lotus set-up so well, it seemed a waste to move to another team and have to start the process all over again. Secondly, Chapman and Phillippe had a new 'secret weapon' car on their drawing board, which would be two wheel drive and much simpler than the 63, which had proved an unmitigated failure. Thirdly, Chapman offered to set up a team, which would be run by Ecclestone, for Jochen in Formula 2, which would be run under the Jochen Rindt Racing banner. However, there is little doubt that the handsome financial package offered by Chapman to encourage Rindt to stay was the final element which probably tipped the balance in favour of Lotus.

It had become clear well before the end of the season that the 63, and four wheel drive generally, was not the answer to Chapman's prayers and the decision was taken to abandon the project entirely. Looking back on that period, Eddie Dennis points out that it was not really surprising that such a complex car was never made to work. 'When you think there were six or seven blokes on the team and we were building four wheel drive cars to start with, and then going back to the 49s and preparing those, and still taking the four wheel drive cars as well ... when you look back, in those days, you just didn't have the time to do development. It was built, loaded into the transporter and the development was done at the circuit. We didn't have the personnel to go and test for two or three days in Spain, or whatever. We used to test in South Africa at the end of the year, but that was paid for by Firestone. Apart from that you might have done an odd quick run at Snetterton or Hethel just to give it a shake-down.'

One of the biggest problems facing Chapman for 1970 was what to do about Graham Hill. Although at one stage there was some doubt about whether, or at least when, he would be fit enough to return to top-flight Grand Prix racing, the friction between him and Rindt had proved counter-productive during 1969 and was not something the Lotus boss was anxious to repeat. Rindt, quite naturally, wanted to avoid it as well. Having proved quite conclusively that he was the faster of the two drivers, he was in a strong position to dictate his terms for 1970 and one of these was that he wouldn't have to drive alongside Hill. In addition, Hill's performances had been, with the exception of Monaco, lacklustre, something which had contrasted starkly with the raw speed and aggression of Rindt. Billy Cowe attributes this to the fact that Hill felt overshadowed and generally unloved. 'Jochen was that much better and his [Hill's] confidence suffered. At times he wasn't as happy within the team as he was at BRM. He'd been with BRMs quite a long time and had quite a lot of say with what went on there and this he would have lost with Colin.' However, Cowe for one was sorry to see him go. 'He wasn't an easy driver to work with, but he was a very nice chap. You couldn't but like the guy, although he was hard work at times!'

Rindt's resistance to having Hill alongside him for another season suited Chapman fine, for he already had his own idea about who Hill should drive for, as Rob Walker recounts with a laugh: 'Colin had been asking me for about two years if I'd take Graham. That was typical Chapman!' With Jo Siffert moving on to March for 1970, Walker was in the market for a driver, but could not afford a top-line or up-and-coming one as they were

Rindt decided to stay put at Team Lotus, tempted by the prospect of a new two-wheel drive car and attractive financial terms.
(Tim Considine)

Siffert and Walker: they understood each other but the Swiss had to move on if he was to further his Grand Prix career. Unfortunately for him, he chose to go to March...
(Ford Motor Company)

Lotus 49

The 49C was the final evolution of the type 49 design. Intended as a stop-gap until the 72 came along, it ended up being used for considerably longer than had been anticipated but still acquitted itself well against new-for-1970 designs. (Classic Team Lotus Collection)

simply too expensive. As a man who had won countless Grand Prix, two World Championships and knew all about getting his car to the finish to score points, Hill was the ideal choice. He would also be driving a car he was extremely familiar with, so it was like home-from-home for the Londoner. With Hill on the team it was hoped that it might also help attract a new sponsor, the wealthy stockbroker Jack Durlacher, who had been a partner of Rob Walker's in the team since 1966, having pulled out at the end of the 1969 season.

Without doubt, the edge seemed to have gone from Hill's driving during 1969 and he had gone from being a regular potential race winner to a regular points scorer but rarely challenging for victory, Monaco and Monza being the obvious exceptions. Perhaps the burden of having to try and compete with such a blindingly fast team-mate had worn him down.

The identity of the replacement for Hill was not decided until late in the year. In October, Colin Chapman had summoned a young Brazilian driver by the name of Emerson Fittipaldi, to Hethel, to discuss his meteoric progress during the 1969 season. The year had seen him go from a Merlyn Formula Ford car to a Lotus 59 Formula 3 car, in which he proceeded to trounce the opposition and win the Lombank Championship (there was more than one Formula 3 championship series), winning almost every race he finished by the end of the season. Such a clear margin of superiority was almost unheard of, in what was a series renowned for its hotly-contested races and evenly-matched cars. It attracted the attention of the Lotus boss, who was keen to help the 22 year-old Brazilian with an eye to the future.

The disarmingly honest Fittipaldi recalls that first meeting with a laugh: 'He called me to his office and I was shaking. It was unbelievable that it had happened to me. I was face-to-face with Colin Chapman! He asked me to drive for him in Formula 1. Actually, the first one to invite me to do Formula 1 was Frank Williams, just one month before Colin. And I said "Colin, I don't have the experience. I've just finished the Formula 3 Championship but I would love to do Formula 2".'

So, Fittipaldi signed a contract with Lotus Components before returning to Brazil for the winter. The programme laid down for him included all the European Trophy Formula 2 races in a Jim Russell-administered, Lotus Components-owned Lotus 59B, plus some international Formula 3 races and the possibility of several sports car races in a modified Lotus 62.

Chapman's eventual choice for number two did not come as a surprise to seasoned Formula 1 observers.

Colour Gallery 2

Double trouble: high wings front and rear were used for the 1969 South African Grand Prix, but the front wing was discarded soon after since it offered no perceptible advantage. This is Graham Hill, who finished second in the race. The height of the rear wing had to be lowered after struts failed during practice. (Len Konings/Robert Young)

For the 1969 Spanish Grand Prix, the works cars emerged with the biggest rear wings ever seen on a Formula 1 car. It was all to end in disaster ... This is Rindt shortly before his accident - the beginnings of a kink in the wing can just be seen. (Phipps Photographic)

With the exception of his victory at Monaco, Hill's 1969 season was a disappointment. His Championship aspirations evaporated with a run of poor form in the last half of the year, culminating in his accident at Watkins Glen. The lowest point was reached at the British Grand Prix, shown here, when Hill had to drive a car borrowed from Jo Bonnier. (Tony Walsh)

Introduced in June 1969, the Lotus 63 was Chapman's intended replacement for the ageing 49. Unfortunately, it did not meet expectations and so the 49s were pressed into service for almost another year … This shot shows the front driveshaft, which passed just above the driver's ankles, front differential unit, which distributed drive from the main driveshaft running to the driver's left side, and reversed DFV with gearbox under the seat back. (Tony Walsh)

Colour Gallery 2

The way it was: Herbie Blash pours a churn of fuel into Rindt's 49 (in the paddock at Silverstone during the 1969 British Grand Prix meeting). Unfortunately, due to the Austrian's flat-out battle with Jackie Stewart, consumption was higher than expected and the luckless Rindt had to make a stop for extra fuel before the end, having already pitted once to have an errant wing endplate removed. (Tony Walsh)

Jo Bonnier finally took delivery of his 49B in time for the 1969 German Grand Prix at the Nürburgring, but his relationship with the car would be brief. (Ed McDonough)

Lotus 49

Victory at last: all season Rindt had been looking as if he would notch up his first win of 1969. He finally did it in the penultimate Championship race at Watkins Glen.
(Phipps Photographic)

Introduced for the 1970 season, the 49C was the final evolution of the 49 with revised suspension geometry and uprights and smaller 13 inch diameter front wheels. (Keith Booker)

Final fling: Rindt's sensational victory at Monaco in 1970 was probably his, and the 49's, finest hour. Note the multi-layer wing borrowed from the 72. (Phipps Photographic)

Lotus 49

Emerson Fittipaldi made his Formula 1 debut at the wheel of a 49C at the 1970 British Grand Prix at Brands Hatch. He acquitted himself well, staying out of trouble to come home eighth. (Colin Taylor Productions)

Pete Lovely in R11 with its 'strange African nation' paint-job, pictured in practice for the 1970 British Grand Prix. (Colin Taylor Productions)

Colour Gallery 2

Another gritty performance from Graham Hill saw him take a well-deserved sixth place in the 1970 British Grand Prix, having come through from the back row of the grid in Rob Walker's 49C. (Colin Taylor Productions)

The final curtain: the Austrian Grand Prix of 1970 marked the last appearance of a works 49 in World Championship competition after a career spanning four different seasons. (Thomas Rohracher)

Lotus 49

From March 1969 onwards, Love ran 49 R3 in the South African Driver's Championship with high front and rear wings, as well as nose fins. However, the compressed air system to make the wings pivot used at the Grand Prix was dropped due to poor reliability, in favour of a more conventional cable-operated system. Love is pictured here on his way to victory in the Rand Autumn Trophy at Kyalami in April 1969.
(Adri Bezuidenhout)

Right – Love's car looked distinctive in its Gunston Cigarettes colour scheme of orange, brown and gold. Here it's pictured after the ban on high wings in international racing was introduced in 1969. Love's lead mechanic, Gordon Jones, is working by the front wheel.
(Adri Bezuidenhout)

Below – For 1970, the first year in which Charlton raced his 49C, it was painted in this rather odd-looking combination of colours. Major sponsorship was still a little way off, a large number of smaller suppliers being responsible for keeping the car on the tracks.
(Adri Bezuidenhout)

Right – For 1971, new South African Driver's Champion Charlton managed to attract sponsorship from the giant United Tobacco company, through its Lucky Strike brand of cigarettes. Consequently, the team and its cars' colours changed to red, white and gold. This is R8 as driven by Peter de Klerk during the latter part of the 1971 season after Charlton took delivery of his Lotus 72. (Jannie van Aswegen)

After its 1967 British Grand Prix practice crash, R1 was rebuilt with a larger oval hatch, offering access to the pedals and the fuel filler, and a strengthening baffle was added in the forward tub in place of the original brace. The path of this baffle can be clearly seen by the line of rivets running down the tub from the front of the air-deflector windscreen, underneath the rocker arm fairing and down to and behind the lower front radius arm mounting point. This modification was later carried out on R2 and these two cars were the only ones to receive such treatment. Longer front rocker arm fairings were also added at the same time. This shot was taken during the 1968 Tasman series. (Ford Motor Company)

Taken in January 1968 during the Tasman series, this picture of R2 shows the separate filler cap has been retained. However, a strengthening baffle has been added, as have longer rocker arm fairings, and a new, overlapping skin has been sewn back on the forward section of the tub. (Ford Motor Company)

A classic side profile, taken in the pits at Monza, showing how R2 finished the 1968 season. (Aldo Zana)

When R3 arrived back in the UK in January 1981 after having been acquired from John Love by Michael Lavers, it was painted a rather gruesome powder blue and in quite a sorry state. However, all the basic components of the car were there, including the ZF box. (Michael Lavers)

Colour Gallery 2

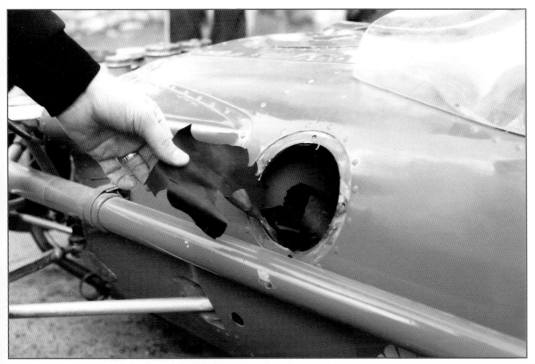

Left - When Doug Nye took delivery of R3 in 1981 in order to arrange its restoration for new owner Mr Hayashi, the FPT rubber fuel cells were perished and came out in handfuls like dried leaves. (Doug Nye)

The ex-Love chassis, shown here in 1981 before its restoration, is unique in that it is the only surviving 49 still in original specification, including the triangular oil tank, fuel pump embedded in the side of the tub and early (originally troublesome) lower rear radius arm mounts. (Doug Nye)

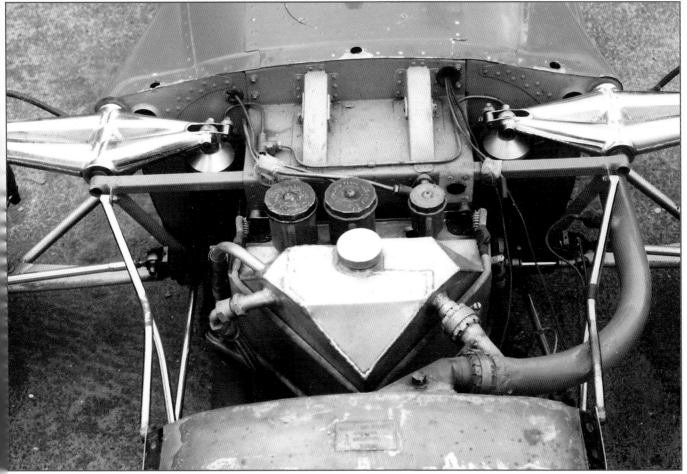

Lotus 49

Opposite top - The Lotus 49 with a 100 per cent winning record! This is Jim Clark in R4 during the only race it started, the South African Grand Prix of 1968. The car was destroyed by fire in March of that year. (Ford Motor Company)

Now owned by the National Motor Museum in the UK, R3 – shown here at the 1997 Goodwood Festival of Speed – still makes regular appearances at historic and other events throughout the country. (Nigel Snowdon)

Opposite bottom - Now owned by the French collector Jacques Setton, R6 has - sadly - not been seen in public for many years and is currently understood to be in storage in Switzerland. It is pictured here during the late 1980s at the time of the preparation of a book on the Setton Collection. (Alberto Martinez)

Colour Gallery 2

189

Lotus 49

Now resplendent in Gold Leaf Team Lotus colours once more, R8 appears regularly at historic race meetings in Australia. This is Colin Bond seen on his way to victory at the Eastern Creek circuit near Sydney in 1992. (John Dawson-Damer)

The Chapman family, through Classic Team Lotus, still owns R10. It is a regular feature at historic events in the UK and Europe, including the Goodwood Festival of Speed, where it is pictured in 1997, with Joaquin Folch at the wheel. (Nigel Snowdon)

Colour Gallery 2

Unique: after the rebuild of R1 in July 1967, Jim Clark's R2 was the only 49 to have a separate fuel filler and access hatch in the scuttle - all other 49s were built with an oval hatch which covered the fuel filler. This unique feature is found today on Pete Lovely's R11 pictured here in 1998, one of a number of attributes which the author believes proves conclusively that this tub is Clark's 1967 Zandvoort winner. (Pete Lovely)

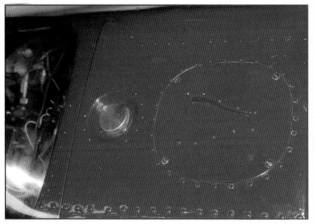

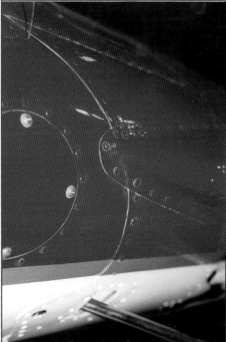

Right - Unique: the modification shown, where a new skin was stitched on overlapping the existing one, following the addition of a strengthening baffle in the forward tub, was made only to two cars. These were R1, which was later resurrected as R9 and then written off by Rindt at Barcelona in 1969 and R2, driven by Clark in 1967, Oliver in 1968 and then sold to Pete Lovely, who photographed this particular feature of his car in 1998. (Pete Lovely)

Below - R11 as it is today, at the 1998 Monterey Festival. (William Taylor)

The Rob Walker car, prior to being auctioned in July 1999. It is shown in a hybrid 49B/C specification, with the high wing from 1968 and early 1969 but the smaller front wheels and revised uprights that were introduced for the 1970 season. (Classic Cars)

The Donington Grand Prix Collection is home for R12, pictured here in 1998. At some stage in the years since its first appearance, it has gained a 49B rear end, with steel crossmember, instead of tubular subframe.
(Michael Oliver)

The main change on the 49C was the introduction of smaller 13 inch front wheels and new fabricated front uprights. (Classic Team Lotus Collection)

(... continued from page 176) John Miles had made a good impression on Chapman and the rest of the team with his efforts in the recalcitrant four wheel drive 63, and so it seemed natural for him to make the step up from occasional third car driver to number two driver for 1970. His run of victories in Formula 3 and his versatility (he had also driven the Lotus 62 to several notable race and class wins in 1969) demonstrated that he had the potential to make it to the top and, by this stage, he had also begun to garner a reputation as a competent test driver. Not only that, but he actually enjoyed test driving, something which was an anathema to many of the emerging generation of Grand Prix stars, including Rindt: 'If I had any success in racing it was due to testing and due to - although I wish I knew then as much as I know now - understanding, or trying to get an understanding of, why things made cars go faster. A lot of drivers weren't interested in all that, just in racing.'

While all this was going on, the race shop was a hive of activity. With the failure of the four wheel drive car, Chapman and Phillippe took the decision towards the end of the 1969 season to build a more conventional two wheel drive replacement for the 49 for 1970. Work on this car – designated the Type 72 - did not start until late in the year and it was obvious that it would be necessary to use the 49s for at least the first couple of races of the season. Consequently, a new specification evolved known as the 49C, the main purpose of which was to take advantage of the new breed of 13-inch front tyres being developed by Firestone and other tyre suppliers.

Switching from the 49B specification 15 inch wheels and tyres required alterations to the front suspension geometry, and the manufacture of new front uprights and wheels. The design of these modifications was entrusted to Mike Pilbeam, Chapman and Phillippe being too occupied with the 72 to have more than a supervisory role in this particular project. Pilbeam would later go on to work for BRM and then found his own racing car manufacturing company, which still trades successfully today. Also working on the revisions with him at the time was Geoff Ferris, who was another who went on to greater things, making his name with the Penske team, where he still works.

Once the fabricated front uprights and new four-spoke wheels had been produced, they were fitted to Rindt's regular mount, 49B/R6, and the car was shipped to Kyalami for some winter tyre testing for Firestone, which took place early in December 1969, with both Rindt and Miles doing the driving. Apart from a brief shakedown at Hethel before packing the car up, this was also the first real opportunity for Miles to get some time behind the wheel of a 49, as well as becoming familiar with the realities of life as a lowly-paid number two driver: 'I remember it as being unbelievably expensive, because in those days I had to pay my own hotel bills! I enjoyed myself actually. I liked nothing better than being left alone to grind round - the thing that Jochen hated - and finding out what made the car go quicker.' His initial reactions were favourable, particularly bearing in mind that his sole experience of a Formula 1 car to date had been the Type 63: 'The thing about the car was by that time it was pretty de-bugged. Like most of Colin's stuff it took some time to get so it didn't fall apart! By this time the 49 was pretty robust and for me it was a 100 per cent reliable car, unlike the 72!'

As well as being able to familiarise himself with the 49C, the testing at Kyalami also enabled Miles to spend a bit of time with his new team-mate, away from the highly-charged atmosphere of a Grand Prix weekend: 'He was a bit like Bernie actually. They got on very well - Bernie was his manager. Jochen was a very aggressive character and he was thought of by some as being quite rude and short with people. But I always found him very fair to deal with. He was out in Lausanne, so he was in the jet-set really but on the occasions we did meet he gave me some good advice. I got a b********g from him once for getting in his way at Monaco! But generally he was OK. He wasn't exactly supportive, 'arm-round-the-shoulder' but at the same time - I wasn't a threat to him - he didn't go out of his way to be unpleasant. He had quite a good sense of humour but there were no shades of grey with him.'

The new modifications and the smaller diameter wheels and tyres worked well. In addition, the new set-up also virtually cured the age-old problem inherent in the 49 design, that of instability under braking. This was most memorably characterised by the antics of Jo Siffert in the Walker car, whose weaving under braking was sufficient to encourage many rivals to concede a corner...

The other works 49, chassis 49B/R10, was also rebuilt to 49C specification. This was the car Hill had crashed at Watkins Glen. Although it had been quite badly damaged, it was repairable, so damaged sections were reskinned and components replaced to restore it to A1 condition. While all this activity was taking place, the car that Jo Bonnier had crashed at the Gold Cup meeting at Oulton Park in August was being rebuilt, prior to being delivered to the South African, Dave Charlton, for use in the South African Drivers' Championship. After the fabricators had repaired the tub, the car was assembled in late October and the first few days of November by Charlton's former mechanic Richie Bray, who then left to spend time in his native South Africa, working on David Piper's Porsche 917 in the Kyalami 9 hours.

Although this car is described in some Lotus history books as being delivered in 49C specification, this seems unlikely, as the new front uprights were not actually drawn by Mike Pilbeam until mid-to-late November (*i.e.* after Bray had completed building the car up) and would almost certainly have needed to be tested in the winter Firestone tyre tests before being fitted to any car delivered to a customer. Although Charlton is insistent that the car was delivered on 13 inch wheels (*i.e.* in C spec) from new, the most likely explanation is that the 'C' spec parts were sent out separately by the works after the car had been despatched. Certainly, by the second round of the South African Drivers' Championship at the end of January 1970, the car was running the new uprights and 13 inch wheels. Therefore, unusually, a privateer was the first to race a Lotus 49C, since the works team did not take part in its first event until March 1970.

The final twist to what had been a year full of turmoil was awaiting Chief Mechanic Leo Wybrott when he returned from the Kyalami testing. Racing Manager Dick Scammell sacked him, along with a number of other long-serving mechanics, including Dennis London, Arthur 'Butty' Birchall and Billy Cowe, while Bob Dance decided it was time to leave, too. It was a case of out with the old, in with the new, a kind of symbolic purging of the Old Guard as the team moved into a new decade. With their departure went a great deal of experience, but also perhaps a groundswell of resentment which had built up throughout the 1969 season due to the demands placed on them and the hours they were expected to work. Cowe sums this up by saying 'We were probably getting a bit "bolshie" and a bit too big for our boots!' Although Jim Endruweit attempted to intervene by speaking on the mechanics' behalf directly to Chapman, it was to no avail. Wybrott was quickly snapped up by McLaren (and went on to have a tremendously successful period with them which brought many wins and two World Championship titles), while Dance moved to March and thence to the STP-McNamara operation, which was running a semi-works March for Mario Andretti.

Cowe, who had been with the team since 1964, and had worked on the Formula 1 cars of the likes of Jimmy Clark, Pete Arundell and Mike Spence before Hill, was sorry that his association with Chapman had come to an end, for he had great admiration for the Lotus boss. 'He was a very hard taskmaster. It depended on the day a bit really – on what he had on his mind and how well things were going. If things were going well he was the best guy under the sun, but if they started going wrong it wasn't the best idea to be too close! But very charismatic: a guy you would do anything for. He could talk you into doing anything, charm the birds out of the trees and more. He'd get you to do no end of work, and you felt happy doing it. In fact, you were disappointed if he didn't ask you for more sometimes!'

The final change that took place in readiness for the 1970 season was the confirmation of the appointment of Peter Warr as Team Lotus Competitions Manager in place of Andrew Ferguson, who had resigned midway through 1969 following a series of disagreements with Chapman. With a fresh team, a new car and a settled driver line-up, Team Lotus entered the new season full of optimism.

Chapter 20
The Swan Song

Towards the end of February it was confirmed – although many had considered it merely a formality – that Players would continue to sponsor Team Lotus through its Gold Leaf cigarettes brand. The good news for Players was that the restrictions on the size of advertising decals had also been relaxed, so that the Gold Leaf lettering on the cars could now be significantly larger than it had been before. Aside from the Formula 1 team, the main thrust of Gold Leaf Team Lotus was to be in Formula 3, with the Formula 2 team to be run as a semi-works operation by Jochen Rindt Racing and not sponsored by Gold Leaf.

After the comings and goings of the winter, the mechanics were reshuffled, with Gordon Huckle, who had worked on Clark's car during the 1967 season, returning as Chief Mechanic. Eddie Dennis continued as lead mechanic on Rindt's car having established a good rapport with the Austrian, assisted by Herbie Blash. Dave 'Beaky' Sims was in charge of Miles' car, assisted by Derek 'Joe 90' Mower, while Graeme 'Rabbit' Bartels was responsible for the third car when it ran.

At about the same time as the Gold Leaf announcement, Graham Hill was having his first drive in a Formula 1 car since his Watkins Glen accident. This consisted of half a dozen laps at the wheel of the Rob Walker 49C at Kyalami, the session being cut short when the fuel pump drive broke. As he was still unable to walk without the aid of a stick and was in considerable discomfort still, there was some doubt as to whether he would be able to endure the rigours of a Grand Prix weekend. As a result, Rob Walker had asked the popular Lancastrian, Brian Redman, to stand by to take over the drive, although this might have been something of a baptism of fire for Redman because he had never sat in the car, let alone driven it before going out to South Africa.

The Lotus 49 presence at the opening round of the World Championship in early March 1970 was as large as it had been the previous year, with five cars entered. In addition to the two works cars of Rindt (49C/R6) and Miles (49C/R10) there was the Walker car for Hill, Charlton's recently acquired 49C/R8 and Love's by now elderly 49/R3. Hill's car differed only slightly to the works models in that it had smaller master cylinders fitted in order to reduce the braking effort required in a bid to reduce the strain on Hill's legs and feet. It also had the team's own 'pannier' fuel tanks outrigged on the side of the car, which had been used several times during 1969 to extend fuel capacity. All the 49Cs present had an extra modification which was to have twin oil coolers fitted on top of the oil tank under the rear wing, a measure designed to combat the high ambient temperatures found in South Africa at that time of year.

While the first Grand Prix of a new season is always of great interest, this event was particularly fascinating because there was a real glut of new cars on the grid. These included the upstart new March team, which equalled the Lotus entry with five cars, including one for the reigning World Champion Jackie Stewart. Other teams with new cars included Ferrari, (also with a new flat-12 engine), BRM, Brabham, Matra and De Tomaso. In fact, Gold Leaf Team Lotus was the only major team that did not manage to get its new car ready in time.

John Miles in R10 on his way to a first points finish in the South African Grand Prix at Kyalami. Note the twin rear oil coolers used for races in hot climates and the vent for cockpit air. (Len Konings/Robert Young Library)

195

Lotus 49

Practice for the race took place on the afternoons of Wednesday, Thursday and Friday, with the race scheduled for Saturday. The lack of a new car did not seem to be a problem for Rindt, who was second quickest on Wednesday, fourth quickest on Thursday (the fastest of the three sessions overall) and then set the best time in the final session on Friday to line up fourth on the grid. Miles was taking his time getting accustomed to the 49C and lined up 14th, one place behind local hero Dave Charlton, who was using his local knowledge as well as the fact that he had a car to the same specification as the works team to his advantage. Despite the pain he was obviously in, the courageous Hill still managed to set a time which put him 19th on the grid, while Love in 49/R3 (making its third consecutive South African Grand Prix appearance) scraped in on the back row in the penultimate place on the grid. Love was hampered by an engine failure on the first day, as a result of which he missed the second day's practice and only returned for the final session. In this last practice, Redman was given a few laps in the Walker car in order to qualify as a reserve driver should Hill decide that he wasn't up to racing.

Race day dawned and Hill resolved to race, much to Redman's disappointment: 'I was taken out to drive in case Graham couldn't manage - but being Graham, there was no way he was NOT going to manage!' On the first lap of the race, Rindt collided with Brabham, was launched into the air and spun at the first corner. When the field came by at the end of the lap, he was languishing down in 16th place and driving furiously to make up lost ground. What followed was a superb recovery, which took him up to 9th place by lap 13, but rather petered out after that. However, the natural rate of attrition in the hot conditions, and it being the first race of the season with all the consequent new-car niggles, saw him move up to fifth place by lap 60 (three-quarters distance) a position he held until the 72nd lap, when he coasted in to retire with a blown engine.

Miles had been troubled from the start by a fuel leak, which meant that he was sitting in a pool of petrol. However, he got on with the job and managed to pass the number two Tyrrell car of Johnny Servoz-Gavin and then the privately run 49C of Charlton. Like his team-mate he then benefited from the retirements of others (including, ironically, Rindt's) to end the race in fifth, his first ever finish in the points. Hill put in probably the grittiest drive

True grit: despite being barely able to walk, Graham Hill climbed aboard Rob Walker's 49C in South Africa and drove to a heroic sixth place finish. Here he is pictured ahead of Henri Pescarolo (Matra), Rolf Stommelen (Brabham) and Piers Courage (De Tomaso). (Nigel Snowdon)

The Swan Song

of his career. From his position of 19th on the grid he worked his way up through the field, although some measure of the attrition rate is the fact that he passed only three cars during the entire race. This is not to detract from the strength of character it must have taken to get to the finish, for his constant discomfort was obvious. His courage was rewarded by a sixth place finish and a well-deserved point.

Local hero Charlton ran ahead of both works cars while Rindt was recovering from his spin and showed Miles the way round his 'home' circuit until the works driver got the better of him on the 49th lap. By three-quarters distance he was up to 7th and looked all set for a finish in the points until his left rear Firestone threw a tread. When he pitted to get the wheel changed, his mechanics were unable to remove the offending item and, when they did manage to do so, the engine refused to start, forcing his retirement. This is an episode Charlton still recalls ruefully even today, for it is not often that a local driver had the opportunity to demonstrate his talents in a car which was in the latest possible specification. 'We had Firestone tyres in those days, and they weren't slicks. They had a funny old tread on them. Five laps from the end, the left rear began to vibrate and it blistered and that was it. We just didn't have the equipment to do a quick pit stop.' It was a sad end to an impressive drive against better-resourced works opposition. The other 'local' driver, Rhodesian John Love, drove a steady if unspectacular race into eighth place, two laps down on the winner, Jack Brabham. What this race clearly illustrated was that what the 49Cs might have lacked in out and out speed (for the basic design was nearly three years old) it made up for in reliability and there is little doubt that, by this stage, it was a reliable and bug-free car.

The *Daily Mail* Race of Champions was the next event on the teams' itinerary. Although they had hoped to have the 72 ready for this race, it was still not finished, so a singleton 49C was entered for Rindt, Miles' car not having returned from South Africa in time. Hill was entered in the Walker car, while Pete Lovely returned to the place where he took delivery of his 49B as the first part of his 1970 Tour of Europe, his car still in B spec. For a number of reasons, he had decided not to update his car to 49C spec. As it turned out, this proved to be something of a false economy. 'The reason that I didn't was that it was going to cost me US$1800 more [approximately £750 then, or £6600 at late 90s values] to buy all of the kit – the uprights and the rocking levers and the wheels – and that was a big chunk of money for me. So I talked to my friend Bruce Harre who was the Firestone Chief Development Engineer and he said that he would continue to find and supply me with the 15 inch diameter front tyres. The problem was that they had destroyed all the moulds and stuff, I mean they just got rid of them, scrapped them. Everybody else had gone to 13 inches, so the main production of Formula 1 tyres for the fronts of the cars, Goodyear and Firestone, went down to 13 inches. So he had to dig out some moulds that were done for Le Mans cars and sports cars and they actually built tyres specifically for my car! They were supposedly made with a softer compound, softer than what the sports cars would have run, but we still didn't have the proper grip and stuff.'

The 49Cs, both of which were back in their normal single oil cooler trim, were not overly impressive in practice, Rindt taking fourth place on the grid and Hill eighth. The Englishman was still suffering from a damaged nerve in his right foot, which made moving that foot off the throttle and onto the brake a rather slow process. Lovely was the slowest of the entries, some five seconds behind pole-man Jackie Stewart's March.

Rindt had a rather uneventful and lonely race, benefiting from the retirement of Oliver's BRM, which had run a strong second and then, right at the end, leader Jack Brabham's pit stop to change a black box. As a result, he ended up a rather flattering second. Hill came in fifth, finishing with only four gears after having lost second

Local hero Dave Charlton thrilled his home crowd by running ahead of the works car of Miles for much of the race while the Englishman was coming to terms with sitting in a pool of fuel. Miles steadily moved up through the field and passed Charlton on the 49th lap out of 80, while Charlton was forced out with an engine that refused to start after a pit stop. (Len Konings/Robert Young Library)

197

A lone 49C appeared for Jochen Rindt in the Race of Champions at Brands Hatch. This shot clearly shows the new fabricated front uprights and hubs designed to take the new size 13 inch front wheels. (Steve Wyatt/ Quilter House)

Despite suffering from a damaged nerve in his right foot, Hill drove the Rob Walker car with great gusto at the Race of Champions, finishing fifth. (Nigel Snowdon)

Pete Lovely began his 1970 'European Tour' at the Race of Champions but finished up in the bank at Paddock early in the race. (Ford Motor Company)

gear on lap six out of 50. Lovely's race was altogether more eventful, ending up in the armco at Paddock Bend due, as the driver candidly put it, to 'brain fade'. However, it was also partially to do with the quality of the tyres he was supplied with. 'It understeered off the road and we busted the radiator, the nose and the wing so we had to get all that patched up before we could go on and do the rest of our 1970 thing.'

While the nose and wings were relatively easy to replace, the radiator was a more expensive proposition, as Lovely recounts. 'Well, it would have been had we bought stuff from Team Lotus but we went to the radiator people, Serck, and took them the busted radiator. The guy said "Well, we can't do this. The way we build these radiators is to Lotus's drawings. We can't supply it to you direct, you have to buy it through Lotus". And we're saying "But gee, we know their prices are going to end up being more expensive than if you just build it and hand it to us". And the guy said "Well, maybe we could modify it a little bit" So they built the core out of two separate pieces with a space down the centre. It wasn't exactly to the Lotus drawings so they sold it to us considerably cheaper than had we bought it from Lotus! We were on a shoestring and we had to do it the best way we could. So, it was expensive, but it didn't kill us.'

With a non-Championship race at Salzburgring over the Easter weekend cancelled due to bad weather, the next Formula 1 race on the calendar was Round 2 of the World Championship, the Spanish Grand Prix. After the previous year's race at Barcelona, this had returned to the tight and twisty, but apparently safer, confines of the Jarama circuit.

On the Wednesday before the Easter weekend, the Type 72 had finally been unveiled to the press, John Miles undertaking the initial shakedown test. Placed alongside the 49, it was considerably sleeker, the siting of dual hip radiators alongside the driver on either side of the cockpit enabling a wedge shape reminiscent of the 1968 Indy Type 56s, or the abortive Type 57/58 car, to be achieved. This resulted in a car with a smaller frontal area than the 49, but which produced about as much downforce as the 49s had with the massive high wings.

Four cars were taken to Jarama, with two 72s joining the regular cars, 49C/R6 - which was the works team spare - and 49C/10. Disappointed not to have taken delivery of their 72, the Walker team were once again reliant on their faithful old 49C/R7. Chassis 10 had been entered under the banner of Garvey Team Lotus for Alex Soler-Roig, a good friend of Rindt and the son of a wealthy surgeon who had looked after the Austrian in the wake of his huge accident in the Spanish Grand Prix the previous year. The drive was arranged in order to try and raise some extra money for the team, as Peter Warr explains. 'It was only the third year of the John Player sponsorship, which had started off as a fairly insignificant sum of money - although it was important in the total scheme of things - and we were short of money to do

The Swan Song

everything we had to do. Then we got caught up in the very expensive modifications to the 72s. It would be wrong to say they cost as much as the cars did in the first place, but they were very expensive and we needed some money.

'So the way in which we achieved that was Colin sent me out to Spain to do a deal with Soler-Roig, who got his funding through the state-owned fuel company, Repsol and Garvey sherry. It was a deal that started out being fairly straightforward but then became complex as we swapped backwards and forwards between the 72s and the 49s and poor old Soler-Roig more often than not finished up as tail-end-Charlie. The deal originally was for a 49, but that it would eventually be a 72. Then the 72s didn't perform and had to be modified so it all became a bit complex.'

A crash for Rindt in private practice in the 72 caused a rethink on the use of solid brake discs and resulted in ventilated ones being fitted to his car for when official practice began. At this stage, these modifications hadn't been completed on Miles' car, so he had to practice in 49C/R6 on the Friday. For Saturday he was back in the 72 but was unable to get up to pace quick enough and did not qualify. Soler-Roig was rather out of his depth and, although he was only a few tenths slower than Hill, he also failed to qualify. Hill only made the grid by virtue of his being a graded driver, his practice thwarted by an accident, which damaged the rear of the car. Miles had turned in a time faster than Hill, but fell foul of the rule which guaranteed a start for ten graded drivers plus the six fastest non-graded drivers. Rindt slowly got up to speed, but the cars were rolling too much in the corners and suffering from electrical gremlins, which were affecting the engines. Eventually, he lined up on the outside of the third row, in eighth position on the grid.

A possible reprieve for the four non-qualifiers on the day of the race was turned down at the last minute and, having assembled on the grid, they suffered the frustration of being pushed away again and back into the paddock. As a result, only two Lotuses started, and only one of these – Hill's Walker car – was a 49. The race was rather dull, with Stewart leading from start to finish and never really being challenged after the retirement of South African Grand Prix winner Brabham at two-thirds distance. While Rindt went out early with engine maladies, Hill drove a superb race from the back end of the grid to finish fourth. At one stage, he had even managed to get ahead of both Bruce McLaren and Mario Andretti - who finished second and third - but problems with his legs resulted in him falling back towards the end. However, he still claimed another three points to add to the one he had scored at Kyalami and, at this stage, was the best-placed Lotus driver in the Championship standings!

Only a week later, many of the teams that had been present at Jarama assembled at Silverstone for the Daily Express/GKN International Trophy race. The entry was open to Formula 1 and Formula 5000 cars, giving the first real opportunity that year to compare the performance of the two types of car. Gold Leaf Team Lotus brought a pair of 72s along for Rindt and Miles, as well as a 49C as a spare 'just in case'. The only 49s in the field were the Walker 49C and Pete Lovely's 49B. Hill's car was resplendent in new decals advertising the food firm, Brooke Bond Oxo and the team had been renamed Brooke Bond Oxo Racing with Rob Walker as a result, while Lovely's car had been repaired after its Brands shunt.

Hill ended up seventh quickest, on the outside of row two, while Lovely was down in 20th, albeit ahead of the works 72 of Miles and only on the row behind Rindt. The two Gold Leaf cars were well down because they had arrived too late to get any meaningful practice in on the Friday (sound familiar?), and Saturday's practice was wet.

The race was held over two heats, the winner being decided on aggregate times. In the first, neither of the 49s had a particular good time, Hill coming in to the pits to investigate a loose steering column and Lovely stopping to have his goggles cleaned after following a particularly oily Formula 5000 car. They eventually finished two and

Wedge wonder: Chapman and Phillippe's replacement for the 49 was the type 72. Retaining the proven DFV engine, the car's hip radiators allowed the wedge shape which had first been used on the 1968 Indy cars to be revived, with devastating effect. (Ford Motor Company)

Lotus 49

No April Fool: revealed to the press on April 1st 1970, the Lotus 72 was a stunning looking car. Angular where the 49 had been rounded, it bore little resemblance to its predecessor in technical terms either, with torsion bar suspension and full anti-squat and anti-dive. (Ford Motor Company)

Out with the old and in with the new: the 49 looks positively bulbous next to the sleek 72 but it would still win a Grand Prix in 1970... (Ford Motor Company)

The Swan Song

three laps down respectively on the winner, Amon. In the second heat, Hill tigered to a strong fourth, while Lovely again finished three laps down. Overall, they were a disappointing 9th and 13th places, reflecting their problems in the first leg. Meanwhile, the 72s had another difficult race, with unpredictable handling troubling both drivers. Rindt eventually retired in the second leg with ignition trouble after finishing a distant 5th in the first heat, while Miles finished 11th in the first heat and retired with a broken throttle cable in the second.

The problems that were being experienced with the new car effectively extended the life of the 49Cs further – much more than had been intended when they were originally modified. Graham Hill tried a 72 after the Silverstone meeting and his views concurred with those of Rindt and Miles – the car was basically fast, with good roadholding and braking capabilities, but it was hard to drive on the limit. Consequently, a decision was made to modify the 72s by removing the anti-squat built into the rear suspension and to modify suspension mounting points on one of the cars in time for the next race, the Monaco Grand Prix.

The work on the 72s was not completed in time for Monaco, so the team decided to take the two 49Cs with them to use in the race, with a 72 as a spare car. The only modification to the cars was the addition of the Lotus 72 'tri-wing', which had been fitted to all the 49Cs including the Rob Walker car, while the twin oil cooler set up was also fitted on all three cars.

The pressure was really on Miles following his non-qualification in Spain. There was the additional problem that he had never driven on the Monaco track, even in Formula 3. 'I found Monaco was a bit of a culture shock. If you've never been to a street circuit before of that nature - although I had been to Pau in Formula 2 - you've got so much else to think about besides hammering round somewhere you are reasonably familiar with. Initially, you tend to get rather tense because you are not quite sure ... you need such a rhythm there to be so precise and yet so relaxed, so confident to be able to find the extra two or three seconds you need. You get very tied up with other cars in qualifying. You can't overtake, there is so much traffic. It was all a bit overwhelming, really'. These comments closely mirror those of Jackie Oliver about his experience in 1968, showing what a difficult circuit it was to come to terms with.

Miles' problems were accentuated by the fact that the track could only accommodate 16 starters but there were 21 entries. With the ten seeded drivers guaranteed a start, this left 11 drivers fighting for the six remaining spots on the grid and it was always going to be difficult with unseeded drivers of the calibre of Pescarolo, McLaren, Courage, Siffert and Oliver.

Meanwhile, Rindt had been going very well in the first practice on Thursday before his engine blew, forcing him to sit out the rest of the session. On the Friday, he did just a few laps in the rain to make sure the DFV was working OK and then sat out the rest of the session. Dave Sims recalls he was not keen on going out in the rain. 'He didn't want to practice in the wet, thought it was a waste of time.' Although the third session on Saturday was dry, overnight changes to camber angles rendered his car far

On his way to fourth place, Hill splashes through the river of water and fire extinguishant which flowed across the track after the fiery accident which engulfed the Ferrari of Ickx and the BRM of Oliver in the Spanish Grand Prix. (Ted Walker/Ferret Fotographics)

more difficult to drive and he was two seconds slower than he had been on Thursday. Nonetheless, he had qualified, albeit down in an unaccustomed eighth place on the fourth row. On Saturday, Hill had set a time that would put him twelfth on the grid, on the sixth row. However, soon after this, he had a monumental accident on the climb from St. Devote after hitting a damp patch and this badly damaged the front of the tub and wrote off the suspension, rendering the car *hors de combat* for the race.

Salvation was at hand in the form of Miles, who struggled in the final unseeded drivers' qualifying session and did not make the grid. This was particularly galling for the Englishman, for he had lapped somewhat quicker than the BRM of Rodriguez, who was in last place on the grid, but was a seeded driver and therefore guaranteed a start. A second consecutive non-qualification was just the kind of set-back Miles didn't want. 'I was really, really depressed, I must say. I couldn't get to grips properly with the circuit and I was sort of caught really because I didn't want to make a mistake and plant the car, and on the other hand I'd got to kind of judge it, I needed a few

Hill settled well into the Walker team, striking up an instant rapport with team principal Rob Walker. Considering they had an outdated car, theirs was a reasonably successful season, particularly the first half, when they invariably finished in the points. (Ted Walker/Ferret Fotographics)

Lotus 49

The tension mounts: Second placed man Denny Hulme speeds by while Eddie Dennis holds out the pits board to Rindt telling him that he is only two seconds behind the New Zealander. Mechanic Dave Sims (left) urges the Austrian on while Herbie Blash (right) looks on impassively. (Dave Sims Collection)

results under my belt. It couldn't have come at a worse time really. What I needed was a couple more races in the 49, and reliable finishes to get the confidence. So it was a pretty low point, actually.'

As soon as it became clear that Miles was not going to be starting the race, Rob Walker approached Chapman to ask if he could borrow 49C/R10 for Hill to use the next day, which the Lotus boss duly agreed to do. It was a nice touch of irony, for this meant he would have driven the same chassis at the Monaco Grand Prix for three consecutive years, since R10 was the 1968 car R5, which had been renumbered at the end of that year. This set in train a frantic period of work on the part of the Walker mechanics, who had to re-spray the car Walker Blue and set it up to fit Hill's requirements in time for the race the next day.

Come race day, Rindt was in a bit of a sulk. Fed up that the new 'wonder-car' which was supposed to be making him World Champion wasn't performing to expectations and disappointed with his lowly grid position, he was not in a good frame of mind for the race. It could not have been further from his mind that he was about to drive what was, without doubt, his greatest ever race and would go down in the history books as one of the all-time great Grand Prix races. Hill, in the hastily repainted R10, was forced to start from the back of the grid, as he was racing a car that he had not practised in.

Pole-man Stewart jumped into the lead at the start and it looked as if it would be another repeat of his dominant Jarama performance. Rindt, having got the better of his fellow fourth-row man Pescarolo, found himself passed by the shrill V12 Matra on the third lap. The natural rate of attrition from the Monaco circuit ensured that he moved up the order without actually overtaking anyone, to the point where he was fifth by lap 30. It was around this stage that he finally seemed to wake up to the fact that he had a very good-handling car and that he was not yet out of contention for a win, or at the very least a top three finish. He passed Pescarolo on lap 36 into the Gasometer hairpin and within five laps was past Hulme - who was suffering from having lost first gear - setting a new lap record which was one and a half seconds faster than he had been in practice. This elevated him to third place, behind Amon's March and Brabham, who had inherited the lead when Stewart's engine went sick. With only 20 of the 80 laps left, Amon went out when a bolt dropped out of his rear suspension.

Suddenly, there was only one man – Brabham – lying between Rindt and victory. The Austrian sensed there was the potential for victory as Monaco was punishing on both cars and drivers and tired drivers made mistakes ... While he was certainly very experienced and canny, Brabham, at 44, was the oldest man in the field and not immune to a slip-up towards the end of a long race. From 13 seconds with 20 laps to go, Rindt whittled the Australian's advantage down to 10 seconds with 10 to go. In these final ten laps, Rindt redoubled his efforts and, on some laps, was taking two seconds out of Brabham's lead. With four laps to go, the leader came upon Siffert's fuel-starved March, which was weaving through Casino Square in a bid to pick up the remaining dregs from the tanks, and suddenly his lead could be measured in car lengths.

As they went into the final lap the gap was only 1.5 seconds but still it seemed certain Brabham would hang

The Swan Song

Closing in: 'Black Jack' heads along the promenade towards the Gasometer hairpin for the last time as Rindt hurls his Lotus out of Tabac in pursuit. (Phipps Photographic)

on. As they came down the promenade towards the Gasometer hairpin for the last time, the Lotus was closer but not close enough to make a serious overtaking manoeuvre. The final distraction was the cars of Peterson and Courage, who were keeping well to the left in order not to get in the way of the battling duo. This forced Brabham onto the inside line for the hairpin and, distracted, he left his braking too late, locked up all his wheels and slid straight on into the straw bales.

Rindt could hardly believe his luck and crossed the line to take a sensational victory. Even the man with the chequered flag was caught by surprise, for he was looking for Brabham and did not even acknowledge Rindt as he passed! To put Rindt's charge in the last few laps into perspective, he set a new lap record on the final lap of 1.23.2, which was eight-tenths of a second faster than Stewart's pole time and two seconds faster than the existing record. Not bad for an obsolete car ... This had also preserved Gold Leaf Team Lotus's amazing 100 per cent record in Monte Carlo, having won the races in 1968 (the first year of Players' involvement) and 1969.

To many of those present, Rindt's drive confirmed him as the fastest driver in Grand Prix racing, even if Stewart might be considered the top driver overall. 'Personally, this is the best race I saw him do' recalls Jabby Crombac. 'There is no doubt about that. Jochen was a very moody guy. He was upset that the 72 wasn't going as well as he was hoping, so he was brooding, he was unhappy. He didn't qualify well, he was in the middle of the pack at the start. Suddenly, he realised he had his chance, and then he showed what he was really worth.' Peter Warr also has fond memories of that day. 'That was classic Rindt in the sense that he was at the end of a fairly long, deep low where everything that he'd hoped for in the season hadn't transpired and was showing no signs of doing so.

'He was pretty lacklustre in qualifying and very lacklustre at the start of the race. Then people started dropping out and, all of sudden, he just switched on and

thought "Christ! I can do something here" and just started driving like the Rindt we all knew, and just put up an absolutely sensational drive.' Despite his understandable disappointment at having thrown the race away, Brabham concedes that his rival did well to win that day. 'Rindt did a fantastic job there really, didn't he? He was a bloody good driver, Jochen. Him and that car went together. They really suited one another.'

Rindt's victory also put him right back in contention in the Championship race, for points leader Stewart had failed to score and the Austrian was now joint third behind Brabham and Stewart. All the drama of the final lap overshadowed another excellent and consistent drive from Graham Hill, who was unable to repeat his winning feats of the previous two years but nonetheless drove to a tremendously satisfying fifth place, given the trials and tribulations of the weekend.

In the hastily repainted R10 borrowed from the works team after Miles' non-qualification, Hill dives into Mirabeau on his way to fifth place. (Ford Motor Company)

Chapter 21
A New Star Rises

A four-week gap between the Monaco Grand Prix and the Belgian round of the Championship gave Lotus some much-needed breathing space to try and re-work the 72s. It also gave Chapman the opportunity to offer a Formula 1 test drive to Emerson Fittipaldi, who had adapted to Formula 2 impressively during 1970, even if he had not yet scored an outright win. Emerson's first test is, like his first meeting with Colin Chapman, still firmly etched in his mind. 'It was at Silverstone. We didn't have the chicanes and it was quite a fast track. I flew there with Colin in his plane. When we arrived, Jochen Rindt was testing the Lotus 72. He was quite upset with Colin and me because Colin stopped him testing and said "Jochen, go back into the 49, do ten laps, see how the car is and then we'll give it to Emerson". He was complaining a little bit but then he drove the car.

'When I was doing my very first few laps in Formula 1 ever with the Lotus 49, it was like a dream come true. I'd seen the car for so many years being driven by great Champions all over the world and there I was driving one myself. Then I was very pleased, Jochen got very much involved with my test. To my surprise, on the third or fourth lap, when I was picking up speed, Jochen came on to the pit wall and held out the time board for me. You can imagine how I felt. I came in after five or six laps and Colin said "How's the car doing?" And I said "Well, the car's understeering just about everywhere." And then Jochen smiled, looked at Colin and then looked at me again. They were squatting down by the cockpit, Jochen on one side, Colin on the other. Then Jochen said "I know how to change the car: just use the throttle, then you will have oversteer!" And, of course, he was right! So you know it was a very exciting moment in my career.'

With the likelihood that he would have a spare 49C available, Chapman offered the young Brazilian the chance to step up to Formula 1 at the Dutch Grand Prix in June but, showing admirable self-restraint, Emerson declined, saying that he did not feel ready for it yet.

After the 1969 race had been cancelled on safety grounds, the Belgian Grand Prix returned to the ultra-high speed Spa-Francorchamps circuits in the Ardennes region. The circuit was now surrounded by armco barrier and a flexible start time had been agreed with the organisers, in the case of rain, to prevent a re-occurrence of the first lap in 1966, when a cloudburst caused many accidents and eliminated a number of cars. Furthermore, a chicane had been added at Malmedy corner, to slow cars round the corner and onto the long Masta straight.

Gold Leaf Team Lotus took two 72s along, both much revised, plus 49C/R6 as a third car. As well as Rindt and Miles, a semi-works entry had been made for Alex Soler-Roig, under the World Wide Racing banner. He was to get the spare car, whichever that turned out to be after Rindt and Miles had made their choice. The other 49 present was the Walker car, which had been completely rebuilt following its Monaco accident and which was only just finished in time for the Belgian trip.

On the first day of practice, Rindt had problems with a seized hub which broke the lower wishbone mounting on the 72B and switched to the 49C for the rest of the weekend, not satisfied that the new car was either raceworthy or safe. Miles had been using the 49C to familiarise himself with the circuit – another one he had no previous experience of – but took over the 72B after it had been repaired, which was for the first Saturday session. Soler-Roig didn't get a run at all as he was waiting for the first heavily-revised 72C to be completed in the paddock. Hill was struggling to get up to speed in the Walker car, spending some time running without nose fins or a rear wing in an effort to increase straightline speed. Although this gave him the top speed through the trap just before the Masta kink at 182mph (294kph), the trade-off in terms of lost downforce on the fast sweeps that characterise the Spa circuit did not appear to be worth it and he was well off the pace.

In the first Saturday session, Rindt set the second

quickest time before engine problems set in with his 49C. Fortunately, there was sufficient time between the 2.30pm finish of the first session and the 6pm start of the second to enable an engine change to be effected and the car was ready with a fresh unit when the track was opened. Miles went slightly slower in the 72B than he had done in the 49C, while Hill – with nose fins and rear wing back on the car – was slightly further up the order and actually quicker than the Lotus number two. In the Saturday evening session, Rindt concentrated on making sure his car was 'just right' for the next day's race, while both Miles and Hill improved, the former by some seven seconds. Clearly there was nothing wrong with the 49C round Spa either, because there was one on the front row, just as there had been in 1967. However, either the Walker car wasn't quite right after its Monaco shunt, or the driver wasn't, because it was on the back row of the grid … Soler-Roig only managed two slow laps in the new 72, so was not permitted to start because he had not completed the mandatory five practice laps.

In the race, Rindt made a terrific start to lead the field up through Eau Rouge and along the Kemmel straight, hotly pursued by his fellow front row men, Amon and Stewart. However, at the end of the first lap, the two Marches had got in front of the Lotus and on lap three Rodriguez's BRM had taken Rindt, followed on the next lap by Ickx's Ferrari. After another two laps, Brabham had demoted him a further place to sixth and it seemed as if the Lotus was a little out of breath. He did manage to regain some ground when he re-passed Ickx and then Brabham overshot the chicane, letting the Lotus past, on consecutive laps. However, at this point a piston failure intervened and put the Austrian out of the race. The other 49C of Hill was running last but one, ahead of Miles, who was having all sorts of problems with the 72. Eventually, the Gold Leaf car lost pressure from both rear tyres and was retired, while Hill continued in last place until his engine blew. A truly stupendous race ensued without any Lotus interest, Rodriguez scoring a historic victory for BRM, 1.1 seconds ahead of Amon after 245 miles of wheel-to-wheel racing.

For the Dutch Grand Prix three weeks later, the Lotus 72 had finally been sorted and, although there were plans to run Soler-Roig in a 49C, these were abandoned to concentrate on the new cars. Thus, the only 49s entered were the Walker car for Hill – his 72 not having materialised yet, due to the problems encountered by the works cars – and Pete Lovely's 49B. With 20 places on the grid and 24 entries, it was clear that there were again going to be several disappointed runners by Saturday evening. As it transpired, only Hill's guaranteed start from being a graded driver saved him from non-qualification – perhaps a little unjust on the non-graded driver de Adamich, who was faster, but those were the rules.

Lovely never looked like qualifying, although his was a gallant effort given the fact that his car was effectively in 1968/69 spec with a 1968 specification engine installed. 'After we'd done our best to qualify and we didn't make it, Rob Walker came over and said "Pete, don't feel bad because your car probably isn't in as fine a tune or set-up as my car was a year ago when Seppi drove it and you were faster here today than Seppi was last year". And of course he was in the race, I don't know where he was, he was like eighth or tenth on the [1969] grid. So, he really made me feel good. I should have trusted him because Rob Walker is a very trustworthy person, but I did go look it up in the books and found that he was not just joshing me to make me feel better! He really meant it.' This story amply illustrates the pace of technological change in Formula 1 at the time and the extent to which tyre development particularly was having an influence on lap times.

Another factor which made things more difficult in 1970 was the sheer number of entrants trying to get into each race. In the 1969 Dutch Grand Prix, there was a place on the grid for all entrants, but in many races in 1970, including the Dutch round, this was not the case. 'What we really should have done was stayed in Europe in 1969 because we could have gotten in to all the races, because they were short of entries' says Lovely. 'But we came home and did our thing here and then we decided to go back in 1970, and that was when March came on the scene with their own works cars and customer cars and who knows what else. There was myself and my wife and one mechanic and a Volkswagen pick-up truck to transport it around, we just weren't quite up to the speed that we should have been. I felt pretty bad about all that'.

So Hill's was the only 49C in the race and, once again, he didn't exactly set the race alight, coming into the pits from 14th place on the 22nd lap complaining of bad handling, rejoining six laps down and finishing in 12th and last place. This was the race which marked the turnaround in Rindt and Lotus's fortunes, the Austrian romping to a comfortable victory, while Miles ran for a long time in the points before being pipped in the closing laps for the final point and finishing seventh. However, Rindt's joy at winning was completely overshadowed by the death of his close friend, Piers Courage, in a fiery accident at the wheel of Frank Williams' de Tomaso.

A fortnight later virtually all the same protagonists assembled at the Charade circuit near Clermont-Ferrand in France. After missing Zandvoort, Soler-Roig was again entered under the World Wide Racing banner at the wheel of a 49C, chassis R6. Lovely rolled up with his car on the back of his VW truck, while the Walker team were once again relying on their trusty 49C, their 72 still being some way off completion. In fact, it transpired that the works intended to build a new 72C for Miles and then convert Miles' 72B to C spec before selling it to Walker. All this would take some time, so it was apparent that several more races with the 49C would be necessary – not quite what Hill or Walker had envisaged when they had joined forces or when they had originally placed their order for a 72 …

In practice, Hill qualified last again, almost three seconds behind the next slowest qualifier and only 24 hundredths of a second quicker than little Silvio Moser in his bulky Bellasi 'DFV kit car'. Soler-Roig was another non-qualifier although, as Peter Warr explains, there was a good reason for him being off the pace. 'Because of the total state of panic in getting the 72s ready, his car still had Spa ratios in it. Eventually he came in to the pits and said "I think there's a problem with the car, I can't get fifth gear." So 'Rabbit' Bartels leant into the cockpit and

Next generation: Emerson Fittipaldi made his debut in a third Gold Leaf Team Lotus entry, 49C/R10, at the British Grand Prix. Within ten weeks he would be the team's lead driver... (Ford Motor Company)

said "Well go out and get a tow!" It had the Spa top gear in for Clermont ...' Along with Soler-Roig and Moser the other non-qualifier was Pete Lovely – who found the 51 corners and five miles to the lap rather daunting.

While Rindt went on to win his second consecutive race, profiting largely from other people's retirements and only passing one car during the entire race, Hill had another rather uninspired drive into tenth place.

On his return from Clermont-Ferrand, Chapman was forced to act fast in order to retain the services of Fittipaldi, for that excellent judge of new talent, Frank Williams, had again shown interest in signing Emerson to drive the de Tomaso. The Lotus boss immediately stepped in and got the Brazilian's name on a contract and arranged for him to drive in the next Championship round, which would be the team's home race, the British Grand Prix at Brands Hatch.

Fittipaldi was entered for the British Grand Prix in the ex-Hill chassis 49B/R10, which the Londoner had driven at Monaco after his own car was damaged. Its specification was little changed to that of Monaco, except for the fact that there was no need in the cool British conditions, to run a twin oil cooler set-up, so the regular single cooler arrangement was re-adopted. The other two 49s in this race were Hill's Walker car – the 72 not now destined to appear until Austria in mid-August at the earliest – and Lovely's old 49B. This time, Lovely was guaranteed a start, as the RAC was not placing any restriction on the number of cars which could start the race. Perhaps a sound reflection on their age relative to the other cars in the field, all three 49s lined up on the back row of the grid, with Fittipaldi emerging fastest of the trio, three-tenths quicker than Hill, with Lovely nearly two seconds further back. After the ease with which he'd adapted to the 49C at Silverstone, the Kent circuit was a rather more taxing challenge for the Brazilian. 'Over the bumps at Brands Hatch it was much more difficult than driving around Silverstone, where it was very smooth. It was not an easy car to drive over the bumps, to me at least. I remember I was surrounded by my heroes. Everyone was ahead of me, except Graham, who started next to me!'

Buoyed no doubt by being on home ground and a track he knew well, Hill made an absolutely stupendous start and came round at the end of the first lap in 14th place, having made up eight places. He was soon in the thick of a midfield battle and, after Ickx, Miles, Pescarolo and Siffert had all pitted, was up to eighth place by lap 25 of 80. After passing an ailing Peterson in his March, he moved up to seventh but then was demoted by a flying Surtees, having his first outing in the first Formula 1 car to bear him name, the TS7. When both Surtees and Stewart and then Oliver were forced to retire, Hill was up to a tremendous fifth place. However, with eleven laps to go, Chris Amon got past him and he was left to collect just a single point. It was a stirring drive and probably his best that season, aside from the sheer bravery aspect of his sixth place in South Africa at the beginning of the year.

Fittipaldi drove a sensible race, passing one or two other cars, but otherwise making sure he kept out of the way when being lapped by the leaders and generally impressing onlookers with his 'old head on young shoulders' approach. After running seventh for a while having got ahead of Cevert in the number two Tyrrell March, Emerson was passed again by the Frenchman just before three-quarters distance and finished a satisfied eighth, despite a split exhaust manifold which made his engine sound rougher than it really was.

Lovely had a fairly uneventful race until he was forced to pit twice, which put him right to the back of the order. 'My first pit stop was to cool the driver off! When I went into the pits I said "Water, water!" So the girls poured out a glass of water... Carlo [his mechanic] knew what I meant. He grabbed the watering can, and he just watered me down and then pushed me out. Away I went. When he was giving me the push to leave, the right rear tyre was very soft. So the next time I came around, he was holding out the sign for me to come back in. I said to myself 'What's the matter with him, doesn't he remember I was just there?!' So, I thought 'Well, I'd better go in and see what this is all about'. So in the meantime, we didn't have our spares out in the racing pit, we had them back in the paddock. The thinking was, jeez, if you ever

A New Star Rises

An attentive Fittipaldi listens to advice. (Ford Motor Company)

had to stop for a tyre change in those days, well you were out of it anyway.

'Carlo went over and Bette Hill was sitting on top of a stack of tyres, timing Graham in Rob Walker's car. So Carlo said "Bette, get up" and so she stood up, she's got her chart in her hand and her watches and all that stuff and he grabbed one of the rain tyres. He knew they wouldn't be needing a rain tyre but at least if it had air in it, well, it would be better than what we had. When I stopped, well, he's changing the wheel and I'm looking at it, you know, "What the heck's going on?" He came and warned me and, of course, by then I knew it was a puncture and he said "Look out, it's got a rain tyre" so I went out and actually it gripped pretty good because the track was so oily by that time, a rain tyre worked good!' Lovely continued to finish tenth and last, some 11 laps down but delighted to have driven in, and finished, a 1970 Grand Prix at last.

It was a good weekend for Lotus, for once again Rindt benefited from Jack Brabham's misfortune as the Australian ran out of fuel on the final lap, due apparently to his mechanic setting the mixture to run slightly too rich! The amazed Austrian once again inherited the lead within sight of the flag to snatch an unexpected victory.

With Lovely having headed back to the USA after the British Grand Prix, only two 49s attempted to qualify for the German Prix, which had been moved from its spiritual home at the Nürburging to the featureless circuit at Hockenheim. With its flat-out blinds down the straights, punctuated only by chicanes halfway down each one and the twisty stadium section through the start and finish area, this was a circuit not ideally suited to the 49 which, not surprisingly, was by this stage beginning to show its age. Consequently, it was a very impressive performance for Fittipaldi to qualify 49C/R10 in 13th place. As the weekend progressed, the Brazilian went faster and faster, ending up only 2.5 seconds slower than pole-man Ickx in the flat-12 Ferrari. 'At Hockenheim, it was a very smooth track and the car, compared to Brands Hatch, was much easier to drive. I was more familiar with it and I could just attack.' Hill, on the other hand, was at the back end of the grid once more, only Beltoise's Matra being slower. This was likely to be the last race for the venerable R7, as Rob Walker felt that it was simply no longer a competitive proposition.

The race saw Cosworth-powered cars decimated, although the engine in Rindt's Lotus hung together to give him his fourth straight win and set him up as the one most likely to become World Champion. Fittipaldi found himself in the midst of a midfield battle from which he emerged in fourth place after all the retirements. 'I remember the first few laps, braking for the chicane, there was a big line [of cars] and I saw Pedro Rodriguez in the BRM passing three or four cars, locking his wheels under braking. And I said "Well, that's the way to go." I remember it gave me a lot of courage to see Rodriguez overtaking twice at the chicane and then I started to be more aggressive as the race settled, and I started passing people, it was very exciting. I was very happy with my performance there.' He maintained this position to the finish, the last car not to be lapped by the flying Rindt, to claim three Championship points in only

Lotus 49

First fruits: in only his second Grand Prix, Fittipaldi drove a superb race in an outdated car to finish fourth in the German GP. Here he flings his 49C through one of the corners in the stadium section of Hockenheim, pursued by Rolf Stommelen (Brabham, obscured), John Surtees (Surtees), Henri Pescarolo (Matra) and Denny Hulme (McLaren). (Nigel Snowdon)

his second ever Grand Prix start. In so doing, he had truly arrived, marking himself down as a potential Champion of the future. In complete contrast, Hill tooled round towards the back of the field and then retired with 12 laps to go and shortly after he had been lapped by Rindt, due to crankshaft failure in his DFV.

The Austrian Grand Prix at the new Österreichring circuit near Zeltweg, was to mark the end of an era in Formula 1 racing, for it was here that the Lotus 49 made its last ever Grand Prix start. Once again, after his impressive display at Hockenheim, Fittipaldi was entered at the wheel of 49C/R10. His was the only 49 present, Rob Walker having stuck to his word and stayed at home awaiting delivery of his 72.

Fittipaldi qualified well, in 16th place out of a grid of 24 cars, despite problems with his throttle linkage. After running a strong 12th in the early stages, he started to experience problems with his car's fuel mixture after 10 of the race's 60 laps, which slowed him considerably. Then, with only seven laps to go, he was forced to pit for more fuel because the problem had increased consumption dramatically. This firmly demoted him to 15th and last place and he finished a disappointing five laps down on the winner, Ickx.

No 49s were entered for the next event, the Italian Grand Prix at Monza, since a full complement of three 72s was available to the works team and the Rob Walker team had finally taken delivery of its 72. This should have been the meeting at which Rindt triumphantly wrapped up his first ever World Championship title. However, it all went tragically wrong in practice when his car speared left into the barriers on the approach to the Parabolica corner and Rindt died from injuries sustained in the crash. It was a terrible irony that Rindt had waited so long for a car capable of taking him to the title, and yet he never lived to know that Chapman's 72 did ultimately do that. It was perhaps fitting that, four weeks later, Emerson Fittipaldi won the US Grand Prix at Watkins Glen in a Lotus 72 – only his fourth race in a Formula 1 car. In doing so, he made it mathematically impossible for the

A New Star Rises

Ferrari driver, Jacky Ickx, to win the Championship and Rindt became the only person to date ever to have won the World Drivers' Championship posthumously.

Although Pete Lovely entered his 49B for the US Grand Prix, he failed to qualify, so Fittipaldi's race in Austria marked the end of the road for the Grand Prix career of the Lotus 49. However, a works-entered 49 did make two further starts in non-Championship Formula 1 races in 1971. As a precursor to holding a fully-blown round of the World Championship in 1972, Argentina hosted a non-Championship Grand Prix event in January 1971. In order to generate local interest, rides were brokered for a couple of promising South American drivers, including one at Gold Leaf Team Lotus. Originally, the plan had been to take the Lotus 56B turbine, plus two 72s, but at the last minute it was decided that the turbine was not yet raceworthy, so it was left at Hethel and the Monaco Grand Prix-winning 49B/R6 resurrected once more.

The recipient of the drive was none other than Emerson Fittipaldi's elder brother, Wilson. It was a baptism of fire for the Brazilian, for he had never experienced anything more powerful than a Formula 2 car and that had only consisted of five laps round Snetterton in his brother's car. However, he drove sensibly, qualifying ninth of the ten Formula 1 cars. In the first heat, he finished eighth, on the same lap as the winner, while in the second he was running a strong fifth ahead of Jo Siffert when his engine let go on the 21st of the 50 laps. He was classified ninth overall but, more importantly, had made a positive impression on a number of leading Grand Prix teams, something which would eventually lead to a World Championship drive, although not with Lotus.

The day after the Argentinian event, Wilson's brother Emerson made a 'guest' appearance at the wheel of R6, which was arranged by the organisers of a Formula 3 race at the Interlagos track near his birthplace of Sao Paulo. This was the first time a modern-day Formula 1 car had run in Brazil and was intended to create interest in a possible Brazilian Grand Prix which was being mooted to take place in the next few years. Getting the car there involved a bizarre and sometimes hair-raising trip, according to Dave Sims. 'They sent me and Joe 90 [Derek Mower]. Peter Warr told me "You're going to Brazil tomorrow." I said "What?" We had no visas for Brazil, nothing. The 49 went in this DC3 Dakota, loaded with oranges, just literally loose oranges and me and Joe 90 had to sit on the floor, and one of the crew was drinking brandy or wine out of a bottle!' After an exhausting and sometimes harrowing journey, it was amazing that the mechanics were able to focus on doing their jobs. 'The objective was for Emerson to break the lap record and he did. We ran it so low that it bottomed out and wore through the radiator, lost water and the engine overheated. But he did it anyway!' Derek Mower recalls that, on their return to the UK, Chapman 'went beserk because it was a development engine or a relatively new one at least and he [Fittipaldi] didn't see the signs and it seized.'

Tragic irony: Rindt's Monaco victory in the 49C kick-started his 1970 title challenge. When the Lotus 72 finally came good, it enabled Colin Chapman to deliver him a richly-deserved Championship title. Sadly, the Austrian did not live to see this day, for he was killed in practice for the Italian Grand Prix and was crowned the sport's only posthumous World Champion to date. (Ford Motor Company)

The old faithful R6 took its final bow at Oulton Park, for the non-Championship Rothmans International Trophy held on Good Friday, April 1971. The driver on this occasion was the Formula 3 driver, Tony Trimmer, who had won the Formula 3 race at the same Monaco Grand Prix meeting as the one in which his Oulton mount had scored its sensational victory in the hands of Jochen Rindt. After getting down to a reasonable time quite early on in the Thursday practice session, Trimmer blotted his copy book by crashing the car after cresting the rise at Clay Hill. The rear of the car, particularly the suspension, was quite badly damaged by the impact with the bank, with the result that the hard-pressed Lotus mechanics had to resort to cannibalising 49B/R10 which, fortunately, was on display nearby in Bradford at Normands Garage. The resulting rebuild was an overnight job, but they managed to get Trimmer onto the grid for the race. After starting 10th out of the twelve cars, he dropped down the order due to a pit stop on the second lap to cure a sticking throttle, while a further stop to adjust the mixture in the latter part of the race lost him further time. However, with the rate of attrition in the race, he still finished in sixth place, albeit four laps behind winner Rodriguez.

Although private owners continued to race their cars after this date, the Oulton Park race marked the final appearance of a Team Lotus-entered 49. By this stage, they were well into the development of the already proven Lotus 72 and were also trying to establish whether or not turbine power was an alternative way forward in Formula 1 ... yet another example of Chapman's ability to explore all possible avenues in the search for that all important edge over his rivals.

Chapter 22
The Exiles

Three 49s had active contemporary racing careers following their life as Team Lotus cars. These were chassis numbers 49/R3, 49B/R11 (formerly 49/R2 and 49B/R2) and 49C/R8. Two of these chassis, R3 and R8, ended up in South Africa which, along with Australia and New Zealand, was one of the major markets for obsolete or unwanted Formula 1 cars. This was due to the fact that it had a national series, the South African Drivers' Championship, which catered for Grand Prix cars, along with Formula A/5000 and Formula 2 cars.

In the USA, the arrival of the SCCA Formula A Championship created a third market for these cars, with 3-litre Formula 1 and 5-litre stock-block engined cars being deemed eligible. It was against this backdrop that the American Pete Lovely bought himself a 49, raced it in a few European Formula 1 events, and then took it back with him to the USA in order to compete in the SCCA series, or at least those rounds that were within relatively easy reach of his Tacoma, Washington base.

The first 49 to be sold to a foreign owner was chassis 49/R3. After its virtual overnight creation for the British Grand Prix and subsequent rebuild after its shunt in practice for the German race, R3 became Graham Hill's regular mount for the rest of the 1967 season. He also drove it to second place in South Africa in the opening Championship round of 1968, finishing second to team-mate Jimmy Clark. It was at this meeting that the Rhodesian driver John Love, already a four-times South African Champion, enquired about the possibility of purchasing a 49 to replace his Repco-Brabham BT20. 'I'd raced for Ken Tyrrell in England and I had the opportunity of buying that or a Repco-Brabham. I'd spoken to Colin while he was out in South Africa and I'd been offered the car, and Jack also offered me one of the works Brabhams. I mentioned to Ken that I had the option of buying either one or the other.

'He said to me, "You don't have to make a decision, do you?" When I asked why, he said "It's obvious – you've got to buy the Lotus". And that's how it went, that's when I bought the 49. They took it back to England and then sent it back to me. I don't think I'd quite made a decision or, at that particular stage, raised the finance to buy it. So, it had to go back and then I organised the finances and the sponsors I had in South Africa [the Rhodesian tobacco firm Gunston and Abraham Harpaz's Peco – Performance Equipment Co] and then they shipped it out to Cape Town.

'It was in Team Lotus colours when it arrived in Cape Town. I trailered it back to Johannesburg [a journey of some 1000 miles], got it there on the Wednesday, we practised on the Thursday and the Friday. They took the car away on Friday night and it was back at the track the following day in Team Gunston colours! It was quite a tremendous set up, they worked from about 6 o'clock in the evening, when we'd finished with the car,

On Ken Tyrrell's recommendation, John Love bought the best car money could buy when he bought Lotus 49 R3 from the works in February 1968. Before its first race, it had to be rapidly resprayed in the colours of the team's sponsor, Gunston cigarettes. (John Love Collection)

and took it away and it was back at the circuit the following morning at 12 o'clock with a new paint job!' The new colours consisted of orange with a brown stripe down the side with 'Team Gunston' on it and Peco underneath on the orange, while the brown stripes were joined together by a brown band round the nose.

As might have been expected for a car which had quite conclusively proved itself to be the fastest Formula 1 car on the Grand Prix grids, there was very little that Love needed to do to set it up for the first race. 'The only thing we found that was really wrong was that it wandered on braking and we just reset all the bump steer and stuff like that on the front and it was tremendous.' The only notable change to the car's specification was a switch from Firestone rubber to Dunlop, the company with which Love was contracted.

Love produced a totally brilliant performance in his first race with the new car, the Rand Autumn Trophy at Kyalami, which was the second round of the South African Formula 1 Championship. He qualified on pole position, more than a second quicker than his nearest rival, Basil van Rooyen in the STP-sponsored Repco-Brabham BT24/1 and led from start to finish. However, because he had jumped the start, he was given a one minute penalty so that, although he finished well clear on the road, he actually had to drive very hard to make up the time he had been penalized. By the finish, he had turned the deficit into a 15 second lead over Dave Charlton's Repco-Brabham and set a fastest lap 1.7 seconds inside Jim Clark's outright record, set in the South African Grand Prix at the beginning of the year.

He repeated his dominant Kyalami drive in the Coronation 100 race, held at the Roy Hesketh circuit in Pietermaritzburg, winning both heats and setting a new lap record once again. His only opposition came from the two 1967 Brabhams but neither finished both heats, Love's Team Gunston team-mate, Tingle, running out of fuel in the first and van Rooyen's retiring in the second with engine problems.

Van Rooyen turned the tables on Love in the next round of the Championship, the Players Gold Leaf Trade Fair 100, at the Bulawayo circuit in Rhodesia. Love, in what was effectively his 'home' race, suffered from a faulty spark plug, which caused his engine to run on only seven of its eight cylinders for most of the race. Nonetheless, he finished second to the red STP car and still managed to share a new lap record with Van Rooyen.

With his DFV back on song again, Love was unbeatable in the Republic Trophy race over 100 miles at Kyalami on June 1st. Although van Rooyen chased him hard, it was in vain as his Repco engine lacked the horsepower of his rival's Cosworth. These two adversaries renewed their battle at the Roy Hesketh circuit in Pietermaritzburg three weeks later, for the Natal Winter Trophy. This time van Rooyen emerged victorious, but only after Love had retired with a broken brake pipe when leading. At this stage in the Championship, after six of the ten rounds, Love led with 33 points to the 29 of Jackie Pretorius, who had been fast and consistent at the wheel of a Formula A Lola T140, and the 24 of van Rooyen.

However, another comfortable victory from pole position in the Border 100 at the East London circuit at the beginning of July, changed this, despite van Rooyen becoming the first runner in the Championship to adopt a high wing, as was the fashion in Formula 1 at that time. Love followed up his East London performance with a start to finish win in the Rand Winter Trophy at Kyalami in early August, to clinch his fifth successive Championship with two rounds still to go, van Rooyen again trailing in second.

The penultimate round of the series was the False Bay 100 at the Killarney circuit near Cape Town. This saw Love – who had already secured pole position – suffer a rare moment of brain fade during a wet practice session on the morning of the race, which saw him spin off and damage the car too badly to be repaired in time for the race, so he non-started.

The Championship season ended on a disappointing note for Love at Kyalami in October with the Rand Spring Trophy Race. After qualifying on pole position, he sailed into his customary lead but was forced to retire on the 11th of the 40 laps when a front suspension ball-joint broke. This race marked the emergence of Dave Charlton as a front-runner, his team Scuderia Scribante having taken delivery of a brand new Formula A/5000 Lola T140, which he placed second on the grid alongside Love and then proceeded to drive to victory following his rival's demise. It was also the first occasion when Love tried a rear wing, mounted on the chassis rather than the suspension. However, he – like his rival van Rooyen – decided against using it in the race.

The final race of the year was the non-Championship Rhodesian Grand Prix, the second race of the year held at Bulawayo, Love's home circuit. This time he did use his rear wing, which was very similar in size and position to the ones used by McLaren towards the end of the 1968 Grand Prix season. It was trimmed out at the front by a beard spoiler of the style experimented with by Lotus at the Race of Champions early that year and adopted by a number of other teams at varying stages of the season. The Rhodesian made amends for his failure to win at the same circuit in May, although only after Charlton had spun away a first lap lead and then, when catching Love once more, had been forced to pit to have a bucket of water thrown over him to combat the excessive heat! Van Rooyen missed the meeting altogether, being in England attempting to buy a McLaren Formula 1 car with which to try and get on terms with Love in 1969 …

At the end of that year, Love took the decision to convert his car to take a Hewland gearbox, as the works Gold Leaf team had done with the introduction of the 49B in May. This was not for reasons of reliability, for the ZF had by 1968 been proved to be an exceptionally reliable piece of equipment, but for greater flexibility in gearing. 'The Hewland box was so easy to change gear ratios', says Love. 'With the ZFs I had two boxes and four ratios, so it was difficult to get the right ratios in the box for any particular circuit. With the Hewland you could change any particular gear so you could gear for the circuit. The ZF box was a fixed ratio, all you could do was change the ratio on the crownwheel and pinion. But it [the ZF] was a good, nice, reliable box, it was fine.' A Hewland FG400 box was fitted and this stayed on the car

Lotus 49

Love, now with a high wing and twin oil coolers mounted at the back of the car, leads Basil van Rooyen in his ex-works McLaren M7A at Killarney before he retired. (John Love Collection)

for the rest of the time Love owned it.

Following current design trends in Formula 1, Love and his team added a high suspension mounted rear wing, with nose fins at the front of the car to balance out the effect of the wing, while they also introduced a rather ugly twin rear oil cooler arrangement over the gearbox. The reason for this was that in the warm climate and the high altitudes found in South Africa, the standard oil and water cooling arrangements simply weren't up to the job. To compensate, two oil coolers were added, although the oil tank in the nose was retained in contrast to the revisions introduced by Lotus for 1968, which saw it moved to the back of the car over the gearbox. In place of the Lotus-issue Serck combined oil/water radiator, a locally sourced water-only radiator was fitted with a core almost twice the size of the original. These tweaks proved very effective and prevented any further problems in this area.

The season-opening race of the 1969 Championship was the Cape South Easter Trophy, at Killarney. Van Rooyen was back, having procured an ex-works McLaren M7A, complete with wing, although not as large, or as high, as that on the Lotus. In practice he and Love traded fastest laps, the Gunston car eventually proving the quicker of the two by 0.3 seconds. At the start, van Rooyen got the jump on Love, who followed in the McLaren's wheeltracks for eight laps until he managed to squeeze past. After that, he pulled away easily as the McLaren faded with brake trouble and looked to be heading for an easy win when, at around half-distance, a driveshaft universal joint broke, gifting the win to van Rooyen. The only consolation for the Gunston driver was that he set a new fastest lap, almost four seconds inside the outright record.

The second round also went to van Rooyen. Although Love qualified on pole, he was unable to convert this into a win and came home second. By this stage, the Gunston car had sprouted a front wing and both front and rear wings were using the feathering system, described in Chapter 15, which was proving hard to get right. The car had also grown a more solid-looking roll-over bar, which it needed to have fitted to fall in line with new FIA regulations so that it could compete in the South African Grand Prix.

Over in the United States, Pete Lovely had arrived back from his English racing trip, complete with his newly-acquired 49B/R11 but with one vital exception: 'When we shipped it back to the States, Pan-American Airways lost the rear wing! So, for the first half of the season in Formula A, I didn't have a whole lot of success because of the fact that I didn't have a rear wing. I continued to run with the front wings, but with them flattened out quite a bit so that I wouldn't get too much understeer. I didn't want to run it totally wingless, it wasn't supposed to be that way, on that particular car. We did Riverside with it. Gee, it went real good down the long straightaway, but it was a little bit slow on the corners!'

The Riverside Continental Grand Prix was the first round of the SCCA's Continental Championship, which allowed for stock-block engines of up to five litres or racing engines of up to three litres, the latter category being permitted 26 gallons of fuel compared to 30 for the stock-blocks. Lovely's 49B was the only Formula 1 car entered, the rest of the field being made up of Formula A/5000 cars from the likes of Eagle, McLaren and Lola. A major attraction was the prize money or 'purse', the winner taking home US$5550 – approximately £1500 at 1969 exchange rates or £14,000 at late 90s values. This was almost five times more than they would get for coming home first in a British Formula 5000 race. In practice, despite his lack of wings, Lovely recorded the sixth fastest time and he continued in this vein to finish in sixth place, albeit two laps behind the winner, John Cannon, in an Eagle-Chevy.

Back in South Africa, Love finally scored his first victory of the 1969 season at Kyalami towards the end of April. By this stage Team Gunston had put aside the complex compressed air system for feathering the wings which had first appeared at the South African Grand Prix and replaced it with a more simple and reliable spring-loaded system with a cable linked to the cockpit pedal, similar to that found on the works Lotus cars at the time. The pattern of the race was similar to the Killarney round, with van Rooyen taking an early lead, Love biding his time and then passing the McLaren, although this time he did not manage to do so until around half-distance and, even then, only due to confusion when lapping back-markers. Nonetheless, once he was past, he pulled out a 10 second margin on his rival, which he held to the finish, despite the drain plug dropping out of the diff and the unit losing all its oil as a result.

The second round of the SCCA Championship took place two weeks after Riverside, at the Laguna Seca Raceway in California. Having qualified a strong third in practice, only just over a second away from pole, Lovely proceeded – much to his amazement – to run out of fuel in the race, when running well up the order. The reason for this was linked to the fact that, between the Riverside and Laguna Seca races, Lovely had sent his DFV engine back to Cosworth for a rebuild. 'For a long time I ran the thing on full-rich because I didn't know how to adjust it. When we sent the engine back for a rebuild, I think they sent the metering unit back to Lucas. The datum pin, which screws into the side of the little thing where you adjust the fuel mixture, is the thing that keeps it from turning once you adjust it. There are six notches. You can go from full rich to full lean. Well, the datum pin was not tightened when they re-assembled it, and it came unscrewed at Laguna Seca, so that it went on, not only to full rich but all the way around to even more than full rich, and then it ran out of fuel!

'So, everybody said "Oh yeah, that Pete Lovely, he's so cheap he won't put enough gas in the car!" When you are going from six and a half miles to the gallon to about four miles to the gallon and you don't really want to carry any more fuel than you absolutely have to, why, with three laps to go in the race, well it just ran out …'

It was around this time that Lovely and his wife decided to get their car resprayed, with an unusual twist. 'We were trying to be nationalistic about the whole thing so we were going to do white with a blue stripe. It ended up that my wife decided that just plain white wasn't too exciting so she wanted it to be pearlescent white! In those days it was a little bit difficult to come up with a pearlescent colour. I mean, at least in our area. And we had a terrible time.

'We finally found some guy that would do it and the way they accomplished it was with fish scales! They took fish scales and mixed it in with the paints. I don't know how they ever managed to blow it through a paint spray gun but … anyway, they did. Then I said "I'd like a darker blue, not just a baby blue." He shot a darker blue for the stripe and, unless you got it in brilliant sunlight, it looked black. So there were people who wondered: 'What national colours were white and black?' That must be some obscure African country or something like that! The paint job was so expensive – it cost me almost U$350! [around £150 in 1969 but approaching £1400/US$2200 at late 90s values]. I wasn't going to repaint it, I just left it that way, because the important things were buying gas for my Volkswagen pick-up truck that we transported it around on, and to get to and from the races and spending - wasting - money on fancy paint jobs was just not my idea of the way to go.'

Lovely also found a solution to his lack of rear wing. 'Some guy in California who was an aerodynamicist and had worked for aircraft companies and stuff came up with this wing. He was marketing them to the guys with the Formula A cars, the Chevy powered ones. So he took some measurements and he built the wing for me, so at least I finally got a pseudo-wing on and it helped a lot, it nailed the back end, which was pretty loose without it.'

Back in South Africa, John Love's next race in the Gunston car was at home in Bulawayo, on the Kumalo circuit. This event, the Bulawayo 100, was not part of the South African Driver's Championship calendar, so a weaker than normal field attended and Love ran out an easy victor. However, it was notable in that it was the last race where high suspension-mounted wings were permitted, since the following weekend the FIA banned such devices and the SADC organisers conformed to this decision. Love's mechanics subsequently replaced it with a chassis mounted version. Three weeks later at the South African Republic Trophy meeting at Kyalami, the Championship tide seemed to be turning in Love's favour, for van Rooyen had an enormous accident in practice after a wheel came off his McLaren and the car was extensively damaged, beyond immediate repair. However, although the Rhodesian easily took pole, he did not convert this to a win as expected, a variety of troubles dropping him to a lowly eighth place at the flag.

Living in Tacoma, Washington, on the Western side of the USA, meant that Pete Lovely couldn't take in all the SCCA Continental Championship races on the calendar, due to the distances involved. 'I didn't do the

Love drove to a comfortable victory from pole in the Rand Autumn Trophy at Kyalami. His car sported high wings at front and rear and a pedal/cable-controlled feathering system in place of the compressed air system used in the South African Grand Prix. (Len Konings/Robert Young Library)

whole thing because I couldn't afford to travel to the East. The United States is a big place and 3000 miles each way is a long way to go from my part of the world to the East Coast. We just didn't have the time to be able to do it, so we just concentrated on the ones that were within reach for us.' Consequently, he had a seven week gap between the race at Laguna Seca and his next start at Sears Point International Raceway, 25 miles north of San Francisco. After qualifying seventh, just over two seconds away from pole, Lovely was forced to retire from the race with broken suspension.

After his troubles at Kyalami, Love also failed to score in the next round of the SADC, the Natal Winter Trophy, held towards the end of June at Roy Hesketh. This was after Love had again put the car on pole but then collided with the Lola T142 Formula A/5000 car of John McNicol. With refreshing candour, he admits that he was entirely to blame for the coming together. 'I tried to go on the inside of the corner and our wheels locked and I ran over him, our wheels went over wheels and when it landed it broke the front suspension. So we had to get a new rocker and a few bits and pieces. That was just a stupid bloody move on my part, I think!'

However, Love was back on track for the Border 100 at East London two weeks later and stormed to pole position followed by an easy victory. He repeated this feat in the non-Championship race at the Lourenco Marques track in Mozambique, but was in trouble yet again at the Rand Winter Trophy meeting at Kyalami, where he took pole only for his engine to blow, forcing him to non-start as he had no spare. This left McNicol, who had consistently finished well up in the points all season, to go on and win the race, and consolidate his Championship lead.

After another seven week break, Lovely re-entered the SCCA Championship fray at the Donnybrooke circuit in Minnesota, known today as the Brainerd International Raceway. He really got the 49B wound up, planting it on the second row of the grid, alongside the highly rated John Cannon and only 0.3 seconds slower than the 'King' of British Formula 5000 at that time, Peter Gethin. However, another Briton, David Hobbs, overshadowed everybody by taking pole by a clear 1.5 seconds. In the race, which was divided into two heats, Lovely was thwarted by another fuel-related drama. 'I can't remember what the lengths of the heats were, maybe fifty miles, maybe longer. My wife and I had taken the car there, just the two of us. When we unloaded it [journalist] Pete Lyons was there, first time I met him. He helped us get it down off the truck because we had to take it through scrutineering. And then we didn't have a pit crew.

'A couple of guys came over and said "Did we need any help?" and my wife talked to them and the next thing I know, well, I've got a pit crew! I ran the first heat and it was hot back there and I wasn't in particularly good shape physically. I'm several years older now, but I'm more fit now than I was then! So they said "Go lay down in the truck, put some ice on your head and cool off and get comfortable and we'll service the car. And how much fuel do you want in?" I told them to put in four cans of five gallons or something like that. I just forgot about what happened. It turned out that they only put in three when we needed four. I was doing quite well in the race, probably in second place, conceivably I could have gotten into first in that second heat, and I ran out of fuel … which was pretty nuts.'

Despite running out of fuel, Lovely was still classified 10th overall, but it was a disappointing end to a weekend which promised so much more. This was his final SCCA event of the year, although he went on to contest the three North American World Championship Grand Prix in Canada, the USA and Mexico in September and October. Eventually, the mystery of his missing wing was solved. 'After about six or eight months, Pan-Am called me up and said "We found something. It was part of a shipment that you …" Well, it turned out to be the original wing. Because we'd broken it down with the struts in a bag along with the wing, they'd laid it on a top shelf some place and nobody ever found it. Then one day somebody got up on a ladder and looked and said "Oh, there's something else!" So we finally got the original wing back.' Ironically, it was too late by then because high wings had been banned in FIA events and would be outlawed in SCCA races for 1970 and all that remained for Lovely that year were the three FIA Championship races, which required a low, chassis-mounted rear wing.

In South Africa, the Gunston car returned, its engine rebuilt, two weeks after the Rand Winter Trophy meeting, for the False Bay 100 at Killarney. This time there were no problems, and Love romped to a comfortable victory. When he took pole for the next round, on

In mid-May 1969, high wings were banned, so Love reverted to a chassis-mounted wing in line with the new regulations. Note the sturdy mandatory roll-over hoop and very non-standard exhaust system which gave it a claimed extra 15bhp. (John Love Collection)

Pete Lovely (in wrap-around shades), wife Nevele and helper struggle to unload his lavishly-resprayed 49B/R11 from the back of his VW truck in the paddock at Donnybrooke. Photographer Pete Lyons ended up putting away his camera and helping to unload the car... (Pete Lyons)

his home Kumalo track at Bulawayo, things seemed to be going in his favour. However, he had not counted on a dramatic change of weather, which saw a very cold day and the track soaked by a prolonged period of drizzle. Damp conditions presented problems for the competitors, since wet tyres were hard to come by. This was because Rhodesia was subject to sanctions by the UK and USA. For the race, Love opted to start on his own hand-cut 'semi-wets' but these were worse than useless and he spun on lap 2 and then five more times within two laps. Eventually, he came home a dejected sixth, while Dave Charlton, who had chosen dry tyres for his Lola T142, ran out the winner as the conditions eased.

This left just one race in the Championship, the Rand Spring Trophy at Kyalami, a circuit Love knew so well and usually performed well at. He was only four points behind McNicol in the standings and the mathematics were quite simple: he had to win, and his rival had to finish lower than second – if he finished in second he could not be overhauled. Love again put his car on pole, as he had done in every race that season, and shot off into the lead. McNicol did everything he had to do, getting his car up into second place but, inexplicably, he crashed, throwing away the title. Love went on to win and was crowned Champion for an incredible sixth consecutive year.

The fitting of wings and nose fins had given the two-year old Gunston car a new lease of life for the latter part of 1968 and into 1969. However, it was clear from the way that van Rooyen had given Love a run for his money in the early part of the season, that it would be possible for someone else to come along with more modern equipment and regularly beat the Champion in 1970. Someone who had already come to this conclusion towards the end of 1969 was Dave Charlton. Together with his long-time patron, Aldo Scribante, he hatched a plan, which he hoped would see the great Love finally vanquished and the Championship go to someone else for the first time since 1963.

After the final race of the 1969 season, Charlton travelled to England in a bid to source a more competitive car than his existing Lola Formula A/5000. Phone calls to the major Grand Prix teams revealed that Lotus Components – the customer arm of Lotus - had a 49B that they were looking to sell. This was the car that Jo Bonnier had crashed heavily at Oulton Park in the Gold Cup meeting, chassis 49B/R8. Charlton takes up the story: 'We needed a new motor car. I went and scratched around in England and the only car available was that car, which Chapman had sold to Bonnier, who'd crashed it at Oulton Park and he'd also repossessed it because Bonnier hadn't paid. So it was Chapman with whom I dealt. And, it was like brand new, when they fixed it up. I just got the lads there at Hethel to cobble it together very smartly.'

When the car was rebuilt, the jigs were set up for 49Bs, so it was built to this specification, rather than the earlier Tasman layout in which it had originally been built in the latter part of 1968. The work undertaken was quite comprehensive, because it involved replacing the rear bulkhead and reworking the cutouts for the lower rear radius arm mounting points from their deep 'Tasman' format to the shallower design found on conventional

49Bs. Doing this also alleviated fuel capacity problems which had been caused by the higher cutouts leaving room for a smaller than average bag tank behind the seat. Following this work, the car then had new side-skins fitted and the damaged rocker arm fairing on the left side of the car was replaced. As Charlton recalls, the decision to undertake such a comprehensive rebuild was partly made by Lotus and partly dictated by his own wish to have a 49 built to the latest specification. 'It was a combined decision. We didn't want to walk in and buy an outdated motor car. Chapman was very accommodating.'

In October 1969, once the fabrication work had been completed and the car sprayed, Charlton's former mechanic Richie Bray, who had worked during 1969 for Paul Hawkins and JW-Gulf, moved temporarily to Hethel to build the car up in an area adjacent to the Formula 1 race shop. This he did over a two or three week period. Charlton was very keen to have the most up-to-date car possible, so he asked for the car to be built to the same specification as the works cars being prepared at the time. He was also keen to get his hands on the car in time for the first round of the SADC, which was to take place on January 10th 1970 at the Killarney. To do this, he had to resort to some rather unorthodox methods in order both to procure the latest specification 'goodies' and to get them quickly enough. 'When I first saw it, it was a bare monocoque on the trestles! The monocoque work had already been done, it was a question of putting the car together. And I had a mate over there, who sort of supervised the whole thing. He gave the lads a few bob on the side, to do the job in a shorter time than normal. And they did an extremely good job.' The lure of hard cash must have worked wonders, for the car was indeed ready in time for it to be shipped out to Cape Town for the first race of the season!

The car was actually paid for, and raced under the name of, Aldo Scribante, who ran the Scuderia Scribante team which had fielded Charlton in the previous two seasons of the SADC. The final cost was a very reasonable £20,000 (US$48,000 at 1969 prices, £187,000/ US$300,000 at late 90s values) including two Cosworth DFV engines, which at the time were selling for £7500 each!

For the 1970 season in South Africa, Love's car was fitted with a smaller, chassis-mounted wing on the rear and a standard nose cone and fins arrangement at the front. In addition, the Gunston team – which now included Pieter de Klerk in an ex-works Brabham BT26 - secured a brace of 9 series engines, DFV 944 replacing Love's venerable 1967 specification DFV (No. 705) which had been in the car since he bought it and the other engine going into the Brabham. This meant that, for the first time, Love would enjoy the luxury of having a spare engine available. With only one engine in the previous two years, it had been it impractical to have it serviced by Cosworth during the season. Fortunately, as Love points out, 'The Cossie motor was a tremendously reliable motor. I said to Ken [Tyrrell] that I had done about nine or ten races [without a rebuild] and he said that was bloody unbelievable! I didn't have a spare motor so I didn't have a chance to send it back to England.' Aside from this the basic specification of the car was unchanged compared to 1969, as Love recounts. 'It wasn't easy for us. We raced on a shoestring, we didn't have a lot of money. And the Gunston backers said "OK, that's fine, you've been fairly successful, just leave it as it is".'

With Killarney being in Cape Town, Charlton's car did not have far to travel for its inaugural outing on South African soil. This first race offered few clues as to the likely impact of Charlton's 49C on the series, for he was still getting used to the car. Even before the car turned a wheel there was drama, as Charlton remembers. 'The first thing that went wrong was when we put petrol in it and the petrol ran out again because the bag tank under my backside was leaking. We had to haul that out and I think we borrowed one from John Love, nobody down there could fix it. That was the way it was in those days.' In practice, he qualified third, while Love took his customary pole. In the race itself, Charlton jumped into the lead but was passed early on by Love, who went on to take a comfortable victory. Charlton eventually retired when his battery cable sheared. On the face of it, nothing had changed...

In the next round, the Highveld 100 at Kyalami, there was the first indication that the balance of power was beginning to shift. Although Charlton qualified second on the grid alongside Love, who had taken pole once more, he got his 49C ahead of the Gunston car and drove to a comfortable victory, Love trailing in a distant second.

After having a miserable South African Grand Prix meeting, where he was out-qualified by both Charlton and de Klerk and then finished a lowly eighth, two laps down on the winner, Love was then beaten to pole for the next race, the Coronation 100, at the Roy Hesketh circuit near Pietermaritzburg. Once again it was Charlton who deposed him, and proceeded to defeat him easily in the race. The writing was on the wall...

At the beginning of May, Love took the plunge and ordered a new car. Looking back on it now, he wishes that he'd taken the opportunity to buy a 49B when he had the chance. 'We were talking about it and then they [Lotus] didn't stick around too long, there were a lot of people looking for that sort of vehicle. By the time we got in there it was too late, and we ended up with a March by mistake! Ken [Tyrrell] had one which Jackie Stewart was driving and said "Well, it's not too bad a motor car" and so the sponsors gee'd one of those up, but it was bad news, it wasn't a great motor car. We had the chance of a Lotus 72 but they thought at that stage it was going to be difficult to maintain it from a distance. We went the wrong direction, we bought a March and then obviously after that, we went even further in the wrong direction by buying a Surtees! Which wasn't a great motor car, but then we bought that on the strength of John's run in the 71 [South African] Grand Prix, which he could have won but he had a problem towards the end.'

Due to component shortages over in England, Love's March was not ready in time for the next round of the Championship, the Republic Trophy at Kyalami, so he was forced to rely on his trusty 49 for one last time. In practice he qualified third to put him on the front row alongside pole-man Charlton - albeit a second slower than the 49C - and de Klerk's BT26. However, disaster

struck Charlton for, when they went to start up and move forward from the dummy grid to the actual grid, his car refused to fire up and the race started with the pole position spot vacant. His mechanics had to wait until after the start before pushing the car into the pits, where an extra 12 volts of power did the trick and the engine burst into life. By this stage, de Klerk had almost completed two laps, so the task for Charlton seemed an impossible one. Nonetheless, he proceeded to drive like a man possessed and was quickly scything his way through the field.

At the front of the race, de Klerk was pulling out a gradual advantage over his Gunston team-mate Love, who seemed unable to match the Brabham for pace. By half distance, Charlton had pulled himself up to sixth place and, on lap 28 of 40, broke the lap record set by John Surtees in the Grand Prix earlier in the year, demonstrating just how hard he was driving. A lap later, Love was in the pits with a loose wheel nut. The problem took four laps to sort, pushing him down to sixth, where he stayed for the remainder of the race. Meanwhile, Charlton broke the lap record again on lap 35 as he moved into second position. He was still a lap behind de Klerk and this is the way things stayed until the finish. However, his tigering drive brought an excellent reward, for his second put him into the lead of the Championship, two points ahead of Love, who only scored a single point in this race.

This was a disappointing end to the Rhodesian's time with the 49, which he had raced at the top level since March 1968. During this time, he had entered 28 races, including two South African Grand Prix (1969 and 1970), started 26 of them, and won 13. He had also taken an incredible 23 pole positions. On this basis, in terms of pure wins, his 49 was the most successful of all, although obviously the quality of the opposition needs to be borne in mind before any attempts are made to compare this with works cars which raced in World Championship Grand Prix. Today, Love recalls the car with great affection. 'For me at that particular stage, it was brilliant. It was a good motor car. I'd driven Coopers prior to that and that Tasman Cooper I bought from Bruce was really a sprint car, because the races in the Tasman series weren't long at all. It was a lot more sophisticated than the Cooper. So it was a nice, good, successful car for me. It was so reliable, a tremendous motor car that didn't give any trouble.'

The first race at the new circuit at Bulawayo, replacing the old one which had been widely regarded as unsafe and unsuitable for the current generation of Formula 1 and Formula A/5000 cars, saw the first appearance of the Love March. Two days before the race, he had sold his 49 to fellow Rhodesian, Peter Parnell, who had made his mark driving an FVA-engined Brabham in the series in 1969. In practice, Charlton put his 49C on pole, with Love in his still-unsorted March alongside, while Parnell managed to qualify a respectable sixth on his first taste of DFV power.

Having secured pole position for the South Africa Republic Day Trophy race at Kyalami, Charlton suffered the frustration of a late start due to a recalcitrant starter motor. He drove the entire race absolutely flat out and managed to salvage a second place for his troubles. (Len Konings/Robert Young Library)

Charlton hurls R8 out of Leeukop Bend during the Rand Winter Trophy at Kyalami, on his way to a convincing victory. (Len Konings/Robert Young Library)

An eventful race ensued. Jackie Pretorius took off like a dragster in his Surtees TS5 Formula A/5000 car but, by the end of the first lap, Charlton was in front. However, on lap five he spun and this allowed Love to close right up and pass him while he was gathering it all together. Charlton quickly regained the lead and looked all set for an easy victory when he suddenly spun off at high speed at the end of the main straight, luckily without hitting anything. This was due to a front radius rod breaking. 'I put the brakes on at the end of the straight and the front wheel tucked underneath the car and it spun. Fortunately, it stayed in one piece, I was lucky.' It transpired that he had clipped some hard rubber marker cones on the inside of one of the corners just prior to his accident, and this probably damaged the radius rod. As a result, he had to sit and watch, while local man Love went on his way in front of a partisan crowd to score a debut win for his March. Parnell ran fifth for a time before being dispossessed by an off-colour de Klerk, but still scored a point in what was his first outing in the car.

With five rounds completed, this was now the halfway stage of the Championship. Drivers could keep their four best results, dropping their fifth. The consistent Love had finished all five races in the points, so had to drop his sixth from the Republic Trophy, while Charlton did not have to drop any, as he had only finished three races. This left Love leading the standings with 30 points, ahead of Charlton on 24. With Love in a new car, it seemed as if Charlton was going to have a fight on his hands if he was to overhaul the Rhodesian to take the title.

What transpired was that the March didn't live up to Love's expectations at all and Charlton scored a whitewash in the second half of the year, winning every Championship and non-Championship event held. At Roy Hesketh in July, the luck was with him for he had to slow dramatically towards the end with a punctured left rear tyre. Fortunately, he had been nearly a lap ahead of his nearest challenger, de Klerk, which enabled him to finish with a 19 seconds cushion still in hand. Meanwhile, Love comprehensively blew his engine up when he was lying second, while Parnell again finished in the points with fourth place. To complete a good day, Charlton also took joint fastest lap with Love in a new record time.

Two weeks later, at the non-Championship race in Mozambique at the Lourenco Marques circuit, he was dominant, taking pole and leading all the way to take an easy win, while Parnell crashed 49/R3 in practice and was unable to effect repairs in time to make the race. When the Championship resumed with the Rand Winter Trophy at Kyalami on August 1st, Charlton picked up where he had left off. He took another pole and led from start to finish, winning despite a 60 second penalty for a jumped start, while his closest rival Love failed to score after crashing in practice. Consistent if not blindingly fast, Parnell collected more points with a fifth place in Love's old 49. This win put Charlton 12 points clear of Love, meaning that the latter had to win the next race at Killarney to stand any chance of defending his title.

The False Bay 100 merely confirmed Charlton's position as the dominant driver in the series. Although Love started his weekend well by gaining pole position, Charlton took the lead at the start and was never headed, Love eventually dropping out with fuel pump troubles. This victory took Charlton's Championship points tally to 51, which meant that even if Love won the remaining two races in the series he would not have been able to surpass the South African's total. After six long years, the reign of King John had ended and a new era in the history of the South African Drivers' Championship was about to begin.

Undaunted by his defeat, Love took pole by 0.4 seconds ahead of Charlton in the penultimate round, the Rhodesian Grand Prix, on his local track at Bulawayo. However, a start-line collision caused him to spin, fortunately without doing serious damage to his car but losing him 15 seconds. Charlton, who was driving on a spare wheel with a worn tyre after a last minute drama on the warm-up lap, drove an intelligent race, doing just enough to keep ahead of the fast-recovering Love, who eventually finished 2.2 seconds behind him in second. Once again, Peter Parnell drove the venerable ex-Love 49 into the points, this time in a solid fourth place. This was actually his and the car's last appearance in competition for, although, he entered the final Championship round, the Rand Spring Trophy, at Kyalami, he did not appear.

Charlton drove another demonstration race at Kyalami, taking pole and leading from lights to flag. The final race of the season was a non-Championship event, the Welkom 100, held at the new Goldfields Raceway in Orange Free State. Love's run of bad luck continued, with another blown engine in practice forcing him to non-start. This contrasted starkly with Charlton, who after being headed for a lap and a half in the early stages, took a comfortable victory, his seventh consecutive win in what was fast becoming an unbeatable combination.

After blowing his engine in the final race of the season, Love was on the lookout for a replacement unit and ended up acquiring his old 49 from Parnell, who would later be killed by terrorists in the Zambese Valley. 'Before that happened I actually bought it back. A little bit of non-payment I suppose! I said to Peter "Look, you haven't paid for it, let me just pay you out and I'll take it back". I wanted the engine'. Love had no plans to run the car in the SADC as it was by now quite outdated and the team already had two more modern Formula 1 cars in the form of the March 701 and the Brabham BT26.

Meanwhile, Pete Lovely had skipped the entire US racing season in order to return to Europe to try and take in the two non-Championship early season British Formula 1 races, plus three Grand Prix in the Netherlands, France and Britain. As mentioned in an earlier chapter, it was a move that wasn't particularly successful, mainly due to the dramatic increase in 1970 in the number of cars trying to get a ride in Grand Prix races, which made it harder to qualify.

Over in South Africa, Charlton's success in the SADC had raised hopes that he might be able attract a sponsor, having raced through 1970 without one, aside from the backing of Scribante and some support from South African Airways and trade suppliers such as Armstrong shock absorbers. 'It was a low-budget operation. We did 2500 miles on the engines. We had an 8 and a 9 series engine. So we used the 8 on the circuits where we needed torque more than top end power and the 9 for places like Kyalami. You changed the pinion bearing in the gearbox in the middle of the season and two sets of dog rings in the whole season. How's that?!' However, despite Charlton being the reigning Champion, new sponsorship was not forthcoming and he started the 1971 season as he had finished 1970, still with no major backer.

Charlton had an inauspicious start to his title defence in the opening round, the Cape South-Easter Trophy at Killarney. After comfortably qualifying on pole, he was forced to retire in the early stages, amazingly with an identical problem to the one he experienced in the same race a year earlier - a broken battery terminal. Arch-rival Love was unable to capitalise on this gift, for he again fell victim to engine troubles, with a broken camshaft, leaving his new Gunston team-mate Jackie Pretorius in the Brabham to win the race.

At Kyalami three weeks later, for the Highveld 100, Charlton was slightly off the pace in practice. He qualified third alongside Paddy Driver's McLaren M10B Formula A/5000 car and pole man Love in the March. A terrific race ensued, with Charlton and Love swapping places constantly and rarely more than a second apart for the full 40 laps. On the final tour, Love tried a banzai outbraking manoeuvre up the inside of Charlton into Crowthorne, the corner at the end of the main straight on the old Kyalami layout, which saw him run wide on the exit. This gave Charlton the breathing space necessary to hold on and take the win a mere 0.8 seconds ahead of his Rhodesian adversary. Pretorius finished third to maintain his Championship lead.

A break of nearly two and a half months followed before the next round of the SADC, during which time Charlton, Love and Pretorius all competed in the South African Grand Prix. Charlton had arranged to drive the second works Brabham and had brought with him sponsorship for this race from United Tobacco's Lucky Strike cigarettes brand. At one stage it had looked as if he would be in line for the permanent number two slot at Brabham but this eventually went to Australian Tim Schenken. None of the South African drivers finished the race, but it had enabled Charlton to make a vital breakthrough in establishing a relationship with United Tobacco.

Pete Lovely's first event of 1971 was in a car which was – albeit in a different specification – just about to enter its fifth season of racing. The non-Championship Questor Grand Prix, held at the newly-opened Ontario Motor Speedway facility in Ontario, California, was an invitation-only event for 30 starters, consisting of 17 Formula 1 and 13 Formula A/5000 cars. Despite the fact that he was a comparatively local driver with a Formula 1 car, Lovely says that he had to work hard to get an entry. 'We had to invite ourselves there. They weren't even going to consider our entry because they wanted US guys in Formula A/5000 cars. I called them up and said "Hey, you've got an American with a Formula 1 car!" And they said "OK, if you can qualify, come on down." So we went.'

Unfortunately, when he arrived, it transpired that, along with two Formula A/5000 cars, Lovely had been made an 'Alternate', which was a reserve who would only get a run if one of the invited cars could not start. During practice, he had a heart-stopping moment when he was going for a faster time towards the end of session and he was thankful for the wide open space that constituted the road circuit loop within the speedway oval. 'We were set to get a better qualifying time because I'm a real slow 'get up to speed' kind of a person. Bruce Harre the Firestone tyre development guy came around. He was doing tyre temperatures and he said to Carlo "Give it another half degree negative in the right rear" and this was with, like, five minutes to go.

'So Carlo grabbed the wrenches and turned it – it was going to be a turn or something in the upper link – the wrong way, so it went to a half degree positive! So I'm going around in the back infield section with Tim Schenken right behind me and there was a fairly fast flat left-hander, with dirt run-off and so-on and it was the right rear that had been changed to positive camber instead of negative. Boy it let loose so fast! Tim came up to me afterwards and said "Wow, you should have been where I was, you should have seen your spin, it was going on and on!"' Although he produced a time in qualifying which was faster than six of the invited drivers, he was second-fastest Alternate, which meant that two of the invited runners would have to non-start for him to get a race. Additionally, there was little chance of the organisers revising the rules and letting him run for, among the six whose times he had beaten were the great Al Unser Snr., his brother Bobby Unser and the legendary A.J. Foyt! Predictably, he didn't get to start...

For the third race in the SADC, Dave Charlton had finally secured the sponsorship deal he longed for and deserved. It came from Lucky Strike, who had obviously been impressed by Charlton's performance in his home Grand Prix, with additional support from the Rotary Watches concern. The team was, from that point on, to be known as Lucky Strike Racing. The narrow and bumpy Roy Hesketh circuit was the venue for this race and, once again, Charlton's main rival was Love. Although the Lucky Strike car took pole by half a second, in the race the Team Gunston March tracked the Lotus very closely in the early stages it seemed as if he was biding his time, waiting for a good opportunity to get past. However, approaching one third distance, Love spun at the end of the main straight. When he rejoined, his engine was misfiring badly and he retired a lap later with a broken fuel feed. Charlton went on to win, moving him into the lead of the Championship standings, just ahead of the consistent Pretorius, who took another third place.

The previous two races had suggested that Charlton and Love were now pretty evenly matched, the latter having got his March well sorted and up to speed. The fourth round of the SADC, the Goldfields Autumn Trophy at the Goldfields Raceway promised a renewal of hostilities between the two and it did not disappoint. In practice, both qualified with an identical time. In the race, they each took their turn to spin out of the lead on the first lap and then they both stormed back through the field in formation, with Love ahead of Charlton. By lap eight of 40, they were back in front and then, a lap later, Charlton spun again, which lost him some ground. Although he closed up again, try as he might, Charlton could not get close enough to challenge for the lead and, eventually, the Gunston car ran out the winner. Nonetheless, Charlton had done enough to maintain his leadership of the SADC, three points ahead of the number two Gunston car of Pretorius and nine points clear of Love. Already, at such an early stage in the proceedings, the prospects of Love regaining his Championship crown were receding.

By this stage, Pete Lovely had embarked on a plan to create a 'special' by grafting the rear end of his 49, including the engine, gearbox, uprights and radius arms, onto the tub of his Formula 2 Lotus 69. The aim was to have the car ready for the Laguna Seca round of the SCCA Continental Championship, which by now enjoyed a prestigious sponsorship from the L&M cigarettes concern. However, all did not go to plan. 'We had intended to take the Lotus 69 with the F1 engine on the back to Laguna Seca but at the last minute we couldn't get the thing to work right. So we threw the 49 all back together again. We had the whole kit off the back end and had it installed, more or less, on the 69. It turned out it was a wiring problem, the electric fuel pump was running backwards – it didn't make fuel pressure, it made zero! So that's the reason why we took the 49 back to Laguna Seca and we intended to run the 69 there to get the bugs out of it before the Grand Prix, because we'd planned to do the Canadian and US Grand Prix with it.'

After qualifying a lowly 14th on the grid, nearly four seconds off pole, Lovely finished a satisfying sixth overall after the two 39 lap heats. This would be the last time the 49 would run in competition, since by the time the Canadian Grand Prix came around, the Lotus special was sorted.

For the non-Championship race event on Love's home track in Bulawayo, Charlton was overshadowed by the Team Gunston car, qualifying in second while Love took pole. However, in the race Charlton took the win, further extending his lead in the title race, the Rhodesian again failing to finish. Although it had been proved fast, probably as quick as Charlton's Lotus, the Team Gunston March had proved chronically unreliable and Love resolved to buy a replacement. Impressed by John Surtees' strong drive in that year's South African Grand Prix in his latest Surtees TS9, he ordered one for himself and the car arrived in time for the fifth round of the SADC, the Republic Trophy at Kyalami in early June.

The Team Gunston driver promptly put his new car on pole, 0.3 seconds ahead of Charlton's 49C, but the handling of the Surtees was not truly sorted and Charlton took the lead after Love spun at around quarter distance. He was never headed, cruising to an easy victory ahead of his nearest Championship rival, Jackie Pretorius, ironically enjoying a considerably better year in terms of results than his Gunston team-mate Love.

With Love having upped the stakes by buying a 1971-spec car, Charlton responded by visiting Lotus between the race at Kyalami and the next round of the Championship at Roy Hesketh, to open discussions about buying a Lotus 72D. He had arranged to drive a 72 at the Dutch Grand Prix in June, but the Australian driver Dave Walker crashed his car in practice and it could not be repaired in time. As the next Grand Prix clashed with the Roy Hesketh race, Charlton had to return home to South Africa

The Roy Hesketh circuit had been widened and resurfaced since the last SADC round there at Easter. In practice, Charlton had once again had to give second best to Love's Surtees, with the next fastest competitor being Pretorius a full second slower than the two front-row men. On race day, unseasonal steady rain put many teams in a spin, some not having wet weather tyres at all, so uncommon was it. Prior to the race, an extra practice session was laid on to allow drivers to acclimatise to the wet conditions.

There was drama among the Lucky Strike crew, for when they went to fit the wet-weather tyres and wheels onto the 49C prior to this 15 minute session, they could not get the left front wheel over the brake caliper! Various attempts were made to solve the problem, including filing down the caliper, which made it pop and a new one had to be fitted, which was also found to be too big! As a result of all this, Charlton missed the entire session and for some 20 minutes afterwards sat impassively in the cockpit with an umbrella over him while his mechanic filed the caliper and tried to get the wheel on. Eventually, the clerk of the course announced that the race would start in five minutes, at which point they gave up and put on a wheel with another tyre compound and Charlton splashed off round the course having had no experience of the wet conditions so far.

At the start Love got the drop on Charlton and quickly pulled out a lead. Charlton was struggling to come to terms with the combination of the conditions and his mix of tyres, and on the second lap coming up Beacon Hill went off in a big way. The impact knocked two wheels off the car, damaging it quite severely. Meanwhile, Love's miserable run of luck continued when he was forced to retire with a broken camshaft while holding a commanding lead. Pretorius went on to win the race, pulling him up to within one point of Charlton in the Championship standings.

The race at Roy Hesketh was to be Charlton's last in the 49C and it was an unfortunate way to end his career with the car, having been so successful during his 18 month tenure in the cockpit. Immediately after the event, Charlton went back to Europe to contest the British Grand Prix in the car he was to buy as a replacement, Lotus 72D/3. The car was then flown home in time to be on the grid for the 25th Anniversary Trophy at Kyalami.

While Charlton was in the UK, his mechanic Ken Howes was busy rebuilding the 49C following its shunt. Now the team had two cars, it would have a better chance to see off the two-pronged attack of Team Gunston if they were both run together, rather than one staying in the paddock as a spare. As a result, Charlton brought in Pieter de Klerk, who had driven for Gunston in 1970, but had been replaced by Jackie Pretorius at the start of 1971. At Kyalami, after qualifying fifth, de Klerk drove a sensible race in an unfamiliar car to come home fourth. It was in this race that the throttle stuck open on Love's Surtees and he had a huge accident, writing the car off. As a result, for subsequent races he was forced to revert to the March 701.

The next two races in the Championship were a huge disappointment to de Klerk and to Lucky Strike Racing. At Killarney, in the False Bay 100, he was eliminated in a start-line shunt having qualified 6th, while Charlton retired with a blown engine, leaving Love to take an easy victory. Then at Bulawayo in the Rhodesian Grand Prix, de Klerk was forced to retire with no oil pressure, again having qualified well up in 5th but was still classified 7th, seven laps behind the winner. Charlton retired once more, leaving Love to win and close to within striking distance in the points standings.

For the final Championship race, the Rand Spring Trophy at Kyalami, de Klerk joined Charlton on the front row, qualifying third. However, in the race he gradually dropped back with throttle trouble and eventually finished right out of the points in seventh. Consolation for Lucky Strike Racing came in the form of Charlton's start to finish victory, which secured him a second successive South African Drivers' Championship title. De Klerk's problems continued in the non-Championship Welkom 100 race at the Goldfields Raceway, where a lengthy pit stop to have a throttle cable repaired cost him 18 laps and then he shunted the car in the final few laps. However, due to the rate of attrition and the small field, he was still classified 6th, while team-mate Charlton scored another victory.

For 1972, Charlton and Lucky Strike Racing continued to run the 49C for de Klerk alongside Charlton's 72D. Being so far away from the Lotus works, Charlton had to resort to novel methods to get the 49C straightened out for the new season, as mechanic Denzil Schultz recalls. 'When I joined them, the car was in pieces because de Klerk had had a fairly minor accident at Welkom. Because of our lack of facility here to work with chromemoly, and the like, and because of the relationship that Charlton had with South African Airways - it was quite a small community - we stripped the car completely and sent the tub off to Airways for them to straighten it. It wasn't bad at all, but it was easier to send the tub to them for their sheet-metal men to repair it. And then I reassembled the car.'

Once the 49C had been rebuilt, invariably it was on, or close to, the pace of the fastest cars, despite the basic design by this stage being nearly five years old! At Killarney and Kyalami for the Cape South Easter Trophy and Highveld 100 respectively, de Klerk qualified third and finished second to Charlton both times.

These turned out to be the only finishes that de Klerk recorded from the seven events he was entered for.

In the Coronation 100 at Roy Hesketh, he retired with gearbox problems having qualified fourth, while at the Goldfields Raceway he had an accident, having again qualified fourth. At the following non-Championship race at Bulawayo, his car was plagued by a misfire, which hampered him in practice, limiting him to seventh on the grid, and then forced him to non-start. The misfire recurred at Kyalami, in the Republic Trophy, resulting in his retirement after qualifying a strong third, and then his engine blew before he was able to do any meaningful practice at the Natal Winter Trophy meeting at Roy Hesketh. This really was the last straw for the frustrated de Klerk, who had a major falling out with Charlton and left the team.

Their disagreements dated back to an engine problem in a race in 1971, when de Klerk had been in a comfortable second position. Exiting a corner he felt something go in the engine, dipped the clutch and killed the ignition. When the team got back to base it was decided to strip the engine and, if possible, to effect repairs locally. Up until that point DFVs had gone back to the UK, but it was becoming very expensive, so Charlton decided to undertake the rebuild himself. On examination it was found that it had just started to run the bearings but there was no damage done. This demonstrated both de Klerk's uncanny empathy for things mechanical and that he was right to switch off when he did.

As there were no dyno facilities to handle these engines, the only way of testing them after a rebuild was to fit them in the car. However, when this engine was reinstalled in the 49 part-way through the 1972 season, it had the misfire which plagued de Klerk for several races. The problem was finally resolved later that year when Charlton entered three European Grand Prix in his Lotus and all the ancillaries of the 'problem' engine, i.e. injection system, black box, distributor, etc., were taken to Race Engine Services in the UK, tested and found to be working OK. Back in South Africa, closer inspection of the engine revealed that the cam timing was out. When corrected it ran perfectly, so this had been the problem all along. What this whole episode served to confirm was the problematic nature of 'do-it-yourself' rebuilds.

A 'gentleman driver' by the name of Meyer Botha was recruited in de Klerk's place. In his first race, the Rand Winter Trophy at Kyalami, he acquitted himself well, taking sixth on the grid and driving steadily to fifth at the finish. However, in the next race at Killarney, the False Bay 100, he had a huge accident from which he was lucky to emerge unscathed, the impact putting a massive dent in the left-side of the tub and writing it off.

So ended the international racing career of the Lotus 49, which had spanned five years and three months. This was a feat that would go on to be repeated by Chapman and Phillippe's replacement for the 49, the Lotus 72, demonstrating the longevity of Lotus designs and illustrating just how ahead of its time the company's design thinking was for its day.

Chapter 23
The Final Analysis

The purpose of this section is to illustrate just how successful the Lotus 49 was during its Grand Prix career, and to assess the reasons why it did so well. For example, was it the car that made the engine, or did the engine make the car?

A straightforward statistical analysis reveals the extent to which the term 'successful' can justifiably be used when describing the Lotus 49. During its career, the 49 clocked up an amazing 12 wins from 41 Grand Prix contested, an admirable win rate of 29%.

The accompanying figures clearly demonstrate that the height of the 49's superiority was in 1967, when it was the only car running the DFV engine, winning 44% of the races it started, and taking pole position every time. In 1968, although the win rate was similar the percentage of poles scored fell as Ferrari got its act together, taking four poles. The four-cam Repco-powered Brabham (twice) and the Honda being the only other cars to get a look in. Despite the fact that the DFV had been made available to both McLaren and Matra for the 1968 season, Lotus 49s were the only Cosworth-powered cars to score pole positions, suggesting that, in terms of pure speed (if not reliability) they were still the fastest DFV-powered cars in the field.

The most successful driver - as shown by the accompanying figures - was Jim Clark, who scored five victories at the wheel of a 49, closely followed by Graham Hill on four. Jochen Rindt and Jo Siffert were the only other drivers to win in one. Clark also had by far the highest win rate, taking victory in half of his starts in a 49.

Graham Hill took the most starts, at 39. Incredibly, there were only two Grand Prix in which a 49 raced when Hill did not drive one – at Mexico in 1969 after his Watkins Glen accident and then in Austria after Rob Walker decided not to go. This was a man inextricably linked with the Lotus 49, from the early tests to putting it on pole for its first race, winning the World Championship in 1968 and then campaigning it like a well-worn slipper almost to the end of its racing career. Hill had lived and breathed the Lotus 49, had nurtured it and developed it and then watched it gradually fade into obsolescence. He truly was 'Mr 49'.

Despite the fact that Graham Hill raced a 49 more times than anyone, his widow Bette still professes to a little frustration that the history books, perhaps, under-

The Lotus 49's record in Grand Prix racing

Year	GPs contested No.	Wins No.	Win rate %	Poles No.	Pole rate %
1967	9	4	44	9	100
1968	12	5	42	5	42
1969	11	2	18	5	45
1970	9	1	11	-	-
Total	41	12	29	19	46

Grand Prix wins in a Lotus 49 by driver

Driver	No. of wins	Win rate %
J. Clark	5	50
G. Hill	4	10
J. Rindt	2	15
J. Siffert	1	5
Total	12	29

state her husband's contribution to the development of the 49 and the DFV engine. 'All the records say Jimmy Clark won the first race with a Ford engine and no-one, but no-one, has ever said it was all down to Graham. He spent hours, days, working with Keith Duckworth, Colin Chapman and Mike Costin getting that engine right and I never felt he got the credit for that, ever.'

What is indisputable is that, without the Ford Cosworth DFV, the Lotus 49 was a fairly basic, uncomplicated chassis. However, it was just the sort of simple bracket (albeit one that broke a lot initially!) that was needed to enable Cosworth and Lotus between them to focus on getting the best out of the DFV in its formative year. By 1968 and 1969 the car had matured into a very well sorted package that had few equals. So, it could be said that the combination of Chapman's engineering genius, Duckworth's brilliance and Ford's and Walter Hayes' foresight in providing the financial resources to make the project happen was indeed the perfect partnership.

Without doubt, the conceptual genius of Colin Chapman was crucial in getting the project off the ground, and then his willingness to push back the technological boundaries ensured that the Lotus 49 stayed ahead of the opposition for most of its racing life. Although the Lotus boss could be abrasive and difficult to work for at times, the high regard in which his mechanics still hold Chapman is indicative of the tremendous loyalty that he engendered among his workforce. Bob Sparshott's comments are typical of this. 'He was inspirational as a leader. He could really come up with some fantastic ideas and he would enthuse you into working. And that's why people gave more than they could give. I mean people collapsed through exhaustion on the team, but they still got up when they'd had a little bit of time to recover, and they'd be off again! They wouldn't do that for many people.

'You knew that, of all the things he got wrong, every now and again, he really would get it right, and you'd be streets ahead of everybody else because they wouldn't do it, they were either frightened or they didn't have the imagination to innovatively design racing cars - they copied. You had more scope in those days, because the rules weren't as tight as they are now. It was hard to develop good engineering skills because you were asked to do things that you thought "Christ! I don't like the look of that ..." but it got results, amazing success.'

Meyer Botha's crash at Killarney was an inauspicious end to the story of the Lotus 49 but, in a way, was very fitting, for it served to underline the structural integrity of the basic design. In an era when drivers perished on a regular basis at the wheel of Formula 1 cars, it was an achievement to be able to say that no-one died at the wheel of a 49 and this fact goes a long way towards rebuffing the critics of Colin Chapman and Lotus who levelled the charge that his cars were fragile. That wasn't to say that a number of drivers didn't have cause to test the soundness of the design along the way, as the accidents to Oliver at Rouen in 1968, Rindt at Barcelona 1969 and Hill at Watkins Glen 1969 bear testimony to, but the important thing is that they survived to fight another day …

Who better to objectively assess the impact of the Lotus 49 than rival drivers and designers around at the time of its introduction and subsequent successful racing career? John Surtees is a man with a unique perspective. The only person to have won World Championship titles on two and four wheels, he drove 3-litre Formula 1 cars of both V8 and V12 configuration from Ferrari, Cooper-Maserati, Honda, BRM, McLaren and his own eponymous marque. He also had first-hand experience of driving for and working with Chapman, for he had raced for Team Lotus in 1960 at the wheel of a Lotus 18. 'Colin Chapman was the foremost creator of Formula 1 cars at that time because he was someone who had been able to drive quickly, and he got his concepts right in that you didn't go along and waste the efficiency of the car with scrubbing off lots of power through corners or making it overweight, or such like. I've often said that the 18 was probably the most competitive car that I ever drove relative to the opposition - in any of my motor racing years.'

Surtees remembers that no-one knew much about the 49 before it arrived at Zandvoort, indicating that Walter Hayes and Colin Chapman must have done a good job of keeping details of their respective parts of the project under wraps. However, he was in no doubt that whatever came out of the partnership would be competitive. 'There was very little information about what had come out. But just because of what Duckworth and his little team had done beforehand with the Formula 2 engines, and things like that, showed that they would come up with something that would be pretty good. By putting Chapman, Ford money and the Cosworth team together, they came up with a package which was another step forward. It was a continuation of the dominance of the English power units. Although there is lots of talk about power characteristics and things, certainly none of the teams I was ever with got anywhere near it.'

Given that the sophisticated four-valve Honda Formula 2 engine had trounced Cosworth's SCA in

Hill: 'Mr 49.' (Ford Motor Company)

Grand Prix starts in a Lotus 49 by driver

Driver	Starts
Graham Hill	39
Jo Siffert	22
Jochen Rindt	13
Jim Clark	10
Jackie Oliver	8
Pete Lovely	4
Emerson Fittipaldi	3
Moises Solana	3
John Love	2
Mario Andretti	2
Richard Attwood	1
Giancarlo Baghetti	1
Jo Bonnier	1
Bill Brack	1
Dave Charlton	1
John Miles	1
Eppie Wietzes	1
Total	113

Lotus 49

Beauty and the Beasts: the contrast between the bulky V12-engined cars and the light V8-motivated Lotus 49s is clearly illustrated by this shot, which shows Surtees' Honda leading the Cooper-Maserati of Jo Siffert as Clark and Hill come up to lap them, pursued by Amon's Ferrari. (Richard Spelberg Collection)

1966, it might have been expected that the Japanese concern would provide some of the stiffest opposition to the DFV in Formula 1. However, the Honda engine that Surtees was using in 1968 always had a significant weight disadvantage which meant that, even if it had a good power output, it was always going to struggle against the lighter DFVs. 'It was taking motor cycle concepts to the extreme, e.g. that you had to have a roller bearing crankshaft - and that meant a centre power take off, and a three shaft gearbox - which meant that we had 200lb of extra weight to carry! Basically, the engine, when it was running correctly and on song – and particularly in the 1967 version – had a good performance but it was held back by silly things like low pressure fuel injection systems.

Whilst not wishing to detract from the obvious superiority of the Lotus 49 and the DFV engine, Surtees believes that the disorganised state of the opposition also played into the hands of the Lotus-Ford-Cosworth combination. 'There weren't any real modern engines, as such, on the scene. Ferrari had thrown their chances of success away with my departure. We were the only ones likely to have challenged with what was there. But you can't go giving a second, a second and a half away on drivers, which is what Ferrari did. They sold the Old Man [Enzo Ferrari] on the idea that they had so much power coming that anybody could drive them! Ferrari should have leap-frogged into a new 3-litre Formula with a car that just disappeared. As it was, of course, they never built the car, because of finance and politics. They went along and did a short-stroke sports car engine for me to start with, which was slower than the 2.4!

'Honda came up with a brave effort but a rather old fashioned one. Against the totally new concept and originality of the [Honda] 1.5 litre car, the 3-litre car was old before its time. That was an old concept and, when I joined them, my faith was all placed in the new car we would get. But that got stopped and we ended up with the disastrous air-cooled thing. BRMs were doing neither one thing nor another. They were just playing around and didn't have their feet on the ground at all. You had the Eagle but, frankly, with the sort of quality that the Weslake was all done to, Dan didn't have a chance of any consistency at all. So Cosworths were in a different league.'

'Repco-Brabham was the most down to earth team, very much on the same sort of lines, the same sort of pragmatic approach as Duckworth had taken. Jack cut a few corners, Ferrari played right into his hands and gave him a couple of Championships! And, of course, the other one that was in was Maserati. If they'd had a budget, and could have made their minds up as to what to do, they had some good potential but no-one had this single objective which Colin Chapman and Duckworth pulled together. The only other one who had his feet on the ground, and who was going for it, was Jack [Brabham]. The others were all their own worst enemies.'

Sir Jack Brabham O.B.E. won the first three litre Formula 1 World Championship title in 1966 in his ultra-reliable Repco-Brabham, and his team-mate Denny Hulme went on to secure the title for the team once again in 1967, despite the clear superiority of, and greater number of wins scored by, the Lotus 49 of Jim Clark. Certainly he rued the day when the 49 hit the tracks. 'Colin coming to light with that car was the biggest bad news we had. It was a very competitive motor car to race against. It was certainly a competitor no matter who was in it! Where Colin shone on that one was that the car and the engine were designed together. Whereas a lot of people – even ourselves - had to design the car around the engine they were using, which in our case was the Repco one. Then when it came to the DFV, Colin had done a much better job of tying the two together. He and Duckworth worked together and they were able to design the car and the engine together, which was the secret of the success of the car.'

Sir Jack does not necessarily subscribe to the popular perception that the chassis was nothing out of the ordinary, as he explains. 'It not only handled very well, the big thing it had over competitors, such as ourselves, was that it had much better traction out of corners. That was the thing I noticed more than anything: that it could come out of a corner and put the power on the road, and

The Final Analysis

that was the big gain that he [Chapman] had made.' The DFV also had a significant power advantage although, as Brabham points out, this could sometimes be to its disadvantage when compared to the driver-friendly characteristics of the Repco engine. 'The DFV was getting on for 40 or 50 more horsepower, but the Repco had a very good torque range and was much easier to drive on short corners. It just lacked the power at the top end.'

Dan Gurney, armed with Weslake V12 power, was another potential threat to the supremacy of the Lotus 49, although the fact that Harry Weslake's engines were all unique, handcrafted pieces of engineering eventually proved too great an obstacle to overcome, there being little or no interchangeability of parts between engines. 'That is the way things were done if you didn't have any real modern machinery. If you were willing to have a go at it anyway, that's what you did. The craftsmen that built the individual engines, they still did a hell of a good job.

'They didn't have Ford Motor Company money behind them and therefore were doing it the old-fashioned way. By dint of tenacity and excellence they were making a competitive engine with old-fashioned machinery. In fact, the cylinder block was made on some old Royal Navy World War I machinery that was purchased almost as scrap, and which was turned into something that had the weight and the rigidity so that they could make accurate machinings! They were totally hand-built. There wasn't any such thing as tape-control. The Weslake group did a fabulous job considering what they were up against and how much money we had to throw at it. In the end they probably received a lot less credit for what they had done than what they deserved.

'It was fundamentally a reliable engine. One of the nice things about the Cosworth – we didn't know it at the time – was that it had a superior scavenging system, and that was one of the main weaknesses of the Weslake engine. It was a good engine, relatively light for a 12 cylinder and it had four valves per cylinder. But it lacked certain things and it could have been a good deal better, but it would have needed a much bigger budget than what we had to work with. The whole effort was underfunded. We had lots of passion, but not much money!'

Gurney believes that the Lotus 49 was '... a technological breakthrough. The engine itself put the fear into me because it had a lot behind it. Even then, the word in the industry was that Duckworth, and so forth, were 'the ones'. At the same time, like the proverbial iceberg, Ford Motor Company had endorsed it and the manufacturing capability, compared to everything else out there, was a whole different ball game. It was a turning point, the dawn of a new era. The Lotus 49 was equal parts Cosworth and equal parts Lotus, a collaboration – one of the first from a design standpoint. It was very, very significant.'

Chris Amon joined Ferrari for 1967 as the junior driver in the team alongside Lorenzo Bandini, Ludovico Scarfiotti and Mike Parkes. By the third race of the season he was the only one of this quartet still racing, Bandini having perished as a result of his Monaco accident, Parkes having an awful crash at Spa and Scarfiotti reportedly too upset by these accidents to continue. Despite apparently being one of the best prepared teams for the new 3-litre

formula, Ferrari failed to capitalise on its position and was undoubtedly ill-prepared for the impact which the combination of the Lotus 49 and the Cosworth DFV would make on Formula 1. As New Zealander Amon points out: 'It certainly had a huge impact in that it sort of changed everybody's ideas of how competitive they were.

'I remember having discussions with Ferrari, particularly with their engine people, and saying "Just watch out when this Cosworth engine comes along, it's probably going to be good because they've done so well in Formula 2 and Formula Junior and Formula 3, and everything else." The reply was "they are going to find Formula 1 a whole lot different!" So I think they were absolutely flabbergasted. I know amongst Ferrari people the feeling was that you needed a V12, anyway. This eight would never do it. It did cause a few problems from the driver's point of view, too. I think people were finding it quite difficult – and not just within Ferrari – to believe that this DFV was so good. When I complained about the fact that we couldn't compete in a straightline, for a while they didn't believe it. Quite a while when I think about it!'

It wasn't just that the DFV engine was quick in a straight line either. As Brabham suggested, the engine also had a significant advantage in terms of torque, as Amon recalls. 'On slower and medium circuits they just had so much better acceleration. On the really quick circuits like Monza and Spa it was a little bit easier to compete with them because the torque wasn't a factor there, it was more of an out-and-out power thing.' Amon – concurring with the opinion expressed earlier by Gurney – also feels that there was one particular aspect of the DFV's design that enabled it to shine. 'One of the secrets of the DFV engine was that Duckworth was really the

Superior: the power and torque of the DFV engine made the 49 unbeatable on its day. (Ford Motor Company)

Catalyst: the Ford Cosworth DFV engine paved the way for British companies to dominate the world motor sports arena, something which continues to this day. (The Dave Friedman Collection)

first guy who ever sorted out oil scavenge problems. A lot of the other engines like the Ferrari and probably the Honda too, whilst they had good power on the bench it didn't actually transmit into the car because the crankcase wasn't scavenging properly and there were quite large power losses.'

From having raced in close company with the 49s, Amon felt that their handling was probably their weakness, particularly with the original cars. 'I don't think the early ones were particularly good, but then it was a brilliant combination with the DFV engine. I always felt through 1967 that my Ferrari appeared to be a better-handling car than the 49, which always looked a bit difficult to drive. Even taking it through to the Tasman series when I drove the Formula 2 car with the 2.4 engine and Jimmy Clark had a 49, again I thought the Ferrari was actually a better car. I had quite a few tussles with Jimmy, and he always looked like he was working hard with it. I think in the early stages it was probably flattered a bit by the fact that it was the only car that was running a DFV engine. They had two very good drivers in Jimmy and Graham, so it had everything going for it …

'Then in 1968, when they put those big wings on, it became a very good car. Chapman was the first one to do the suspension-mounted wings and they had a huge amount of downforce. That seemed to transform the car. By then a lot of people had DFV engines, but it really was the car to beat. So I see the 49 as two stages really. A fairly average car without wings and fairly good one with. Why that should be I don't know, but without wings it never looked a particularly easy car to drive.'

Rival designers of that era also readily acknowledge the significance of the Lotus 49. Ron Tauranac, designer of the all-conquering Championship-winning Brabhams of 1966 and 1967 believes that the DFV engine was the vital ingredient that made the 49 such a successful car. 'The new bit was the DFV engine, then the method of mounting it in the car. It wasn't the first, but it was the first popularly known use of the engine as a stressed unit. When you think about it, it was virtually two of their Formula 2 engines. It wasn't that huge a jump in performance that year [1967]. It had its problems with torsional vibrations in the camshaft and also in the drive train. I think the following year it [the DFV's design quality] became more apparent, because Repco had tried to make a four-cam and it had a lot more torsional vibrations than the DFV ever had! Although individually a lot of the things that were involved in it had been done before, they'd never been put together in such an overall good and successful package. So I would say the Duckworth bit of it was the big deal.'

Robin Herd designed the much-admired McLaren M7As of 1968/69 and then went on to work with Keith Duckworth on the Cosworth four wheel drive Formula 1 car until he left to form March for the 1970 season. A vastly experienced designer of both Grand Prix and Indy cars, he feels the chassis was nothing out of the ordinary, particularly for a Lotus. 'The chassis wasn't really rated as a particularly good one. It was unreliable and it didn't handle well. But what was magnificent was the engine, which was superb. If you'd put it in the spaceframe Brabham or something like that it would have gone a lot quicker, because that was a much better balanced and much better handling car. It was a user-friendly engine and also the packaging of it was superb - it was low, light and short, compared to existing engines, and that helped the performance of the chassis.'

So there it is. The consensus of opinion of those that raced against it and designed rival cars seems to be that, without the engine, the car was good but not revolutionary nor way ahead of the rest of the field. However, the combination of the genius of Chapman, the design practicality of Maurice Phillippe and the supreme driving talents of the likes of Jimmy Clark and Graham Hill ensured that the Lotus 49 was, at its height, invariably ahead of the pack.

I'll leave the final word to Walter Hayes, 'father' of the DFV and the man without whose vision the entire Lotus 49 project would probably never have got off the ground. He argues passionately that the combination of the 49 and the DFV laid the foundations for Britain's current position of strength as the home of the motor racing industry. 'I truly believe that we now have supremacy in world motor racing. If there is a centre of excellence for motor racing it is Britain. You cannot win a GP or a World Championship these days if you haven't got something British in your team. What was it that really created this great dominance in Formula 1? It was the Lotus 49, and then, of course, the availability of the DFV engine.' Certainly not a bad legacy for something which evolved from a sketch on the back of a napkin!

Appendix 1
Chassis By Chassis
The Complete Competition Record Of The Team Lotus Type 49

Chassis 49/R1

The prototype chassis, numbered R1, was completed in early May 1967 according to records kept by Team Lotus mechanic Dick Scammell, and first turned a wheel at Hethel, with Mike Costin and Scammell at the wheel. Subsequently, Graham Hill took over testing duties. Hill was also at the wheel for its first race, at Zandvoort, in the 1967 Dutch Grand Prix and drove it at a further ten race meetings, developing quite an attachment to the chassis. His final appearance in the car was rather poignant, marking the end of one era in Formula One and the beginning of another, since it was at the Spanish GP in 1968, the first Grand Prix since the tragic death of Jim Clark. Hill hurled the faithful prototype around the twisty Jarama track to win the race and give Team Lotus the boost it needed in the wake of his team-mate's demise. With that, he bade farewell to his trusty steed, taking the start of the next race in the new specification 49B, chassis R5.

Five other drivers appeared at the wheel of R1 at race meetings, the most illustrious of these being Clark, who took over the chassis for the 1967 Mexican Grand Prix because it had a fresher engine than his regular mount, R2. He scored a dominant victory in that race and continued with R1 at the non-Championship Madrid Grand Prix, where he was again a comfortable victor. However, for the Tasman series in early 1968 Clark reverted back to R2, having briefly switched to chassis R4 to score his final World Championship Grand Prix victory, at Kyalami in South Africa, on New Year's Day. Hill joined Clark for the last four races in the Tasman series, all in Australia, with R1 in 49T form. This was essentially little different from the Grand Prix specification apart from the fact that it used a 2.5 litre variant of the DFV in order to comply with the Tasman regulations: the engine had been re-christened as the DFW in order to differentiate it.

Eppie Wietzes, Giancarlo Baghetti, Moises Solana and Jackie Oliver all made single appearances at the wheel of R1. In the case of the first three, they drove because Team Lotus fielded a third car for a paying driver, alongside its regular duo of Clark and Hill. None of these drivers were able to match Clark's performances in the car, with Solana making the most significant impression, qualifying 7th on the grid at Watkins Glen on his first acquaintance with a Type 49, putting many highly-rated drivers in the shade. Oliver made just about the most inauspicious start possible to his Team Lotus Grand Prix career when he crashed the car coming out of the tunnel on the first lap of the Monaco Grand Prix and comprehensively damaged it. However, as he was the unfortunate victim of someone else's (Bruce McLaren's) accident, he was not entirely to blame.

Apart from the Monaco shunt, R1 suffered one other major trauma during its racing life and a rather public one at that. In practice for the 1967 British Grand Prix at Silverstone, a mounting for the rear radius arm failed, as a result of a faulty weld, as Hill turned into the pit lane in front of the grandstands and he clouted the

As the prototype, R1 was the test 'mule' used to plug round Snetterton while the second car was being built up. It was this car which experienced the problems of the radius arms kinking the tub and which had to be stripped down and rebuilt with additionally strengthening braces. (Ford Motor Company)

retaining wall, pushing the right front wheel back into the tub. Damage was extensive enough to preclude further use of the chassis for the rest of the weekend, although it returned to service three weeks later at the Nürburgring after Hill had another 'off' in R3.

One of only two cars skinned in 18 gauge Alclad aluminium sheet, the car was modified between the British and German rounds of the 1967 World Championship, with the addition of a new combined oval access hatch and fuel filler as had been incorporated on chassis R3 from new. A new aluminium sheet bulkhead (known as a 'baffle') was added to provide greater torsional integrity at the front of the tub and specifically to give more support for the area where the rocker arm fairings were attached to the tub, since there had been problems with these causing the tub to kink. The position of this baffle is shown on all 49s by a line of rivets arching down the sides of the forward part of the tub, starting at the top of the scuttle just by the leading edge of the air-deflector windscreen, running down underneath the rocker arm fairing and extending to just behind the lower front radius arm mounting point and on to the floor of the tub.

To carry out this modification, the front of the tub had to be cut off, the baffle fitted and then a new skin riveted on, overlapping the original skin and following the line of the new bulkhead. At the same time, revised rocker arm fairings, significantly longer than the originals and to the same dimensions as R3, were fitted in order to spread the loads it transmitted over a wider area and prevent kinking of the tub caused by heavy braking forces. Finally, a small disc-shaped indentation is believed to have been added to the rear face of the bulkhead. Recollections of the various people involved differ as to why this was deemed necessary. The most likely explanation is that the bulkhead bulged under full fuel load, pushing itself against the thin aluminium cover over the nose of the crankshaft. By adding the indentation, this ensured that, in such a situation, the crankshaft nose would not drill through the bulkhead and into the tank, which could have had disastrous results ...

The combination of the distinctive overlapping skin on the front of the car and the oval access hatch make this chassis easily identifiable in subsequent photos and is particularly important in confirming that R1 was indeed later rebuilt as a 49T for the 1969 Tasman series and then as a 49B for the 1969 season, and renumbered as 49B/R9.

Debut:	Dutch GP, 1967
Final race:	Monaco GP, 1968
Race starts:	16
Grand Prix starts:	10
Race wins:	3
Grand Prix wins:	2
Other notes:	Re-numbered as chassis R9, November 1968

Chassis 49/R2, 49B/R2

If ever a Lotus 49 is inextricably linked with the name of Jim Clark, then this is it. It was the car in which he drove to victory on the 49's debut at Zandvoort in June 1967 and, in total, he made 16 starts at the wheel of this particular chassis before his untimely death in April 1968. After his Dutch Grand Prix win, Clark went on to score two more victories in R2 that year, the first in the British Grand Prix at Silverstone and the second, a somewhat fortuitous win, in the US round since he crossed the line with his right rear wheel at a rakish angle and probably would not have been able to complete another lap.

Between the Italian and US Grand Prix of 1967, this chassis received similar modifications to those already carried out (and described earlier in this section) on chassis R1. Although, a larger access hatch was provided, the filler cap remained separate, instead of being included in an integral hatch. This was the only 49 which was modified in this way, all others having the combined access/fuel filler hatch. The aluminum 'baffle' bulkhead was also added. At the same time as these modifications were made, the rocker arm fairings were brought up to the same specification as R1 and R3, in a bid to spread the loads put through them and prevent kinking of the tub in this area. Finally, an indentation is believed to have been added to the rear of the bulkhead at this time for the same reasons as with R1.

Jimmy Clark's mechanic on R2 in 1967, Allan McCall, has fond memories of his year working on the car. He also has a couple of rather valuable souvenirs of it, too. 'I've still got the steering wheel of R2! Halfway through the season, Chunky went through one of his lightening campaigns and we went from the Triumph Herald boss to the first of the six-bolt flanges with a lightened top tube. And as Jimmy's was 'my' car I got to keep the old steering wheel. I've also got the laurels for Zandvoort! It's a little tatty because it sat in my toolbox for about six or eight months! I got hold of it, put a rubber band round it and threw it in the box. Things had no value in those days ...'

The same chassis was then converted (with a 2.5 litre DFW engine replacing the DFV) to 49T specification for the 1968 Tasman Cup and Clark drove it throughout, winning four of the eight rounds, and taking the overall series. On the car's return from Australia, it was loaned out by Colin Chapman to Rob Walker, as a temporary replacement for chassis R4, which had been burnt to a cinder by fire the day before the Race of Champions. Jo Siffert made his bow in R2 in the International Trophy meeting at Silverstone in April, and he took in the next five Grand Prix rounds while the works built up a fresh car to 'B' specification.

Siffert's spell in the car was thoroughly miserable, and must have made him wonder why his team boss had attached so much significance to the switch to Lotus. Apart from an inspired 3rd place on the grid at Monte Carlo, Siffert suffered a constant array of engine and gearbox problems, which led to his retirement in every race with the exception of the French Grand Prix at Rouen, where he finished a lowly 11th after stopping to offer his car to Graham Hill - who had broken down on the return leg from the Nouveau Monde hairpin mid-race. His fortunes were dramatically transformed with the arrival of his 49B (chassis R7), in which he won the

This shot of R2, taken during practice for the German Grand Prix, clearly shows the shaped aluminium shoulder support which was fitted only to Clark's car to stop the Scot moving around in the cockpit. Today, the car owned by Pete Lovely still has the nut-sert holes in the tub where this support was fitted. At this juncture the car still has the small circular access hatch, stubby rocker arm fairings and original strengthening brace in it. (Ted Walker/Ferret Fotographics)

next race, but he must have been glad to see the back of R2, despite its impressive pedigree.

Jackie Oliver inherited the 'hot seat' in R2 for the British Grand Prix. Having virtually written off two chassis already that season - the prototype R1 at Monaco and the first 'pukka' 49B, chassis R6, at Rouen - he was entrusted with R2, which was still in original 49 specification, with the ZF gearbox and oil tank and cooler mounted in the nosecone, save for the addition of 49B nose fins and high rear wing. After qualifying a strong second behind Hill and leading in the early stages of the race, Oliver was eliminated due to what was initially

After its period of loan to the Rob Walker team, R2 returned to the works in time for the British Grand Prix, where it was hurriedly repainted in Gold Leaf colours and a rear wing and nose fins added. However, at this stage it was otherwise a standard 1967-spec 49. (Ford Motor Company)

Between the British and German Grand Prix of 1968, R2 was converted to a pukka 49B. This involved, among other things, the tub being reworked with the cut-outs for the lower rear radius rods, the re-siting of the oil tank and cooler to the rear, swapping the ZF box for a Hewland and adding the swept forward rocker arms. Here the car is shown shortly before the Italian Grand Prix, with the new ducted nose exits. (Classic Team Lotus Collection)

Lotus 49

suspected as crownwheel and pinion failure. However, when the Team Lotus mechanics stripped the car down back at Hethel, the ZF box was intact, and the problem was eventually found to be a broken crankshaft in the engine.

Between the British and German rounds, the car was converted to full 49B specification, using some parts from R6 which Oliver had crashed at Rouen. In order to do this, the tub had to be reworked to accept the different lower rear radius rod pick-up points which were a feature of the 'B' spec. During practice for the US Grand Prix at Watkins Glen, Oliver crashed this car in practice again, damaging the chassis sufficiently to eliminate himself from the rest of the weekend's activities, since the spare was being used (to very good effect, for pole position, to be precise) by Mario Andretti. The final bow for R2 was at the Mexican Grand Prix, where underrated local driver Moises Solana qualified well but retired it early on in the race with a collapsed rear wing. During the winter of 1968, R2 was updated to conform to new regulations being introduced for the 1969 season, and was renumbered as chassis R11.

Debut:	Dutch GP, 1967
Final race:	Mexican GP, 1968
Race starts:	28
Grand Prix starts:	18
Race wins:	7
Grand Prix wins:	3
Other notes:	Re-numbered as chassis R11, November 1968

Chassis 49/R3

This was the chassis which was completed overnight in an effort of Herculean proportions prior to the 1967 British Grand Prix. It differed substantially from the first two Lotus 49s built, in that it was skinned in heavier gauge aluminium sheet (16swg rather than 18swg), had the access hatch and fuel filler in the scuttle combined as one, and the additional internal 'baffle' bulkhead was incorporated in the front of the tub from new, therefore the side skins were one sheet from the rear bulkhead to the front. All subsequent 49s were built in this way. It also had longer fairings behind the front rocker arms, to spread the load and try and prevent kinking from heavy braking forces, and the indentation in the rear face of the bulkhead.

Chassis R3 made a fairytale debut with Hill at the wheel, leading the British Grand Prix from Clark until a bolt fell out of the top link, forcing Hill to pit and have it replaced. His engine later blew up. In the next meeting at the Nürburgring, Hill comprehensively wrecked the car in a practice accident, puncturing the tub near the left front wishbone mounting point.

The car was repaired in time for the next race in Canada, where Hill finished fourth and then, at Monza, he was leading comfortably when the crankshaft let go. In the remaining races that season he scored a couple of 2nd places behind team-mate Jimmy Clark and this order was repeated in the last race for R3 in works colours, in South Africa in 1968. Subsequently, the car was sold to John Love, who campaigned it with considerable success in events in South Africa, Rhodesia and Mozambique until 1970, whereupon he sold it to Peter Parnell. With Parnell defaulting on the payments, Love bought the car back so that he could have a spare DFV engine.

The car languished in a shed at Love's house until London dealer Michael Lavers heard from a colleague in South Africa that it might be for sale. 'I contacted him direct via International Directory Enquiries – he has an exhaust business in Bulawayo. This was around April 1980 and the car finally arrived with me in London in January 1981. It was incredibly original and complete with two ZF gearboxes, masses of spare wheels and a nice big hole in the DFV block! It had grown a nasty rollover hoop, but the chromed original one came with the car.

'The whole package arrived in the most enormous crate, which appeared to have been a family home prior to being used for shipping the car! When I sold it via Doug Nye, Jenks came to inspect it and immediately crawled underneath to find the scrape when Graham Hill dinged the underside of the tub at the Nürburgring in 1967!'

After its overnight build-up, R3 made its debut in the 1967 British Grand Prix, driven by Hill. It led for some time until first a loose bolt and then a blown engine thwarted the Englishman's efforts. Note the very non-standard nose cone, which was the only one available at the time. (Ford Motor Company)

Chassis By Chassis

In only its second event, R3 was crashed heavily, as this picture of the car on the back of the Lotus flat truck shows. The tub was punctured on the left side behind the lower front radius rod mounting point, and it was the marks left by the repairs which allowed it to be identified conclusively as R3 when it returned to England in 1980. (Goddard Picture Library)

John Love won the 1968 and 1969 South African Driver's Championships with R3. Here he is seen with his mechanic Gordon Jones and the 1969 Championship trophy. (John Love Collection)

After an extensive restoration carried out by Nye, Neil Pittaway, Michael Cane, Matt Brown and Alan Charles, the car was delivered to its new owner, the collector Mr Hayashi, in Japan. It subsequently appeared in two historic events at Suzuka and Fuji before being sold to the National Motor Museum in March 1985, where it remains today. The car is regularly demonstrated in the UK. Unfortunately, it was badly damaged in a freak accident caused by a misfire in June 1998, which twisted the tub and broke several key suspension components. However, it has since undergone a superb restoration at the hands of Hall & Fowler, who managed to retain virtually all of the original tub and it remains the only surviving original specification Lotus 49 in the world.

Debut:	British GP, 1967
Final race:	Rhodesian GP, 1970
Race starts:	37
Grand Prix starts:	8
Race wins:	13
Grand Prix wins:	None
Other notes:	Now the only surviving original specification Lotus 49

This is how R3 looked following its extensive restoration - much more like a 'proper' 49 should do! (Goddard Picture Library)

Chassis 49/R4

This chassis had the shortest working life of all the 49s built and, in percentage terms, the most successful race record, with a 100 per cent win rate! The reason for this is that the car was destroyed by fire after it had won its first race. The car was driven to victory by Clark at Kyalami in the season-opening Grand Prix of 1968, following which it was sold to the private Rob Walker/Jack Durlacher Racing team for Jo Siffert to drive. However, Siffert crashed and badly damaged the car in unofficial testing on the Friday prior to the Race of Champions, and the car was promptly totally destroyed by the fire which engulfed Rob Walker's garage the next day.

Debut:	South African GP, 1968
Final race:	South African GP, 1968
Race starts:	1
Grand Prix starts:	1
Race wins:	1
Grand Prix wins:	1
Other notes:	A recreation of this car has been built

Chassis 49/R5, 49B/R5

Built to the same basic design as the other four 49s before it, R5 first ran in December 1967 at Snetterton, in the British Racing Green and yellow colours that all new Lotus racing cars were painted in at that time. It incorporated the same features from new as R3 and R4, in that it was built with an oval access hatch covering the fuel filler, long rocker arm fairings and the internal strengthening baffle, as well as the indentation in the rear face of the tub. It was given its first shakedown run by Graham Hill and it was during this testing that Ford is

Lotus 49

After starting life as an original specification 49, R5 was reworked to become the prototype 49B. Although it was ready in time for, and was taken to, the Spanish Grand Prix of 1968, it did not make its race debut until the next event, the Monaco Grand Prix. (Classic Team Lotus Collection)

believed to have filmed the car for its retrospective documentary film *Nine Days in Summer*, no film crew having been present at the initial runs in May 1967. Clues pointing to this include frost on the ground (Bob Dance's test report records the conditions as frosty) and the fact that the car in the film is running with a rear-mounted oil tank and cooler – something which only appeared in 1968 on chassis R5 and then for all subsequent 49Bs.

It was chassis R5 which was displayed, less engine and with a tubular frame in place of the DFV (the team was short of engines at the time!) at the press announcement of Team Lotus's sponsorship deal with John Player in January 1968. It then underwent a period of testing in Hill's hands (Clark was in tax exile by this stage) before making its first race appearance at the Race of Champions in March. It was an inauspicious debut, the car handling badly in practice and breaking a driveshaft yoke in the race. After this, it was dismantled and the tub was cut and reworked to become the prototype 49B. It was not ready in time for the International Trophy meeting at the end of April and, because Colin Chapman was not present, was not run at the Spanish Grand Prix, instead being kept under wraps in the team's garage.

The first person to drive a 49B in anger was Jackie Oliver, who was being evaluated by the team as a possible replacement driver following Jim Clark's death. He undertook three days of testing in the week following the Spanish Grand Prix which were anything but trouble-free. On the first he suffered electric fuel pump failure at Snetterton, whereupon the team returned to Hethel, where he managed to cover only five miles before a washer jammed in the water pump! Short of engines, the team continued the next day with a 2.5 litre DFW unit, managing 20 miles at Snetterton before a broken right rear damper pick-up curtailed the day's activities. The next day, they finally managed some reasonable mileage, doing 85 miles, again with the DFW. That the day was trouble-free is illustrated by the fact that Bob Dance's records simply state '1.27.6, car OK'. Two days later, and with a fresh DFV installed, Graham Hill took over testing work, running 50 miles at Hethel ('Springs too hard!') and then 45 at Snetterton the next day before the plunge joint on the driveshaft broke.

What this illustrates is that the easy victory scored by the 49B on its debut at Monaco six days later was by no means a foregone conclusion, and most of the team must have been expecting reliability problems given the niggles which had affected the development testing. As it turned out, the car was driven quite sympathetically by Hill and suffered no problems at all in completing the arduous Monaco race.

This was to be Hill's regular chassis for most of the rest of 1968, until Monza when he switched to the new chassis, R6. Thereafter, it was used as a spare car to be driven by a third driver. At Monza, this was the young Italian-American driver Mario Andretti, although when he was refused permission to race, Jackie Oliver took over the car as his own had blown an engine in practice. In Canada, Lotus importer Bill Brack took the wheel, while in the US Grand Prix, Andretti finally made his debut and wowed the Formula 1 establishment by taking pole and keeping pace with Jackie Stewart in the early stages before clutch problems put him out of the running. In Mexico, local hero Moises Solana was due to drive R2 but wasn't satisfied with its performance and demanded a swap to R5, with the result that Jackie Oliver had to give up his seat in the car and revert to his regular mount. However, he had the last laugh for, while Solana's rear wing collapsed early in the race, Oliver went on to score his best result of the season!

This was the final race for the car as R5, as it was subsequently modified to meet the new regulations concerning roll-over hoops and fire extinguisher systems and re-numbered R10 for the 1969 season.

After Graham Hill switched to the newer R6, R5 was used by a series of third drivers in the latter part of 1968. These included Mario Andretti (seen here talking to Hill in the Watkins Glen pits), Bill Brack and Moises Solana. (The Dave Friedman Collection)

Debut:	Race of Champions, 1968
Final race:	Mexican GP, 1968
Race starts:	12
Grand Prix starts:	10
Race wins:	1
Grand Prix wins:	1
Other notes:	Re-numbered as chassis R10, November 1968

Chassis 49B/R6

Chassis R6 was the first 49B to be built from new, R5 having been a modified Mk 1 49. It had a short but eventful life, being the car given to Jackie Oliver as a replacement for the prototype 49, R1, which he had written-off at Monaco. After a steady race to 5th place at Spa, albeit 2 laps behind the winner, Oliver finished a lowly 10th in the Dutch Grand Prix, the last race with the original 49B upswept engine cover and nose fins. For the next race in France, his car and that of Graham Hill sprouted the tallest, widest wings on the grid. During practice, Oliver had a monumental accident caused by a sudden loss of downforce from following another car, and from which he was lucky to emerge unscathed. However, the car was not so lucky, the monocoque being a complete write-off, although it is the author's belief that the tub was rebuilt later in 1968 and formed the basis of the show car built for Ford Motor Company, 49/R12.

Although a car bearing the chassis number R6 appeared for the Italian GP at Monza later that year, this was a completely new car (the eighth in the build sequence) that had assumed R6's chassis plate for customs carnet purposes. In fact, some of the salvageable parts from R6 were subsequently used to complete Rob Walker's 49B-R7 and some were used to complete the replacement 'R6'. This totally new car made an unpromising debut in Italy with Graham Hill driving, for it lost a wheel in the race, causing Hill to crash. However, things improved in the remaining races of the season, for he drove R6 to 4th in Canada, 2nd in the US and 1st in Mexico to clinch a well-deserved second World Driver's Championship title.

This chassis continued to be Hill's regular mount in the 1969 Grand Prix rounds until his crash in the Spanish GP at Montjuich Park, which badly damaged the car and necessitated a switch to the spare 49B, R10, for the next round at Monaco. R6 returned for the Dutch Grand Prix that year, but with Hill's team-mate Jochen Rindt at the wheel. It became the Austrian's regular car for the balance of that season and he had an excellent run in the last few

Graham Hill used R6 from the 1968 Italian Grand Prix onwards until his crash in Barcelona in May 1969. Here he is seen in pits during that disastrous weekend for Team Lotus. (Ford Motor Company)

Jochen Rindt took over R6 from the Dutch Grand Prix of 1969 onwards and it became his regular mount for the rest of time that the works team fielded 49s. He is shown blasting off the grid at Silverstone in the 1969 British Grand Prix, alongside Stewart (Matra) and Hulme (McLaren). (Nigel Snowdon)

Lotus 49

For 1970, R6 was fitted with 13 inch front wheels and new fabricated front uprights. It was driven by Jochen Rindt to a sensational victory, watched by millions of TV viewers worldwide, in that year's Monaco Grand Prix. Here he is rounding Station Hairpin. Note the high downforce rear wing borrowed from the Lotus 72. (Nigel Snowdon)

races of the year, culminating in his first Grand Prix victory in the US Grand Prix at Watkins Glen.

While waiting for the new Lotus 72 to arrive in 1970, Rindt piloted his 1969 chassis, brought up to 49C specification with 13in diameter front wheels, in the South African Grand Prix and the non-Championship Race of Champions. When the 72 did not come up to initial expectations in the Spanish Grand Prix at Jarama and the International Trophy at Silverstone, Jochen reverted to his faithful R6 for the Monaco Grand Prix. There he drove what is generally considered to be the race of his life to catch Jack Brabham, in the closing laps, and harry him into a mistake at the final bend, leaving a delighted Rindt to take the chequered flag in a design now almost three years old. His swan song in the car was in Belgium, at Spa-Francorchamps, where he led briefly before being sidelined by piston failure.

The car was advertised for sale in *Autosport* magazine in November 1970, along with a Type 63 four wheel drive car, a Type 62 sports/GT car, two Type 69s and two 59As. It clearly did not sell since the car was used in 1971 to give a few relatively inexperienced drivers a first taster of Formula One. It made its final bow at the non-Championship Rothmans International meeting in April 1971, driven by Tony Trimmer. Even this meeting was eventful, Trimmer bending the car in practice and the Team Lotus mechanics having to make a mercy dash to Normands Garage in Bradford, where chassis R10 conveniently happened to be on display, for spares to rebuild the car. After a pitstop to cure a sticking throttle and adjust the fuel settings, Trimmer had a steady drive to finish 6th, albeit four laps behind the winner, to conclude the chassis's racing life.

In November 1971, a deal was struck with Rob Lamplough to buy the car, as part of a package which included two Lotus 64 Indy cars from the abortive 1969 Indy campaign, some spares for them, and one Lotus 63 four wheel drive car from 1969. In his book *Team Lotus: The Indianapolis Years* (Patrick Stephens Ltd., 1996), former Team Lotus Competitions Manager Andrew Ferguson tells how Chapman had come across these cars and 'issued the instruction to get rid of them. With Colin it was always best to leave matters until he saw the items again in order to establish that he had not changed his mind, but on this occasion someone [Peter Warr] had carried out his order immediately; a buyer [Lamplough] had been found and a deposit paid.' Once Chapman found out what had been done, he tried to stop the sale going through, but it was too late. However, Lamplough had to resort to the courts to get his car in the end and it took some six years for the matter to be resolved.

In the meantime, in June 1972, Team Lotus loaned the car to the National Motor Museum at Beaulieu. Lamplough finally officially assumed ownership of the car in October 1977, but it remained in the museum until 1985, when it went to Paris for an exhibition. It was never to return, for the French Pioneer hi-fi importer Jacques Setton acquired it from Lamplough while it was there, to add to his already stunning collection. Sadly, since the break up of the collection, the car has been in storage in a bonded warehouse in Lausanne, Switzerland, and has not been seen publicly for some years.

Debut:	Belgian GP 1968 ('R6/1'), Italian GP 1968 ('R6/2')
Final race:	Rothmans International Trophy, 1971
Race starts:	24
Grand Prix starts:	19
Race wins:	3
Grand Prix wins:	3
Other notes:	Two cars given same chassis plate for carnet purposes

Chassis 49B/R7

This chassis was ordered by Rob Walker to replace 49/4, which had been destroyed by fire in March 1968. Originally, it had been hoped that the car would have been ready for the Dutch or French Grand Prix, but in the end it was only finished on the eve of the British Grand Prix in July. The car differed from the works machines in that it had smaller nose fins and a 'customer specification' rear wing which was noticeably lower than on the cars of Hill and Oliver. Otherwise it was a standard 49B, constructed in 16 gauge aluminium sheet, with the large oval access hatch with integral filler cap in the scuttle.

It was a fairytale debut, for Siffert won after virtually a race-long battle with the Ferrari of Chris Amon, and he was consistently on the pace – often quicker than the works cars – for the rest of the season, culminating in a scintillating practice lap in Mexico which garnered him pole position. In the race, Siffert probably would have won had the throttle cable not pulled out of its socket, forcing him to pit. Even so, he stormed back through the field, setting a new lap record and finishing sixth.

In 1969, the car was one of the most immaculately turned out on the grid, distinctive because of the highly polished cam covers on its DFV, while it also had its fire extinguisher bottle mounted on the roll-hoop support, over the injection trumpets of the DFV, instead of behind the oil tank as on the works cars. Siffert's season

Chassis By Chassis

Graham Hill, seen here during the South African Grand Prix at Kyalami, took over the wheel of R7 in 1970. Although he finished in the points a number of times, this was as a result of the car's reliability rather than its performance. The German Grand Prix was the car's last start, the team preferring to wait until its 72 arrived before it did any more races. (Ford Motor Company)

Fae left: The Rob Walker team took delivery of R7 just in time for the 1968 British Grand Prix, finishing the build of the car in the van on the way down from Hethel and in the paddock! Siffert repaid the dedication of the team and its mechanics by winning first time out. (Ford Motor Company)

US$560,000 at late 1990s values).

Debut:	British GP, 1968
Final race:	German GP, 1970
Race starts:	28
Grand Prix starts:	24
Race wins:	1
Grand Prix wins:	1
Other notes:	Currently in C spec but with a high wing

Chassis 49T/R8, 49B/R8, 49C/R8

tailed off after some strong results in the first half of the year, with a string of retirements, including a heavy shunt into the trees at the Nürburgring when the suspension broke, an accident which could have had far worse consequences than it actually did. The tub was repaired but, for 1970, the Swiss left the team, to be replaced by Graham Hill. During the winter, R7 was, like the works cars, brought up to 49C specification, with 13 inch front wheels and fabricated front uprights. The Englishman put in several gritty performances in what was, by now, a slightly outdated design, although he blotted his copybook by comprehensively pranging the car beyond immediate repair at Monaco, which forced the team to borrow a car for the race from the works.

By mid-1970, the car was really beginning to show its age, to the extent that, after the German Grand Prix, Rob Walker decided that he wasn't going to race the 49 again, and would wait until his 72 was delivered. His 49 evoked such strong memories of happy days that Walker could not bear to part with it. It was placed in the ownership of the Walker Family Trust and was on permanent loan to the Donington Grand Prix Collection, where it had been 'semi-converted' back to its high wing configuration, albeit still with 13 inch wheels and 49C front uprights. During the late 1990s it was dismantled and was used as a template for the recreation of R4. In 1999, the decision was taken to sell the car and it came under the hammer at H&H Auctions in July. The buyer was the well-known British historic racer, Geoff Farmer and he intends to race the car when it has been refettled. The price paid for the car was £350,000 (approximately

This chassis, the ninth constructed, was built in October 1968 expressly for use in the Tasman series. As with its sister car, R9, it differed from the regular 49Bs because it was a hybrid design utilising some aspects of the 49B specification and some of the original 49. Features carried over from the original 49 were the oil tank still being mounted in the nose, the use of a combined oil/water radiator, the original front rocker arms at 90 degrees to the tub rather than swept forward, the ZF gearbox and the old-style rear suspension hung on the 'fir-tree' brackets bolted to the DFV. The main 49B feature used was the use of cutouts in the lower rear part of the tub to locate the lower rear radius arms. However, these were considerably deeper than the standard 49B specification, because with the old-style suspension, the radius rod mounting was much higher up the side of the tub. As the cars were using high wings, they were mounted on the tops of the uprights using the same method as had been used for Jackie Oliver's R2 when it was hastily converted to a part 49/part 49B spec for the 1968 British Grand Prix. However, by now it was utilising the wing-feathering system as pioneered in Mexico at the end of 1968.

Having driven a 49B for most of the year in the World Championship, Graham Hill was not overly impressed with his Tasman mount. For the first race at Pukekohe, he had a fuel leak on the startline and ending up sitting in a plastic bag in a pool of fuel, a legacy of the long sea journey to New Zealand having caused the empty rubber bag fuel tanks to perish. The Englishman's mood blackened further when, after bending R9 beyond immediate repair, Rindt had a pukka 49B chassis (R10)

Lotus 49

Constructed specifically for the 1969 Tasman series, R8 was basically a 49 with 49B-style cut-outs for the lower rear radius arms, a high wing and nose fins. This is Hill at Wigram Airfield, Christchurch, where he finished second behind teammate Rindt. (Ted Walker/Ferret Fotographics)

After travelling back by sea from the Tasman series, R8 had to be retrieved from under a pile of rotting fruit on a freighter and fettled up for Richard Attwood to drive at the 1969 Monaco Grand Prix. He is seen here on his way to fourth place. (Ted Walker/Ferret Fotographics)

sent out to him, which he used to good effect for the rest of the series while Hill had to struggle on with the old hybrid R8! Hill's Tasman trip was a rather troubled one all in all, his best results being a brace of second places at Christchurch and Invercargill. He certainly seemed to have more than his fair share of problems, something which was probably attributable to the outdated nature of the car, which one mechanic from a rival team was heard to describe as 'a museum piece'!

As it was not required by the works team for the World Championship, R8 was sent home by sea, while Rindt's 49B was flown home so that the team could have a spare car available at short notice. However, when both R9 (which had been rebuilt following its Tasman shunt) and R6 were crashed at Barcelona, R8 had to be hastily retrieved from under a pile of rotting fruit in a freighter at Southampton, and prepared for use in the Monaco Grand Prix. Richard Attwood was chosen to drive the car as Rindt was still recovering from his Barcelona injuries. Literally all that the mechanics had time to do was to install a fresh DFV in place of the 2.5 litre Tasman DFW unit and fit a 1969 specification roll-over hoop and a fire extinguisher system. For the first practice session (as they had not yet been banned) a high wing was also fitted to the car, but for subsequent practice sessions and the race the car ran 'naked' without any wings after they had been outlawed by the CSI. Given that the car was almost in 1967 specification save for the different radius arm mountings, Attwood acquitted himself well, finishing a creditable fourth in the race.

After Monaco, the car went back to the factory at Hethel and was converted to full 49B specification, which involved switching to the B-type rear suspension mountings with the two crossmembers bolted to the back of the DFV, the addition of a Hewland gearbox, re-siting the oil tank over the gearbox and re-drilling the locating points for the lower rear radius arms so that they were further down the side of the tub. The Swedish driver Jo Bonnier had agreed to buy the car, but the work was only finished just before the British Grand Prix and it arrived in Bonnier's van still resplendent in Gold Leaf Team Lotus colours. However, after Hill and Rindt refused to drive the Type 63 four wheel drive cars, it was hastily requisitioned by Chapman and prepared for Graham Hill's use, Bonnier switching to a four wheel drive. Hill drove a poor race and ended up having to stop for fuel before the end, eventually finishing seventh. This was a legacy of the higher cutouts reducing fuel capacity of the car (which was already tight at Silverstone) and the team was forced to run an auxiliary tank in the nose where the oil tank had been located. Even this was not enough …

After this race, Bonnier finally took delivery of the car and had it resprayed so that it was in dark red with a white stripe down the middle, the colours of Scuderia Filipinetti, under whose name he had an arrangement to enter his cars in 1969. The car appeared like this at the German Grand Prix, where he retired with a fuel leak (a familiar story!) and then he crashed it heavily in practice for the Gold Cup at Oulton Park, when the front suspension broke. Bonnier subsequently returned the car to Lotus saying he wanted nothing more to do with it.

Given the fuel capacity problems with the car and the fact that R8 was quite heavily damaged by its Oulton shunt, it was decided to tear it down and rebuild it more comprehensively to a full 49B specification. This work was carried out between mid-August and around October/November 1969 by the fabrication shop within Team Lotus. The work involved replacing the entire rear bulkhead, both side skins of the tub and fitting shallower B-type cutouts and radius rod mountings. While this work was going on, the South African, Dave Charlton, visited the works, saw the car, and bought it. After the fabrication work was completed, it was rebuilt by Charlton's friend Richie Bray and shipped out to South Africa in time for the first round of the South African Drivers' Championship on January 10th, 1970. 49C parts - 13 inch wheels and new fabricated uprights - were sent out for the second round of the SADC, so it was, by a margin of over one month, the first 49C to race in competition, well before the works cars.

Charlton campaigned the car successfully in South Africa, winning the 1970 SADC and driving it in six of the ten rounds in 1971 before switching to a Lotus 72,

By the time of the 1969 British Grand Prix, R8 had been converted to full 49B spec, although it retained the very deep 'Tasman-style' cut-outs for the lower rear radius arms. The car was sold to Jo Bonnier but was borrowed back for Hill to use in the race after he declined to drive the Lotus 63 four-wheel drive car. (Ted Walker/Ferret Fotographics)

winning the title once again that year. In his last drive in the car, Charlton had quite a heavy accident which ripped two wheels off one side of the car. After it was rebuilt, Pieter de Klerk then took over driving duties for almost the next year until a disagreement with Charlton - over engine misfires and preparation - led to him leaving the team. After that, a 'gentleman driver' by the name of Meyer Botha drove the car in two races, but in the second, at Killarney near Cape Town in August 1972, he had a huge accident in the race and ended up hitting a pole. While Botha was lucky to escape relatively unharmed, the car was less fortunate, a huge dent at the point of impact on the left side of the tub effectively writing it off. Charlton eventually sold the car in Decem-

Bonnier's second outing with R8 ended in a nasty practice shunt at the Oulton Park Gold Cup meeting, after which the car was returned to the works for a complete rebuild and upgrade to 49C specification for the 1970 season. (Ted Walker/Ferret Fotographics)

Lotus 49

In late 1969, R8 was sold to the South African driver Dave Charlton, who raced it in 1970 (seen here in his home Grand Prix at Kyalami) and the first part of 1971, before he took delivery of a Lotus 72. (Ford Motor Company)

The damage inflicted on R8 when Meyer Botha crashed at Killarney was significant, as this picture shows. (John Dawson-Damer)

The restoration of R8 was made possible by the fact that the skin from the undamaged side of the car could be used as a template for a new skin to replace the damaged one. (John Dawson-Damer)

ber 1975 to the Australian collector the Hon. John Dawson-Damer.

As the pictures illustrate, the car was in quite a mess when it arrived in Australia; the left side skin was mangled beyond repair so the right side was removed and used as a template for a new left side by Alan Standfield, the Sydney fabricator contracted by Dawson-Damer to carry out the rebuild. The painstaking reconstruction was completed in 1982 and the car has appeared regularly since at historic race meetings in Australia. The 49 is now part of a superb collection which includes a Lotus 79, 63, 39, 25 and 18.

Debut:	New Zealand GP, 1969
Final race:	False Bay 100, 1972
Race starts:	42
Grand Prix starts:	4
Race wins:	13
Grand Prix wins:	None
Other notes:	-

Chassis 49T/R9, 49B/R9

This car was first built up for the 1969 Tasman series, being completed in November/December 1968, after its sister chassis, R8. However, unlike R8, this was not an all-new car, but the prototype 49 chassis R1 rebuilt. It had been heavily damaged in Jackie Oliver's 1968 Monaco Grand Prix shunt but was reskinned and converted to the same hybrid 49/49B specification as R8. Therefore, it had the oil tank in the nose, a combined oil/water radiator, front rocker arms at 90 degrees to the tub rather than swept forward, a ZF 5DS12 gearbox, the old-style suspension attached to the fir-tree and deep cutouts in the rear of the tub for the radius rod mountings. It also had a high rear wing mounted on the original specification

Chassis By Chassis

Built up from the wreckage of R1, R9 was the sister car on the 1969 Tasman series to Hill's R8. Rindt rolled the car in the second round at Levin and it had to be flown back to Hethel for repairs, reappearing in time for the Race of Champions at Brands Hatch in March of that year. Here Rindt is seen at Pukekohe in New Zealand, contemplating the year ahead. (Euan Sarginson)

49 rear uprights, but fitted with the feathering arrangement controlled by a fourth pedal in the cockpit as first seen at the end of the 1968 Grand Prix season in Mexico.

It's possible to say with certainty that this car evolved from R1 because it had a combination of distinguishing features which only one Lotus 49 ever had. These were the internal strengthening baffle which was added retrospectively at the front of the tub, resulting in an overlapping skin, as described in the chassis history of R1, and an oval access hatch/fuel filler cover in the scuttle. The only other car to have the strengthening baffle added after it had been built (and therefore the overlapping skin) was R2, but this tub still retained the separate access hatch and fuel filler, and does so to this day in its later guise as R11. Close examination of pictures of R9 taken during the first part of the 1969 season (including shots in the pits at Barcelona) quite clearly show this second skin and large oval access hatch, confirming that this was the car which was descended from R1.

In its guise of R9, this chassis had a very short competitive career, lasting just six races. In the Tasman series Rindt managed to hit an earth bank and turn it over in only its second race, at Levin. The damage was too bad to effect a repair in New Zealand, so the tub was flown back to the UK and a replacement one airfreighted out.

The tub was rebuilt at the factory in time to be Rindt's car for the opening World Championship round of the season, the South African Grand Prix at Kyalami. Rindt then drove the car in two non-Championship events, the Race of Champions at Brands Hatch and the International Trophy at Silverstone where, after early problems, he scythed through the field with ease to finish second. The Austrian's final drive in the car could well have had fatal consequences. As he crested a brow on the Montjuich Park circuit which was hosting that year's Spanish Grand Prix, his car's wing collapsed, sending him spinning into the barriers. He then hit his team-mate's car, which had suffered an identical problem a few laps earlier and was at the side of the track, causing R9 to turn over and also bending the tub into a banana shape. Mercifully, Rindt escaped without serious injury. However, the same could not be said about the tub, which was literally thrown into the back of the transporter, taken back to Hethel and then thrown into a skip, thereby becoming the only type 49 tub which was deemed beyond repair.

Barcelona marked the final appearance of R9. Its ancestry can be traced back to R1 by the presence of the overlapping skin running down from the front of the air-deflector windscreen to the tip of the rocker arm fairing and down round the back of the lower front radius rod mounting point. This modification was carried out in July 1967 when strengthening baffles were inserted in the tub. (Ford Motor Company)

Debut:	New Zealand GP, 1969
Final race:	Spanish GP, 1969
Race starts:	6
Grand Prix starts:	2
Race wins:	None
Grand Prix wins:	None
Other notes:	Formerly chassis 49/R1

Chassis 49B/R10

This chassis came into existence during the winter of 1968, but was not in any way a new car. Instead, it was simply the prototype 49B, R5, renumbered. It is not immediately apparent why this chassis was renumbered as, initially, it was not modified in any way although this was the intention of the works before the new season got

Lotus 49

underway. However, careful analysis of the rivet patterns of this car reveals matches with those of R5 and there is also evidence of rivet holes where the original type 49 lower rear radius arm mounting brackets would have been fixed to the rear bulkhead. Furthermore, there was no other car in the Team Lotus race shop at that point in time which could have been renumbered as R10.

Although it had been intended to retain R10 for use in the 1969 World Championship season, Rindt's crash in the second round of the Tasman series at Levin threw these plans into disarray, since it necessitated the dispatch of a replacement car from Hethel as R9 was too badly damaged to be repaired *in situ*. The only available complete car at the time was R10. At this stage, it was in exactly the same specification as it had been in Mexico when Moises Solana drove it.

Rindt used the B-spec car to great advantage in the remaining Tasman rounds. Although he did not win the series title, he was quite clearly the class of the field. What was more, his car was obviously superior to that of teammate Hill's, and this seems to have sown the seeds of the Englishman's dissatisfaction with the treatment he was receiving, as reigning World Champion, at the hands of Colin Chapman. In the five Tasman races Rindt contested in R10, he won two, finished second in another and retired twice.

After the series finished, R10 was flown back to the UK because Lotus needed it to be available as a spare car in case either of the two regular chassis were damaged. It also had to be updated to the specification required for the 1969 World Championship season with the more robust roll-over hoop and the now mandatory fire extinguisher system.

It was not needed again until it was requisitioned, in the aftermath of the Barcelona *débâcle*, as Hill's car for the Monaco Grand Prix, renewing the Englishman's association with the tub which had brought him victory in the 1968 race. True to form, Hill drove a sensible race, inheriting the lead after both Stewart and Amon had retired, and scoring a popular fifth win in the Principality. Thereafter, R10 became his regular car for the balance of the season, and it was in this car that he had the accident at Watkins Glen in the US Grand Prix which nearly finished his career.

Although the accident resulted in the car somersaulting end-over-end (and throwing its unbelted driver out in the process) mechanics who worked on the car at the time confirm that the damage was not as bad as had been initially feared, and was confined mainly to the four corners and the footbox area. As a result, the tub was stripped down and despatched back to the UK for repairs and was subsequently rebuilt for the 1970 season.

For 1970 R10 was, like the other works 49B, converted to C specification, with 13 inch wheels and new, Mike Pilbeam-designed, fabricated front uprights. John Miles drove the car to a well-deserved fifth place in the opening round in South Africa, after which it was used by the Spaniard Alex Soler-Roig in an unsuccessful attempt to qualify for his home Grand Prix at Jarama. With the continuing problems with the Lotus 72, Miles reverted to R10 for the Monaco Grand Prix but failed to get to grips with the circuit and did not qualify. His loss was Graham Hill's gain, for it enabled the Rob Walker team to borrow R10 for the race, the Englishman having uncharacteristically damaged his car, R7, beyond immediate repair in a practice accident.

The car was hastily resprayed overnight and Hill took the start (from the back of the grid, as he had not practised this car) for the third successive time in this chassis. Although it did not bring him quite the result he had been used to in the previous two years, he did managed to finish fifth to maintain his record of having scored points in every 1970 World Championship round to date.

The car's next appearance was in the British Grand Prix, where it was used to give the young up-and-coming Brazilian driver, Emerson Fittipaldi, his first taste of Formula 1. After driving a sensible, steady race to eighth, Emerson really set tongues wagging in his next appearance, in the German Grand Prix at Hockenheim, where he finished a scintillating fourth. The final outing for this chassis (and, indeed, the last ever appearance of a Lotus

The first appearance of R10 was at Christchurch, the third round of the 1969 Tasman series. However, it was not a new car but R5 renumbered. As it was called up at short notice following Rindt's Levin accident, it had yet to be updated to 1969 specification and so still had the old-style roll-over hoop and no fire extinguisher system. (Euan Sarginson)

49 in a World Championship race) was in the Austrian Grand Prix, where Fittipaldi finished a disappointed 18th after problems with the engine running too rich.

Since that race, the car has remained in the ownership of the Chapman family, firstly within Team Lotus and, more recently, as part of the Classic Team Lotus stable run by Colin Chapman's son, Clive. After a rebuild by former Team Lotus mechanics Eddie Dennis (who was with the team at the time the car was racing) and Chris Dinnage, the car reappeared for the first time in over 25 years at the Monaco Grand Prix Historique in 1997. Driven by Joaquin Folch, it took a most appropriate victory in the race for cars of its era, repeating the wins of 1968 and 1969. It has appeared occasionally in historic race meetings since, including the Goodwood Festival of Speed where, in 1998, it ran with a high wing for the first time since practice for the 1969 Monaco Grand Prix. Apart from having been resprayed, the car is still in incredibly original condition. Amazingly, this includes a smudge of 'Rob Walker' blue paint still visible on the lip of the cockpit testifying to the hasty paint job it received back in May 1970!

Debut:	Lady Wigram Trophy, 1969
Final race:	Austrian GP, 1970
Race starts:	17
Grand Prix starts:	12
Race wins:	3
Grand Prix wins:	1
Other notes:	Formerly chassis 49/R5, 49B/R5

After the Barcelona debacle, R10 returned to front-line duty at the 1969 Monaco Grand Prix, where Hill took it to a popular victory, complete with hastily cobbled-up chassis-mounted wing. (Ford Motor Company)

After the works team switched to 72s part-way through the 1970 season, R10 was used as a third car to bring on the promising Brazilian driver, Emerson Fittipaldi, seen here at Brands Hatch during the British Grand Prix meeting. It was the last of the works-run 49s to start a Grand Prix, at Austria in 1970. (Ford Motor Company)

Chassis 49B/R11

During the winter of 1968, chassis 49B/R2 was updated to 1969 specification with the addition of a more substantial roll-over hoop and a fire extinguisher system and was sold to the American privateer, Pete Lovely, who has owned the car ever since. It was also renumbered as chassis R11. Although this assertion contradicts previously published accounts of the history of the chassis, it is possible to say this with confidence since it is supported by considerable weight of evidence.

The most prominent clue is the unique separate fuel filler and access hatch arrangement that was only ever found on R2 and which can be clearly seen in pictures of Lovely's car shortly after he had taken delivery of it and remain on the car to this day. In addition, it was one of only two cars which were rebuilt during the 1967 season with an extra internal baffle, as described in the chassis history of R2. The overlapping skin following the path of the baffle is again quite clearly visible in pictures of Lovely's car. The only other car with this feature, chassis R1 (which was later renumbered R9) was destroyed in Rindt's 1969 Barcelona crash.

A large number of other equally significant factors link Lovely's R11 to R2. When Jimmy Clark first drove the 49 at Zandvoort, he reported that the sheer acceleration provided by the DFV was pushing him back in his seat and upwards. Being smaller in stature than his teammate Hill, he was also moving about in the cockpit and had to be 'wedged' in with various bits of padding to make him comfortable. In those days drivers did not wear seatbelts, so Frank, one of the Lotus fabricators who was particularly adept at shaping aluminium, made him up a special shoulder support for subsequent races which held him more firmly in the cockpit, as well as two side pieces which reduced lateral movement. This is clearly visible in overhead shots of Clark's car in 1967 after Zandvoort. Alan McCall, who worked on the car during that year and actually fitted the shoulder support, has inspected Lovely's car recently and reports that the holes for the shoulder support were still there in the cockpit of the car.

Additionally, there was talk at one stage during 1967 of introducing a progressive throttle linkage to counter the sudden surge of power of the DFV at certain revs. As a result, Colin Chapman designed a progressive throttle linkage system which would help him feed in the

Mario Andretti was the first to drive R11 in the 1969 South African Grand Prix, at Kyalami. By this stage it had already been sold to the American privateer Pete Lovely. The separate filler cap identifying it as being the tub of R2 can just be seen on top of the scuttle above the driver's legs. (Nigel Snowdon)

Chassis By Chassis

Lovely's 49B was hastily resprayed in the shade of red used by the Gold Leaf cars and handed over to him in the paddock at the Race of Champions. The distinctive access hatch/fuel filler arrangement from Clark's '67 car is still clearly visible. As with the Rob Walker car, this chassis was delivered with a 'customer' height wing but it still had the feathering system operated by a fourth pedal in the cockpit, which remains in the car today. (Ford Motor Company)

Resprayed in Lovely's own choice of colours, R11 returned to the European racing scene in 1970. He is seen here entering Druids Bend at Brands Hatch during the Race of Champions meeting. (Ford Motor Company)

power more smoothly and this unit was nut-serted into place in the tub, or at least holes were made for it. Drawings for this unit were done in the week following the Dutch Grand Prix and it made its first appearance, according to contemporary reports, at the French Grand Prix. Keith Duckworth contends that the problem was overcome by introducing a progressive linkage at the engine end rather than at the pedal end, so it may have been this which the reports were referring to. Either way, these holes in the tub are still evident in the tub of the car owned by Lovely today. Finally, Lovely's car has a disc-shaped indentation on the rear of the bulkhead, in line with where the nose of the crankshaft would be, which is believed to have been fitted to the Mark I type 49s from R3 onwards and retrospectively fitted to chassis R1 and R2 at the same time as the baffle and access hatch mods were carried out.

While one of these factors on its own might not be convincing enough evidence, combined they add up to, in the author's opinion, conclusive proof that Pete Lovely owns a car which is descended from R2 and, therefore, the 1967 Dutch Grand Prix winner.

Before R11 was actually delivered to Lovely (but after he'd paid for it!) it was used by Gold Leaf Team Lotus one more time, as a third car run for Mario Andretti in the 1969 South African Grand Prix. After this it was hastily re-sprayed, using the red from the Gold Leaf colours, and handed over to the American in the paddock at the Race of Champions. Lovely subsequently contested this race and the International Trophy at Silverstone before returning to the US, where the car was re-sprayed in pearlescent white with a dark blue stripe. Pete ran it in four rounds of the SCCA Formula A Championship, which was open to 3-litre Formula 1 machines as well as 5-litre stock-block Formula 5000-type cars, and then contested the three North American rounds of the World Championship in Canada, the US and Mexico.

In 1970, Lovely returned to Europe to take in the two British non-Championship events and three Grand Prix, but only managed to get a start in the British round, due to oversubscribed fields. He also failed to qualify for his home race at Watkins Glen. In 1971, he tried to get a race in the invitation-only Questor Grand Prix, but was not admitted despite being faster than some of the invited drivers. He then took in a couple of Formula A races before grafting the back end of the 49 onto his Lotus 69 Formula 2 car, in which guise he contested the Canadian and US Grand Prix.

Eventually, Lovely restored both the Lotus 49 and 69 to their original specification. In the 1990s, he reluctantly decided to sell the car and a buyer, the English enthusiast Anthony Mayman, was found. Sadly, before the car could be fully paid for, Mayman died. After a lengthy battle, Lovely regained ownership of the car in 1997 in return for a payment to the Mayman estate, and he still owns it today. What is more, with the car having now been confirmed as such a significant part of Grand Prix history, he has no immediate plans to sell it …

Debut: South African GP, 1969
Final race: Seafair 200, 1971
Race starts: 16
Grand Prix starts: 6
Race wins: None
Grand Prix wins: None
Other notes: Formerly chassis 49/R2, 49B/R2

Chassis 49/R12

This chassis was the only one of the series which was never actually built up for the purpose of racing and, consequently, it never took part in any competitive motor sport, being intended primarily as a static show car for use by Ford Motor Company.

The author believes (although no records exist to corroborate this) that it was based around the original 49B/R6 crashed at Rouen by Jackie Oliver and that it was rebuilt from scratch with new side panels, a new bulkhead - in fact, just about new everything apart from the tub floor. This would have taken place at the same time as the Tasman cars R8 and R9 were built, *i.e.* between October and December 1968, using the Tasman jigs. This explains the 'Tasman' style lower rear radius arm mounting points and cutouts found on R12, despite the fact that it never made the trip Down Under. It even had a ZF box fitted to it, which was still on the car in November 1969 when it appeared at the Jochen Rindt Show, although to make it look more like a 49B, an oil tank was mounted over the gearbox. At some stage since, it has been converted to full 49B specification, with a Hewland box and rear subframe carrying the suspension, but with the mounting points for the rear radius arms unchanged.

The car was delivered to Ford Motor Company in December 1968 and made its first public appearance at the Racing Car Show at Olympia in January 1969. Thereafter, it toured extensively as a promotional vehicle for Ford, before being donated to the Donington Grand Prix Collection at the Donington Park circuit in Derbyshire, England, where it currently resides.

Debut: Not applicable
Final race: Not applicable
Race starts: None
Grand Prix starts: None
Race wins: None
Grand Prix wins: None
Other notes: -

Replicas and recreations

At the time of writing, a replica of the type 49 design was being completed by Hall & Fowler on behalf of Belgian-domiciled collectors John and Peter Morley. It reportedly incorporates the components discarded when chassis 49/R2 was converted from 49 to 49B specification between the British and German rounds of the World Championship in 1968.

A recreation of chassis 49/R4 was completed in 1998 by Hall & Fowler for a consortium led by David McLaughlin, using Walker's chassis R7 as a template. This originally came into being as a project sanctioned by Rob Walker, who authorised his long-time mechanic,

John Chisman, to build it and keep the proceeds should he wish to sell it on. Unfortunately, John died before he could make much progress with it, and his son Jim eventually sold the project to Tom Wheatcroft, from whom David McLaughlin acquired it.

Although scheduled to appear in time for the 1997 Grand Prix de Monaco Historique, with the Belgian driver Jean 'Beurlys' Blaton at the wheel, the monocoque was not finished in time to allow Hall & Fowler to build up the car and, following the withdrawal of Blaton from the consortium, it remained in pieces for some time. It finally made its first public appearance at the Lotus 50th Anniversary Meeting at Brands Hatch in June 1998. Although little was carried over from the original R4, since the vast majority of components were destroyed by the fire at Walker's garage, great efforts were made to source original Lotus 49 components for the car, including the ZF box, uprights, hubs, driveshafts and brakes, and it even features an early specification DFV.

Team Lotus produced a show car for Ford, chassis R12, in time for the Racing Car Show at Olympia in January 1969. A proud Colin Chapman, who opened the show, watches while his six-year old son, Clive, takes the wheel. Today, Clive keeps the Chapman name associated with the Lotus marque by running Classic Team Lotus. (Ted Walker/Ferret Fotographics)

Lotus 49

The Racing Record of Chassis 49/R1

Year	Date	Event	Circuit	Spec	No.	Driver	Race	Notes	Practice
1967	04-Jun	Dutch GP	Zandvoort	49	6	G. Hill	Retired	Camshaft	1st
1967	18-Jun	Belgian GP	Spa-Francorchamps	49	22	G. Hill	Retired	Gearbox	3rd
1967	02-Jul	French GP	Bugatti Crct, Le Mans	49	7	G. Hill	Retired	Gearbox	1st
1967	15-Jul	British GP	Silverstone	49	6	G. Hill	DNS	Crashed in practice	-
1967	06-Aug	German GP	Nürburgring	49	4	G. Hill	Retired	Suspension	13th
1967	27-Aug	Canadian GP	Mosport Park	49	5	E. Wietzes	Retired	Ignition	16th
1967	10-Sep	Italian GP	Monza	49	24	G. Baghetti	Retired	Suspension	17th
1967	01-Oct	US GP	Watkins Glen	49	18	M. Solana	Retired	B/flagged for p/start	7th
1967	22-Oct	Mexican GP	Magdalena Mixhuca	49	5	J. Clark	1st		1st
1967	12-Nov	Spanish GP (NC)	Jarama	49	1	J. Clark	1st		1st
1968	11-Feb	Surfers' Paradise (TC)	Surfers' Paradise	49T	5	G. Hill	2nd		3rd
1968	18-Feb	Warwick Farm Intl (TC)	Warwick Farm	49T	5	G. Hill	2nd		2nd
1968	25-Feb	Australian GP (TC)	Sandown Park	49T	5	G. Hill	3rd		5th
1968	04-Mar	Sth Pacific Trophy (TC)	Longford	49T	5	G. Hill	6th		2nd
1968	27-Apr	Intl Trophy (NC)	Silverstone	49	5	G. Hill	Retired	Fuel line	6th
1968	12-May	Spanish GP	Jarama	49	10	G. Hill	1st		6th
1968	26-May	Monaco GP	Monte Carlo	49	10	J. Oliver	Retired	Accident, sev. damaged	13th

TC = Tasman Cup   NC = Non Championship

The Racing Record of Chassis 49/R2

Year	Date	Event	Circuit	Spec	No.	Driver	Race	Notes	Practice
1967	04-Jun	Dutch GP	Zandvoort	49	5	J. Clark	1st		8th
1967	18-Jun	Belgian GP	Spa-Francorchamps	49	21	J. Clark	6th	-1 lap	1st
1967	02-Jul	French GP	Bugatti Crct, Le Mans	49	6	J. Clark	Retired	Final drive	4th
1967	15-Jul	British GP	Silverstone	49	5	J. Clark	1st		1st
1967	06-Aug	German GP	Nürburgring	49	3	J. Clark	Retired	Suspension	1st
1967	27-Aug	Canadian GP	Mosport Park	49	3	J. Clark	Retired	Ignition	1st
1967	10-Sep	Italian GP	Monza	49	20	J. Clark	3rd		1st
1967	01-Oct	US GP	Watkins Glen	49	5	J. Clark	1st		2nd
1967	22-Oct	Mexican GP	Magdalena Mixhuca	49	18	M. Solana	Retired	Front suspension	9th
1968	06-Jan	New Zealand GP (TC)	Pukekohe	49T	6	J. Clark	Retired	Engine	1st
1968	13-Jan	Levin GP (TC)	Levin	49T	6	J. Clark	Retired	Suspension damage	1st
1968	20-Jan	Lady Wigram Trphy (TC)	Christchurch	49T	6	J. Clark	1st		1st
1968	28-Jan	Teretonga Trophy (TC)	Invercargill	49T	6	J. Clark	2nd		2nd
1968	11-Feb	Surfers' Paradise (TC)	Surfers' Paradise	49T	6	J. Clark	1st		2nd
1968	18-Feb	Warwick Farm Intl (TC)	Warwick Farm	49T	6	J. Clark	1st		1st
1968	25-Feb	Australian GP (TC)	Sandown Park	49T	6	J. Clark	1st		3rd
1968	04-Mar	Sth Pacific Trophy (TC)	Longford	49T	6	J. Clark	5th		1st
1968	27-Apr	Intl Trophy (NC)	Silverstone	49	12	J. Siffert	Retired	Clutch	8th
1968	12-May	Spanish GP	Jarama	49	16	J. Siffert	Retired	Transmission	10th
1968	26-May	Monaco GP	Monte Carlo	49	17	J. Siffert	Retired	Transmission	3rd
1968	09-Jun	Belgian GP	Spa-Francorchamps	49	3	J. Siffert	7th/DNF	Engine	9th
1968	23-Jun	Dutch GP	Zandvoort	49	21	J. Siffert	Retired	Gears	13th
1968	07-Jul	French GP	Rouen-les-Essarts	49	34	J. Siffert	11th		11th
1968	20-Jul	British GP	Brands Hatch	49	9	J. Oliver	Retired	Final drive	2nd
1968	04-Aug	German GP	Nürburgring	49B	21	J. Oliver	11th		13th
1968	17-Aug	Gold Cup (NC)	Oulton Park	49B	3	J. Oliver	3rd		7th
1968	08-Sep	Italian GP	Monza	49B	19	J. Oliver	DNS	Practice only	11th
1968	22-Sep	Canadian GP	Mont Tremblant	49B	4	J. Oliver	Retired	Transmission	9th
1968	06-Oct	US GP	Watkins Glen	49B	11	J. Oliver	DNS	Practice crash (no spare)	16th
1968	03-Nov	Mexican GP	Magdalena Mixhuca	49B	11	J. Oliver	3rd	Practised in R5	14th

TC = Tasman Cup   NC = Non Championship

Chassis By Chassis

The Racing Record of Chassis 49/R3

Year	Date	Event	Circuit	Spec	No.	Driver	Race	Notes	Practice
1967	15-Jul	British GP	Silverstone	49	6	G. Hill	Retired	Engine	2nd
1967	06-Aug	German GP	Nürburgring	49	4	G. Hill	DNS	Crashed in practice	13th
1967	27-Aug	Canadian GP	Mosport Park	49	4	G. Hill	4th		2nd
1967	10-Sep	Italian GP	Monza	49	22	G. Hill	Retired	Broken crankshaft	8th
1967	01-Oct	US GP	Watkins Glen	49	6	G. Hill	2nd		1st
1967	22-Oct	Mexican GP	Magdalena Mixhuca	49	6	G. Hill	Retired	Broken UV joint	4th
1967	12-Nov	Spanish GP (NC)	Jarama	49	2	G. Hill	2nd		2nd
1968	01-Jan	South African GP	Kyalami	49	5	G. Hill	2nd		2nd
1968	30-Mar	Rand Atmn Trophy (SA)	Kyalami	49	1	J. Love	1st		1st
1968	12-Apr	Coronation 100 (SA)	Roy Hesketh	49	1	J. Love	1st		1st
1968	05-May	Bulawayo 100 (NC)	Bulawayo	49	1	J. Love	2nd		1st
1968	01-Jun	S.A. Republic Trphy (SA)	Kyalami	49	1	J. Love	1st		1st
1968	23-Jun	Natal Winter Trphy (SA)	Roy Hesketh	49	1	J. Love	Retired	Brake pipe failure	2nd
1968	08-Jul	Border 100 (SA)	East London	49	1	J. Love	1st		1st
1968	03-Aug	Rand Winter Trphy (SA)	Kyalami	49	1	J. Love	1st		1st
1968	31-Aug	False Bay 100 (SA)	Killarney	49	1	J. Love	DNS	Crashed in practice	1st
1968	05-Oct	Rand Spring Trophy (SA)	Kyalami	49	1	J. Love	Retired	Suspension	1st
1968	01-Dec	Rhodesian GP (SA)	Bulawayo	49	1	J. Love	1st		1st
1969	11-Jan	Cape Sth-Estr Trphy (SA)	Killarney	49	1	J. Love	Retired	Transmission (UV joint)	1st
1969	01-Mar	South African GP	Kyalami	49	16	J. Love	Retired	Ignition	10th
1969	07-Apr	Coronation 100 (SA)	Roy Hesketh	49	1	J. Love	2nd		
1969	26-Apr	Rand Atmn Trphy (SA)	Kyalami	49	1	J. Love	1st		1st
1969	11-May	Bulawayo 100 (NC)	Bulawayo	49	1	J. Love	1st		1st
1969	31-May	S.A. Republic Trphy (SA)	Kyalami	49	1	J. Love	8th		1st
1969	29-Jun	Natal Winter Trphy (SA)	Roy Hesketh	49	1	J. Love	Retired	Accident	1st
1969	14-Jul	Border 100 (SA)	East London	49	1	J. Love	1st		1st
1969	27-Jul	Taca Governador Generale de Mozambique (NC)	Lourenco Marques	49	1	J. Love	1st		1st
1969	09-Aug	Rand Winter Trophy (SA)	Kyalami	49	1	J. Love	DNS	Engine blew in practice	1st
1969	23-Aug	False Bay 100 (SA)	Killarney	49	1	J. Love	1st		1st
1969	14-Sep	Rhodesian GP (SA)	Bulawayo	49	1	J. Love	6th		1st
1969	04-Oct	Rand Spring Trophy (SA)	Kyalami	49	1	J. Love	1st		1st
1970	10-Jan	Cape Sth-Estr Trphy (SA)	Killarney	49	1	J. Love	1st		1st
1970	31-Jan	Highveld 100 (SA)	Kyalami	49	1	J. Love	2nd		1st
1970	07-Mar	South African GP	Kyalami	49	23	J. Love	8th		22nd
1970	30-Mar	Coronation 100 (SA)	Roy Hesketh	49	1	J. Love	2nd		2nd
1970	06-Jun	S.A. Republic Trphy (SA)	Kyalami	49	1	J. Love	6th		3rd
1970	20-Jun	Bulawayo 100 (NC)	Bulawayo	49	3	P. Parnell	6th		7th
1970	05-Jul	Natal Winter Trophy (SA)	Roy Hesketh	49	3	P. Parnell	4th		8th
1970	19-Jul	Taca Governador Generale de Mozambique (NC)	Lourenco Marques	49	3	P. Parnell	DNS	Crashed in practice	-
1970	01-Aug	Rand Winter Trophy (SA)	Kyalami	49	3	P. Parnell	5th		7th
1970	13-Sep	Rhodesian GP (SA)	Bulawayo	49	3	P. Parnell	4th		11th

SA = South African Drivers' Championship   NC = Non Championship

The Racing Record of Chassis 49/R4

Year	Date	Event	Circuit	Spec	No.	Driver	Race	Notes	Practice
1968	01-Jan	South African GP	Kyalami	49	4	J. Clark	1st		Pole
1968	17-Mar	Race of Champions (NC)	Brands Hatch	49	-	J. Siffert	DNS	Crashed in pre-practice Subsequently destroyed by fire	-

NC = Non Championship

Lotus 49

The Racing Record of Chassis 49/R5

Year	Date	Event	Circuit	Spec	No.	Driver	Race	Notes	Practice
1968	17-Mar	Race of Champions (NC)	Brands Hatch	49	4	G. Hill	Retired	Driveshaft	6th
1968	26-May	Monaco GP	Monte Carlo	49B	9	G. Hill	1st		1st
1968	09-Jun	Belgian GP	Spa-Francorchamps	49B	1	G. Hill	Retired	Driveshaft UV joint	14th
1968	23-Jun	Dutch GP	Zandvoort	49B	3	G. Hill	Rtd/9th	Accident	3rd
1968	07-Jul	French GP	Rouen-les-Essarts	49B	12	G. Hill	Retired	Driveshaft	9th
1968	20-Jul	British GP	Brands Hatch	49B	8	G. Hill	Retired	UV joint	1st
1968	04-Aug	German GP	Nürburgring	49D	3	G. Hill	2nd		4th
1968	17-Aug	Gold Cup (NC)	Oulton Park	49B	2	G. Hill	Retired	Final drive	1st
1968	08-Sep	Italian GP	Monza	49B	19	J. Oliver	Retired	Transmission	11th (R2)
1968	22-Sep	Canadian GP	Mont Tremblant	49B	27	B. Brack	Retired	Driveshaft	20th
1968	06-Oct	US GP	Watkins Glen	49B	12	M. Andretti	Retired	Clutch	1st
1968	03-Nov	Mexican GP	Magdalena Mixhuca	49B	12	M. Solana	Retired	Collapsed rear wing	11th (R2)

NC = Non Championship

The Racing Record of Chassis 49/R6

Year	Date	Event	Circuit	Spec	No.	Driver	Race	Notes	Practice
1968	09-Jun	Belgian GP	Spa-Francorchamps	49B	2	J. Oliver	5th	-2 laps	15th
1968	23-Jun	Dutch GP	Zandvoort	49B	4	J. Oliver	10th	Not classified	10th
1968	07-Jul	French GP	Rouen-les-Essarts	49B	14	J. Oliver	DNS	Accident in practice	11th
1968	08-Sep	Italian GP*	Monza	49B	16	G. Hill	Retired	Lost wheel, accident	5th
1968	22-Sep	Canadian GP	Mont Tremblant	49B	3	G. Hill	4th	-4 laps	5th
1968	06-Oct	US GP	Watkins Glen	49B	10	G. Hill	2nd		3rd
1968	03-Nov	Mexican GP	Magdalena Mixhuca	49B	10	G. Hill	1st		3rd
1969	01-Mar	South African GP	Kyalami	49B	1	G. Hill	2nd		7th
1969	16-Mar	Race of Champions (NC)	Brands Hatch	49B	1	G. Hill	2nd		1st
1969	30-Mar	Intl Trophy (NC)	Silverstone	49B	1	G. Hill	7th	-2 laps	10th
1969	04-May	Spanish GP	Montjuich Park	49B	1	G. Hill	Retired	Accident, wing collapsed	3rd
1969	21-Jun	Dutch GP	Zandvoort	49B	2	J. Rindt	Retired	Driveshaft joint	1st
1969	06-Jul	French GP	Clermont-Ferrand	49B	15	J. Rindt	Retired	Driver unwell	3rd
1969	19-Jul	British GP	Silverstone	49B	2	J. Rindt	4th	-1 lap	1st
1969	03-Aug	German GP	Nürburgring	49B	2	J. Rindt	Retired	Ignition	3rd
1969	07-Sep	Italian GP	Monza	49B	4	J. Rindt	2nd		1st
1969	20-Sep	Canadian GP	Mosport Park	49B	2	J. Rindt	3rd		3rd
1969	05-Oct	US GP	Watkins Glen	49B	2	J. Rindt	1st		1st
1969	19-Oct	Mexican GP	Magdalena Mixhuca	49B	2	J. Rindt	Retired	Broken front wishbone	6th
1970	07-Mar	South African GP	Kyalami	49C	9	J. Rindt	13th/DNF	Engine	4th
1970	22-Mar	Race of Champions (NC)	Brands Hatch	49C	8	J. Rindt	2nd		4th
1970	19-Apr	Spanish GP	Jarama	49C	3	J. Rindt	DNS	Spare (raced 72)	-
1970	10-May	Monaco GP	Monte Carlo	49C	3	J. Rindt	1st		8th
1970	07-Jun	Belgian GP	Spa-Francorchamps	49C	20	J. Rindt	Retired	Piston failure	2nd
1970	05-Jul	French GP	Clermont-Ferrand	49C	9	A. Soler-Roig	DNQ		DNQ
1971	24-Jan	Argentine GP (NC)	Buenos Aires	49C	6	W. Fittipaldi	9th	-29 laps	9th
1971	09-Apr	R/mans Intl Trphy (NC)	Oulton Park	49C	27	A. Trimmer	6th	-4 laps	10th

* Effectively a new car for this race   NC = Non Championship

The Racing Record of Chassis 49/R7

Year	Date	Event	Circuit	Spec	No.	Driver	Race	Notes	Practice
1968	20-Jul	British GP	Brands Hatch	49B	22	J. Siffert	1st		4th
1968	04-Aug	German GP	Nürburgring	49B	16	J. Siffert	Retired	Ignition	9th
1968	08-Sep	Italian GP	Monza	49B	20	J. Siffert	Retired	Suspension	9th
1968	22-Sep	Canadian GP	Mont Tremblant	49B	12	J. Siffert	Retired	Oil system	3rd
1968	06-Oct	US GP	Watkins Glen	49B	16	J. Siffert	5th		12th
1968	03-Nov	Mexican GP	Magdalena Mixhuca	49B	16	J. Siffert	6th		1st
1969	01-Mar	South African GP	Kyalami	49B	4	J. Siffert	4th		12th
1969	16-Mar	Race of Champs (NC)	Brands Hatch	49B	14	J. Siffert	4th		3rd
1969	30-Mar	Intl Trophy (NC)	Silverstone	49B	12	J. Siffert	11th		6th
1969	04-May	Spanish GP	Montjuich Park	49B	10	J. Siffert	Retired	Engine	6th
1969	18-May	Monaco GP	Monte Carlo	49B	9	J. Siffert	3rd		5th
1969	21-Jun	Dutch GP	Zandvoort	49B	10	J. Siffert	2nd		10th
1969	06-Jul	French GP	Clermont-Ferrand	49B	3	J. Siffert	9th		9th
1969	19-Jul	British GP	Silverstone	49B	10	J. Siffert	8th		9th
1969	03-Aug	German GP	Nürburgring	49B	11	J. Siffert	Retired	Accdnt - rocker arm u/s	4th
1969	07-Sep	Italian GP	Monza	49B	30	J. Siffert	Rtd/8th	Engine (piston failure)	8th
1969	20-Sep	Canadian GP	Mosport Park	49B	9	J. Siffert	Retired	Driveshaft	5th
1969	05-Oct	US GP	Watkins Glen	49B	10	J. Siffert	Retired	Drive-belt	5th
1969	19-Oct	Mexican GP	Magdalena Mixhuca	49B	10	J. Siffert	Retired	Accident - with Courage	5th
1970	07-Mar	South African GP	Kyalami	49C	11	G. Hill	6th	-1 lap	19th
1970	22-Mar	Race of Champs (NC)	Brands Hatch	49C	9	G. Hill	5th		8th
1970	19-Apr	Spanish GP	Jarama	49C	6	G. Hill	4th	-1 lap	15th
1970	26-Apr	Intl Trophy (NC)	Silverstone	49C	3	G. Hill	9th		7th
1970	10-May	Monaco GP	Monte Carlo	49C	1	G. Hill	DNS	Practice crash - raced R10	16th
1970	07-Jun	Belgian GP	Spa-Francorchamps	49C	23	G. Hill	Retired	Engine	16th
1970	21-Jun	Dutch GP	Zandvoort	49C	15	G. Hill	12th	Not classified	20th
1970	05-Jul	French GP	Clermont-Ferrand	49C	8	G. Hill	10th		20th
1970	18-Jul	British GP	Brands Hatch	49C	14	G. Hill	6th		22nd
1970	02-Aug	German GP	Hockenheim	49C	9	G. Hill	Retired	Engine	20th

NC = Non Championship

Lotus 49

The Racing Record of Chassis 49/R8

Year	Date	Event	Circuit	Spec	No.	Driver	Race	Notes	Practice
1969	04-Jan	New Zealand GP (TC)	Pukekohe	49T	3	G. Hill	Retired	Front suspension failure	3rd
1969	11-Jan	R/mans Intl (TC)	Levin	49T	3	G. Hill	Retired	Transmission	4th
1969	18-Jan	Lady Wigram Trphy (TC)	Christchurch	49T	3	G. Hill	2nd		2nd
1969	25-Jan	Teretonga Trophy (TC)	Invercargill	49T/B	3	G. Hill	2nd		6th
1969	02-Feb	Australian GP (TC)	Lakeside, Queensland	49T/B	3	G. Hill	4th		3rd
1969	09-Feb	International 100 (TC)	Warwick Farm	49T/B	3	G. Hill	NC	Ignition trouble	3rd
1969	16-Feb	International 100 (TC)	Sandown Park	49T/B	3	G. Hill	6th		3rd
1969	18-May	Monaco GP	Monte Carlo	49T	2	R. Attwood	4th		10th
1969	19-Jul	British GP	Silverstone	49B	1	G. Hill	7th		12th
1969	03-Aug	German GP	Nürburgring	49B	16	J. Bonnier	Retired	Fuel tank leaking	14th
1969	16-Aug	Gold Cup (NC)	Oulton Park	49B	9	J. Bonnier	DNS	Accident in practice	3rd
1970	10-Jan	Cape Sth-Estr Trphy (SA)	Killarney	49C	7	D. Charlton	Retired	Electrics - battery cable	3rd
1970	31-Jan	Highveld 100 (SA)	Kyalami	49C	7	D. Charlton	1st		2nd
1970	07-Mar	South African GP	Kyalami	49C	25	D. Charlton	12th	Not run. at finish - eng.	13th
1970	30-Mar	Coronation 100 (SA)	Roy Hesketh	49C	7	D. Charlton	1st		1st
1970	06-Jun	S.A. Republic Trphy (SA)	Kyalami	49C	7	D. Charlton	2nd		1st
1970	20-Jun	Bulawayo 100 (NC)	Bulawayo, Rhodesia	49C	7	D. Charlton	Retired	Accident	1st
1970	05-Jul	Natal Winter Trophy (SA)	Roy Hesketh	49C	7	D. Charlton	1st		1st
1970	19-Jul	Taca Governador Generale de Mozambique (NC)	Lourenco Marques	49C	7	D. Charlton	1st		1st
1970	01-Aug	Rand Winter Trophy (SA)	Kyalami	49C	7	D. Charlton	1st		1st
1970	29-Aug	False Bay 100 (SA)	Killarney	49C	7	D. Charlton	1st		2nd
1970	13-Sep	Rhodesian GP (SA)	Bulawayo	49C	7	D. Charlton	1st		2nd
1970	03-Oct	Rand Spring Trophy (SA)	Kyalami	49C	7	D. Charlton	1st		1st
1970	17-Oct	Welkom 100 (NC)	Goldfields Raceway	49C	7	D. Charlton	1st		1st
1971	09-Jan	Cape Sth-Estr Trphy (SA)	Killarney	49C	1	D. Charlton	Retired	Broken battery terminal	1st
1971	30-Jan	Highveld 100 (SA)	Kyalami	49C	1	D. Charlton	1st		3rd
1971	10-Apr	Coronation 100 (SA)	Roy Hesketh	49C	1	D. Charlton	1st		1st
1971	01-May	G/fields Atmn Trphy (SA)	Goldfields Raceway	49C	1	D. Charlton	2nd		=1st
1971	16-May	Bulawayo 100 (SA - NC)	Bulawayo	49C	1	D. Charlton	1st		2nd
1971	05-Jun	S.A. Republic Trphy (SA)	Kyalami	49C	1	D. Charlton	1st		2nd
1971	04-Jul	Natal Winter Trphy (SA)	Roy Hesketh	49C	1	D. Charlton	Retired	Accident	2nd
1971	07-Aug	25th Anni. Trophy (SA)	Kyalami	49C	4	P. de Klerk	4th		5th
1971	28-Aug	False Bay 100 (SA)	Killarney	49C	4	P. de Klerk	4th		6th
1971	19-Sep	Rhodesian GP (SA)	Bulawayo	49C	4	P. de Klerk	Retired	No oil pressure	5th
1971	10-Oct	Rand Spring Trophy (SA)	Kyalami	49C	4	P. de Klerk	7th	Throttle trouble	3rd
1971	23-Oct	Welkom 100 (NC)	Goldfields Raceway	49C	4	P. de Klerk	6th	-18 laps (accident)	6th
1972	08-Jan	Cape Sth-Estr Trphy (SA)	Killarney	49C	4	P. de Klerk	2nd		3rd
1972	29-Jan	Highveld 100 (SA)	Kyalami	49C	4	P. de Klerk	2nd		3rd
1972	03-Apr	Coronation 100 (SA)	Roy Hesketh	49C	4	P. de Klerk	Retired	Gearbox	4th
1972	22-Apr	G/fields Atmn Trphy (SA)	Goldfields Raceway	49C	4	P. de Klerk	Retired	Accident	4th
1972	14-May	Bulawayo 100 (SA - NC)	Bulawayo	49C	4	P. de Klerk	DNS	Misfire before start	7th
1972	03-Jun	S.A. Republic Trphy (SA)	Kyalami	49C	4	P. de Klerk	Retired	Misfire	3rd
1972	02-Jul	Natal Winter Trophy (SA)	Roy Hesketh	49C	4	P. de Klerk	DNS	Engine - did not practice	-
1972	05-Aug	Rand Winter Trophy (SA)	Kyalami	49C	4	M. Botha	5th		6th
1972	26-Aug	False Bay 100 (SA)	Killarney	49C	4	M. Botha	Retired	Accident (car written-off)	6th

TC = Tasman Cup   SA = South African Drivers' Championship   NC = Non Championship

The Racing Record of Chassis 49/R9

Year	Date	Event	Circuit	Spec	No.	Driver	Race	Notes	Practice
1969	04-Jan	New Zealand GP (TC)	Pukekohe	49T	2	J. Rindt	2nd		2nd
1969	11-Jan	R/mans Intl (TC)	Levin	49T	2	J. Rindt	Retired	Accident - rtd to UK	1st
1969	01-Mar	South African GP	Kyalami	49B	2	J. Rindt	Retired	Engine	2nd
1969	16-Mar	Race of Champions (NC)	Brands Hatch	49B	2	J. Rindt	Retired	Oil pressure	4th
1969	30-Mar	Intl Trophy (NC)	Silverstone	49B	2	J. Rindt	2nd		8th
1969	04-May	Spanish GP	Montjuich Park	49B	2	J. Rindt	Retired	Accident - wing collapsed	1st

TC = Tasman Cup   NC = Non Championship

The Racing Record of Chassis 49/R10

Year	Date	Event	Circuit	Spec	No.	Driver	Race	Notes	Practice
1969	18-Jan	Lady Wigram Trphy (TC)	Christchurch	49B	2	J. Rindt	1st		1st
1969	25-Jan	Teretonga Trophy (TC)	Invercargill	49B	2	J. Rindt	Retired	Broken driveshaft	1st
1969	02-Feb	Australian GP (TC)	Lakeside, Queensland	49B	2	J. Rindt	Retired	Engine	5th
1969	09-Feb	International 100 (TC)	Warwick Farm	49B	2	J. Rindt	1st		1st
1969	16-Feb	International 100 (TC)	Sandown Park	49B	2	J. Rindt	2nd		1st
1969	18-May	Monaco GP	Monte Carlo	49B	1	G. Hill	1st		4th
1969	21-Jun	Dutch GP	Zandvoort	49B	1	G. Hill	7th		3rd
1969	06-Jul	French GP	Clermont-Ferrand	49B	1	G. Hill	6th		8th
1969	03-Aug	German GP	Nürburgring	49B	1	G. Hill	4th		9th
1969	07-Sep	Italian GP	Monza	49B	2	G. Hill	Rtd/9th	Broken R/H driveshaft	9th
1969	20-Sep	Canadian GP	Mosport Park	49B	1	G. Hill	Retired	Camshaft	7th
1969	05-Oct	US GP	Watkins Glen	49B	1	G. Hill	Retired	Accident	4th
1970	07-Mar	South African GP	Kyalami	49C	10	J. Miles	5th		14th
1970	19-Apr	Spanish GP	Jarama	49C	23	A. Soler-Roig	DNQ		DNQ
1970	10-May	Monaco GP	Monte Carlo	49C	2	J. Miles	DNQ		DNQ
1970	10-May	Monaco GP	Monte Carlo	49C	1	G. Hill	5th	Practised R7	-
1970	18-Jul	British GP	Brands Hatch	49C	28	E. Fittipaldi	8th		21st
1970	02-Aug	German GP	Hockenheim	49C	17	E. Fittipaldi	4th		13th
1970	16-Aug	Austrian GP	Osterreichring	49C	8	E. Fittipaldi	15th		16th

TC = Tasman Cup

Lotus 49

The Racing Record of Chassis 49/R11

Year	Date	Event	Circuit	Spec	No.	Driver	Race	Notes	Practice
1969	01-Mar	South African GP	Kyalami	49B	3	M. Andretti	Retired	Transmission	6th
1969	16-Mar	Race of Champions (NC)	Brands Hatch	49B	15	P. Lovely	6th		10th
1969	30-Mar	Intl Trophy (NC)	Silverstone	49B	17	P. Lovely	Retired	Accident on lap 2	12th
1969	20-Apr	R/side Cntl GP (SCCA)	Riverside	49B	15	P. Lovely	6th		6th
1969	04-May	Royal 76 Cntl GP (SCCA)	Laguna Seca	49B	15	P. Lovely	Retired	Ran out of fuel	3rd
1969	22-Jun	Continental 49er (SCCA)	Sears Point	49B	15	P. Lovely	Retired	Suspension	7th
1969	10-Aug	Donnybrooke GP (SCCA)	Donnybrooke	49B	15	P. Lovely	10th		4th
1969	20-Sep	Canadian GP	Mosport Park	49B	25	P. Lovely	7th		16th
1969	05-Oct	US GP	Watkins Glen	49B	21	P. Lovely	Retired	Driveshaft	16th
1969	19-Oct	Mexican GP	Magdalena Mixhuca	49B	21	P. Lovely	9th		16th
1970	22-Mar	Race of Champions (NC)	Brands Hatch	49B	10	P. Lovely	Retired	Accident	12th
1970	26-Apr	Intl Trophy (NC)	Silverstone	49B	17	P. Lovely	13th		20th
1970	21-Jun	Dutch GP	Zandvoort	49B	31	P. Lovely	DNQ		DNQ
1970	05-Jul	French GP	Clermont-Ferrand	49B	25	P. Lovely	DNQ		DNQ
1970	18-Jul	British GP	Brands Hatch	49B	29	P. Lovely	10th	Not classified	23rd
1970	04-Oct	US GP	Watkins Glen	49B	28	P. Lovely			DNQ
1971	28-Mar	Questor GP (NC)	Ontario Speedway	49B	40	P. Lovely	DNS	Practised, but a reserve	DNQ
1971	02-May	L. Seca L&M GP (SCCA)	Laguna Seca	49B	25	P. Lovely	6th		14th
1971	07-Aug	Seafair 200 (USAC)	Seattle Intl Raceway	49B	-	P. Lovely	5th/7th		4th

NC = Non Championship
SCCA = Sports Car Club of America Formula A Championship
USAC = United States Auto Club Formula A series

Appendix 2
Bibliography

Autosport File: Lotus, Temple Press, London, 1988
Colin Chapman: The Man and his Cars, by Gerard ('Jabby') Crombac, Patrick Stephens Ltd, Wellingborough, 1986
Cosworth: The Search For Power, by Graham Robson, Patrick Stephens Ltd, Wellingborough, 1990
Emerson Fittipaldi, by Gordon Kirby, Hazleton Publishing, Richmond, 1990
Jim Clark: The Legend Lives On, by Graham Gauld, Patrick Stephens Ltd, Sparkford, 1989
Jim Clark: Tribute to a Champion, by Eric Dymock, Dove Publishing, Sutton Veny, 1997
Jochen Rindt, by Alan Henry, Hazleton Publishing, Richmond, 1990
John Player Motorsport Yearbook 1973, edited by Barrie Gill, The Queen Anne Press Ltd, London, 1973
Life At The Limit, by Graham Hill, William Kimber, London, 1969
Lotus: A Formula One Team History, by Bruce Grant-Braham, The Crowood Press Ltd, Ramsbury, 1994
Lotus: All the Cars, by Anthony Pritchard, Aston Publications Ltd, Bourne End, 1990

Motor Racing Year 1967-8, edited by the staff of Motor Racing magazine, Knightsbridge Publications, 1967
Springbok Grand Prix, by Robert Young, M.R. Kinsey, Durban, 1970
Team Lotus: The Indianapolis Years, by Andrew Ferguson, Patrick Stephens Ltd, Sparkford, 1996
The 4-Wheel Drives: Racing's Formula for Failure?, by Alan Henry, Macmillan London Ltd, London, 1975
The Autocourse History of the Grand Prix Car 1966-1985, by Doug Nye, Hazleton Publishing, Richmond, 1986
The Grand Prix Who's Who, by Steve Small, Guinness Publishing, London, 1994
The Lotus 49, by David Hodges, Lionel Leventhal, London, 1970
Theme Lotus, by Doug Nye, MRP, London 1978

Magazines referred to for research included: *Motor Sport, Autosport, Motoring News, Autocar, Motor, Autosprint, Auto Motor und Sport, Speedworld International* and *Classic & Sports Car.*

Index

Page numbers in *italics* indicate illustrations

Absalom, Hywel 'Hughie' 128
Allison, Cliff 139
Amon, Chris 39, 48-49, 56, 59-60, *76*, 83-84, 86-87, *90*, 91-92, 102, *107-108*, *108*, 111, 114, 117, 118, 122, 131-132, 134-137, 147, 149, 153, 201, 202, 205-206, 225-226, 234
Andretti, Mario 95, 112-114, *115*, 119-122, *121-122*, 128, 139-140, 158, 166-167, 170-172, 194, 199, 230, 232, *232*, *242*, 244
Arundell, Pete 14, 194
Aston Martin 12
Attwood, Richard 'Dickie' 100, 105, 113, 151, 153-154, *154*, 236, *236*
Austrian GP 1970 208
Autolite spark plugs 20, 43

BMW engines 11
BRM engines 14, 15, *15*, 25
Baghetti, Giancarlo 55-56, 227
Bailey, Len 131
Bandini, Lorenzo 99, 125, 225
Bartels, Graeme 'Rabbit' 195, 206
Belgian GP 1966 14; 1967 41-43, *42*; 1968 101-102; 1970 204-205
Bell, Derek 131, 135-138
Beltoise, Jean-Pierre 96, 103, 153, 163, 170-171, 207
Birchall, Arthur 'Butty' 194
Blash, Michael 'Herbie' 128, 147-148, 150, 153, 158, 173, *179*, 195, *202*
Blaton, Jean 'Beurlys' 245
Bonnier, Jo 125, 161, 163, 166-167, *166*, 169-170, *179*, 194, 236
Botha, Meyer 221, 223, 237
Brabham, Jack 'Black Jack' 37-39, 44-45, 48-49, 54-59, 63, 83, *89-90*, 94, 111, 122, 125, 140, 143-144, 171, *172*, 173-174, 197-199, 202-203, *203*, 205, 207, 210, 224-225, 234
Brack, Bill *78*, 116-119, *116*, 232
Bray, Richie 194, 216, 236
Bridge, Dougie 25, *27*, *33*, *55*, 128, 131, 228
British GP 1967 46-49, *48*; 1968 106-9, *107*; 1969 162-166, *164*; 1970 206-207
Brown, Bill 11, *21*
Brown, David 12
Brown, Matt 231

Canadian GP 1967 53, *54*; 1968 116-118; 1969 170
Cane, Michael 231
Cannon, John 212, 214
Carr, Sid *95*, 101
Cevert, Francois 206
Chamberlain, Jay 139
Champion spark plugs 20
Chapman, Clive 241, *245*
Chapman, Colin 6, 8-13, *8*, 17-19, 23-24, 31, *32-35*, 39, *39*, 41, *41*, *45*, 46-48, *47*, 50-53, *55*, 56, 59-60, *61*, 62-63, *65*, *77*, 82, 85-86, 91, 93, 95, 97-100, 102-103, 105, 109, 112-113, 116, 119, 120-121, 123-128, *123*, *125*, 139-140, 145-147, *145*, *146*, 149-152, 154, 156-158, *156*, 160-161, 163-165, 168, 171, 174, *174*, 176, 193-194, 199, 202, 204, 206, 208-209, *209*, 221, 223-226, 228, 234, 236, 240, 244, *245*
Charles, Alan 231
Charlton, Dave *184*, 194-197, *197*, 211, 215-221, *217-218*, 236-238, *238*

Chisman, John and Jim 245
Clark, Jim 6, 10-11, 14-15, 25, 35-41, *36-40*, 42, 43-44, *44-45*, 45, 47, 49, 51-57, *57*, 58-60, *61*, 62-64, *65-73*, 82-88, *84-90*, 91-92, 94-95, *94*, 98, 100, 129, *185*, *189*, 194, 210, 222-223, 226-228, 230-232, 244
Classic Team Lotus 241
Cleverley, Tony 91, 106, 109, 171
Clifton, Si 20
Collier, Stan 171
Colotti 17
Copp, Harley 13, 39, *41*
Costin, Mike 10-12, 15, *15*, 18, 20, *21*, 31, 33, *33-34*, 46, 136-137, 223, *227*
Cosworth 8, 10-13, 16-17, 43, 64
Courage, Piers 87, *88*, 91, 131, 134-138, 143-144, 154, 160, 165, 170, 201, 203, 205,
Coventry Climax 8, 10-11, 12, 14-15, *27*
Cowe, Billy 128, 149, *149*, 150, 160, 165, 173, 175, 194
Crombac, Gerard 'Jabby' 35-36, 39, 56, 64, 85, 95, 152, 158-9, 203

Dance, Bob *71*, *77*, 82, 93, 95, *95*, 98-101, 104-105, 112-113, 117, 119-125, 128, 158, 162, 164, 165-166, 170, 194, 232
Dawson-Damer, the Hon. John 238
De Dion axles 128
De Havilland 14, 53
De Klerk, Pieter 216-218, 221, 237
Dennis, Eddie *77*, 101, 128, *145*, 150, *159*, *163*, 166, 173, 175, 195, *202*, 241
Dinnage, Chris 241
Donington Grand Prix Collection, The 244
Driver, Paddy 219
Duckett, George 20-21
Duckworth, Keith 6, 9-13, 16-21, *21-22*, 23-24, 27, 33, *33-35*, 38-41, *39*, *41*, 43, *45*, 46, *47*, 49, 52-53, 54-55, 56, *61*, 62, 64-65, 84, 91, 223-226, 244
Duncan, Sir Val 92
Durlacher, Jack 176
Dutch GP 1967 35; 36-40, *37-38*; 1968 102-3; 1969 157-160; 1970 205

Eaton, George 173
Ecclestone, Bernie 152, 174, 175
Elford, Vic 118, 161, 163, 165, 167
Endruweit, Jim 25, 28-29, 47, 82, 95, 98-99, 130, 194

Farmer, Geoff 235
Ferguson, Andrew 82, 95, 151, 194, 234
Ferris, Geoff *24-25*, 193
Fittipaldi, Emerson 176, *182*, 204, 206-209, *206-208*, 241, *241*
Fittipaldi, Wilson 209
Ford, Henry 13
Ford/Cosworth engines
  105E 10
  MAE 10
  SCA 10-11, 224
  FVA 10-11, 13, 16-17, *17*, 20, 51, 64, 128
  FVB 16
  DFV 8-9, 11, 13, 15, 16-23, *20-23*, 24-25, *27-29*, 31, 39, 41, 44, 50-51, 64, *65*, 84, 93, 103, 126-127, 157, 167, 213, 221, 222, 224 226, *225-226*
  DFW 84, 129, 130
Four valve engines 10-11
Foyt, A.J. 219
Franks, Roy 25
French GP 1966 14; 1967 43-45, 46; 1968 104-106; 1969 160-161; 1970 205-206
Frost, Ron 84

254

Index

Gardner, Frank 87, 131, 135, 137
Geoghegan, Leo *88*, 136
German GP 1967 50; 1968 109; 1969 166-167; 1970 207-208
Gethin, Peter 214
Gurney, Dan 37-38, 41, 43, 45, 48, 52-53, 55, 58-59, 63, 117-118, *122*, 125, 225

Hall, Jim 53
Hall, Mike 18
Hall & Fowler 231, 244
Harre, Bruce 197, 219
Hart, Brian 14
Hayashi, Mr 231
Hayes, Walter 9, *11*, 12-13, 15, 17-18, 21, 39, 51-52, 59-60, *61*, 62, 64, 126, 223, 226
Hayward, Brian *24*, *87*
Hemispherical head 11
Hennessy, Sir Patrick 12, 23
Herd, Robin 174, 226
Hewland gearboxes 17, 93, 117, 130, 134-135, *135*, 140, 211-212
Hewland, Mike 93
Hill, Bette *77*, 95, *95*, 97, 111, 126, 207, 222-223
Hill, Graham 6, 8-9, 15, 23, *32-35*, 34-41, *38*, *40*, *42*, 43-44, *45-47*, 47-50, 52-56, *55*, 58-60, *59*, *61*, 62-64, *65-67*, 69, *73-77*, 79-80, 82-84, *83*, *86-88*, 87, 91-92, *92*, 95-104, *95-96*, *100*, 106-109, 112-126, *107*, *125*, 127-128, 131-132, *133*, 134-137, *134-137*, 139-140, 142-144, 147, *147-149*, 149-152, 154-56, *154-155*, 158-163, *158*, 165-167, *165*, *167*, 170-173, *174*, 175, *177-178*, *183*, 194-198, *196*, *198*, 201-203, *201*, *203*, 205-8, 222-223, 226-230, *230*, 232-233, *232-233*, 235-236, *235-237*, 240-241, *241*
Hill, Phil 55
Hobbs, David 49, 115, 214
Holmes, Denis 21
Honda engines 10-11, 223-224
Hopkins, Tex 59, *59*, *61*, 121
Howes, Ken 220-221
Huckle, Gordon *45*, *55*, *61*, 195
Hugenholtz, John 63
Hulme, Denny 38, 44, 48-49, 51-52, 54, 56-57, 59-60, 63, 65, 92, 95, *96-97*, 106, 109, 114-115, 118, 122-125, 131, 142, 161, 163, 170, 172-173, 224

Ickx, Jacky 51, 53, 92, 102, *102*, 105-109, 111, 114-115, 117, 128, 142-144, 167, 170-171, 173, 205-207, 209
Irwin, Chris 49
Italian GP 1966 14; 1967 55; 1968 112-115; 1969 *169*, 170; 1970 208

Jackson, George 20
Jenkinson, Denis 'Jenks' 44, 59, 128, 160, 170, 173, 230
Jones, Gordon 140, *231*
Jones, Roy 18

Knight, Colin 25

Lamplough, Rob 234
Lavers, Michael 230
Lazenby, David 25
Le Mans 12-13, 139
Ligier, Guy 45
London, Dennis 194
Love, John 91, 140, 161, *184*, 195-197, 210-221, *210*, *212-214*, 230, *231*
Lovely, Nevele 140, *215*
Lovely, Pete 139-140, 142-143, *142*, 170-173, *182*, 197-199, *198*, 201, 205-207, 209-210, 212-214, *215*, 218-220, 242, *243*

Lotus racing cars
  Type 16 139,
  Type 18 8, 10, 25, 223, 238
  Type 25 238
  Type 29 14
  Type 33 14, *27*,
  Type 34 14
  Type 38 14
  Type 39 14, 136, 238
  Type 41 116
  Type 42 14
  Type 43 14-15, *15*, 24-25
  Type 48 24, 64,
  Type 49
    Assessment of, by rival drivers and designers 223-226
    Chassis mounted wings on rival cars *102*
    Compared with 72 *200*
    Costings 30
    Design, building and testing of prototype 24-30, *25-30*
    Difference between 1969 49T and 49B 134
    First race of 49B 98-100
    First race 36-40, *65-66*
    First test 31
    First wing fitted 85, *85*
    High wings 1968 *74-75*, 103, *103*, 114
    High wings 1969 *79*, *132*, *135-136*, *141*, *143*, *147-151*, *177*
    Individual chassis histories 227-244
    Last Grand Prix *183*
    Last race in British Racing Green 85
    Origins 8-10
    Pivoting/feathering high wings 123
    Press launch *32*, 65
    Race record of 222-223, 246-252
    Strength of tub 105
    1967, modifications during season 43, *185*
    1968, evolution of 49B specification 92-94, *92*, *99-100*, 112, *112*
    1969, modifications for new season 139
    1969/70, evolution of 49C specification *176-177*, *180*, 193
  Type 56 128, 198
  Type 56B 209
  Type 57/58 128-129, *128-130*, 198
  Type 62 193
  Type 63 129, 139, 144, 155-158, *156-158*, 160, *162*, 168-170, *169*, 171, 175, *178*, 238
  Type 64 128, 144, 158
  Type 72 128, 193, 198, 199, *199-200*, 201, 204, 209, 221
  Type 79 238
Lyons, Pete 214

MIRA 25
Mann, Alan 131
Marelli, Giovanni 86
Mayer, Teddy 159
Mayman, Anthony 244
McCall, Allan 36, 40, 43, *45*, 47-49, 51-52, 56, *58*, 60, *61*, 62-63, 142, 228, 244
McLaren, Bruce 54-56, 65, 87, 96, 99, 102, 107, 114-115, 118, 122, 125, 153, 167, 170, 199, 201, 227
McLaughlin, David 244
McNicol, John 214-215
Melia, Brian 20-21, 43, *45*, *77*
Mexican GP 1966 15; 1967 59; 1968 122-125; 1969 173
Miles, John *156*, 157, 160-163, *162*, 170-171, 173, 176, 194-199, *195*, 201-202, 204-206, 240
Monaco GP 1967 15; 1968 98-101; 1969 153-154; 1970 201-203

Index

Morley, John and Peter 244
Moser, Silvio 205-206
Moss, Stirling *89*
Mower, Derek 128, 195, 209
Murdoch, Jim 128, 160, 173

National Motor Museum 230, 234
Nye, Doug 230

Oliver, Jackie 53, 64, *77*, 91, 98-111, *100-102*, *104*, *106-108*, *110*, 113-114, 117-120, 123-125, 127, 142, 152-153, 163, 201, 206, 223, 227, 229-230, 232-233, 238, 244

Parkes, Mike 55, 225
Parnell, Peter 217-218, 230
Parnell, Tim 166
'Pent-roof' combustion chamber 11
Pescarolo, Henri 201-202
Peterson, Ronnie 203, 206
Pfaff, George 140
Phillippe, Maurice 8, 14, 17, 19, 24-25, *24*, *32-34*, 33, 35, 41, *47*, 48, 53, 91, 109, 120-121, 175, 193, 221, 226
Phipps, David 146
Pilbeam, Mike 193-194, 240
Pittaway, Neil 231
Player, John 'Player's' 85, 127, 195, 199
Porteous, Dale 25, *34*, 47-48, 84, *87*, 101, 129-131, 134, *136*
Porteous, Roger 84, 86, *87*, 129, 131, *136*
Pratt, Bill 20
Pretorius, Jackie 211, 218-219, 221

'Queerbox' 8-9, 82

Rathgeb, Chuck Jnr. 53
Redman, Brian 96, *97*, 195-196
Repco engines 11, 13, 224-226
Rindt, Jochen 38, 56, *80*, 83, 102-103, 107, 109, 111, 114, 117, 122, 127-128, 131-138, *133*, *136-138*, 140-147, *141*, *144*, *146*, 149-153, *150-151*, 155, *157*, 158-164, *163-164*, 166-168, *169*, 170-171, *172-173*, 173-175, *175*, *177*, *180-181*, 193-199, 201-208, *203*, *209*, 222-223, *233-234*, 234, 236, 239-240, *239-240*
Rodriguez, Pedro 43, 96, 102, 118, 122, 125, 142, 144, 173, 201, 205, 207, 209
Rood, Ben 11, 18, *21*
Rudd, Tony 39

Scammell, Dick 25, *27*, 31, *31*, 33, *34*, 37, 40, 43, *45*, 46, 48, 50, *55*, 56, *77*, 82, 100, 146, 149, 154, 163-164, 170, 194, 227
Scarfiotti, Ludovico 99, 225
Schenken, Tim 219
Schlesser, Jo 106
Schultz, Denzil 221
Scribante, Aldo 215-216
Seaman, Trevor *77*, 101, 120, 131-132
Sellens, Dick 133
Servoz-Gavin, Johnny 98-100, 114-115, 125, 196

Setton, Jacques 234
Sherman-Erpe, Paul 18
Siffert, Jo (Guiseppe, 'Seppi') 64, *76-77*, *79*, 91, 95, *96*, 97-100, 102-109, *106*, *108-110*, 112, 114- 115, 117-118, 121, 123-125, 140, 142-143, *143-144*, *146*, 147, 149-150, 153-154, *155*, 159-161, *161*, 163, 167, 170-172, 175, *175*, 194, 201-202, 206, 222-223, 228, 231, 234-235
Sims, Dave 121, 128, 145, 149-150, 152-153, 159, *162*, 163, 173, 195, 201, *202*, 209
Society of Motor Manufacturers and Traders 12
Solana, Moises 58-59, *60*, *62*, 63, 123-124, 227, 230, 232, 240
Soler-Roig, Alex 198-199, 204-206, 240
South African GP 1967 15, 1968 81-84, 1969 140 142, 1970 195 197
Spanish GP 1967 (NC) 63; 1968 95-97; 1969 144-151, *145-151*; 1970 198-199
Sparshott, Bob *77*, *87*, *87*, *95*, 98, 100, 111-112, 118-120, 123-125, 128, 223
Speedwell 10
Spence, Mike 14, 25, 95, 194
Standfield, Alan 238
Stemp, Peter 18
Stewart, Jackie 43, 47, 51, 64-65, *79*, 82-83, 95, 98, 102-103, 106-112, 114-115, 117-118, 121, 123-125, 142-145, 150, 153-154, 158-161, 163-164, *164*, 166-167, 170, *171*, 172, 195, 197, 202-203, 205-206, 232
Surtees, John 41, 49, 51, 56, 83, 96, 99, 100, 102-103, 113-114, 121-122, 124, 127, 206, 217, 220, 223-224,

Tasman series 1968 64-65, 84-91; 1969 131;
Tauranac, Ron 226
Taylor, Henry 20
Terry, Len 14, 25
Trimmer, Tony 209, 234
Tyrrell, Ken 51, 65, 124, 142, 210, 216

United States GP 1966 14; 1967 58, *59-61*; 1968 118-122; 1969 171; 1970 208-209
Unser, Al Snr. 219
Unser, Bobby 113, 219

Valve angles 11, 17, 20
van Rooyen, Basil 140, 211-213, *212*, 215
Villeneuve, Gilles 118

Walker, Dave 220
Walker, Rob 64, 91, 109, *175*, 176, 195, 199, 201, 205, 207-208, 228, 234, 244
Warr, Peter 194, 199, 203, 205, 209, 234
Weslake engines 37, 64, 224-225
Wheatcroft, Tom 245
Wietzes, Eppie *52*, 53-54, 227
Williams, Frank 131, 143, 154, 176, 205-206,
Wybrott, Leo 25, *27*, 84, 86, *87*, *94*, 128-131, *130*, *145*, 153, 156, 160-161, 163, 165, 173, 194

Zahnradfabrik Friedrichshafen (ZF) gearboxes 17, 24, 28-29, *28*, 44, 46-47, 93, 135, 211

Visit Veloce on the Web - www.veloce.co.uk

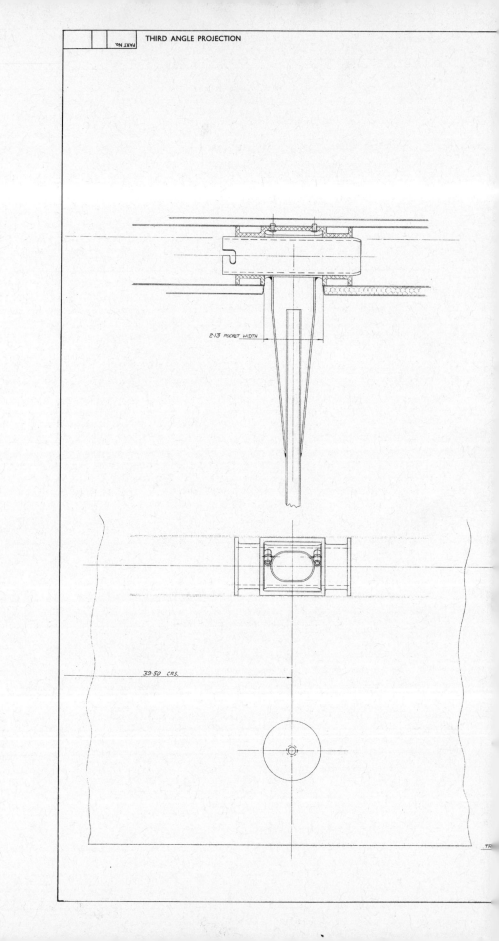